S0-BRY-136

GEOGRAPHY'S INNER WORLDS

GEOGRAPHY'S INNER WORLDS

Pervasive Themes in Contemporary
American Geography

EDITED BY

RONALD F. ABLER

•••

MELVIN G. MARCUS

•••

JUDY M. OLSON

RUTGERS UNIVERSITY PRESS NEW BRUNSWICK, NEW JERSEY

Wingate College Library

Copyright © 1992 by Rutgers, The State University
All rights reserved
Manufactured in the United States of America

Library of Congress Cataloging-in-Publication Data

Geography's inner worlds : pervasive themes in contemporary American
  geography / edited by Ronald F. Abler, Melvin G. Marcus, and Judy M.
  Olson.
      p.   cm. — (Occasional publications of the Association of
  American Geographers ; no. 2)
      Includes bibliographical references and index.
      ISBN 0-8135-1829-6 (cloth) — ISBN 0-8135-1830-X (paper)
      1. Geography—philosophy.   2. Geography—United States.   I. Abler,
  Ronald.   II. Marcus, Melvin G. (Melvin Gerald). 1929–
  III. Olson, Judy M.   IV. Series.
  G70.G0446   1992
  910′.01—dc20   91-43478
                                                                      CIP

British Cataloging-in-Publication information available.

Contents

List of Figures

List of Tables

Preface

Geography's Inner Worlds began as the Survey and Synthesis Project of the Association of American Geographers (AAG). In early 1985, noting that over thirty years had elapsed since a comprehensive survey of the discipline (*American Geography: Inventory and Prospect*) had been undertaken, the AAG Council appointed a task force to evaluate the desirability and feasibility of another overview. The proposed book was to be written primarily for geographers rather than for readers outside the discipline, and it was to focus on what geographers of all specialties and persuasions had in common intellectually.

The task force consisted of Athol Abrahams, State University of New York at Buffalo; Michael Conzen, University of Chicago; John Estes and Michael Goodchild, University of California, Santa Barbara; Chauncy Harris, University of Chicago; Geoffrey Martin, Southern Connecticut State College; John R. Mather, University of Delaware; Janice Monk, University of Arizona; Joel Morrison, U.S. Geological Survey; Christopher Salter, University of Missouri; and Thomas Vale, University of Wisconsin, Madison. Ronald Abler (then AAG vice-president), Sam Natoli (then AAG director for educational affairs), and Risa Palm (then AAG president) were ex officio members of the task force. The task force was asked to: develop a preliminary table of contents; select editors for the book; select an editorial board to provide guidance and assistance to the editors; prepare preliminary proposals for obtaining the funds needed to support preparation of the volume; and evaluate a proposal for a volume entitled *Geography in America* proposed by Gary Gaile of the University of Colorado and Cort Willmott of the University of Delaware.

The task force recommended a volume to be entitled *American Geography: Survey and Synthesis* and submitted a table of contents that it put forth as a general guideline for authors and editors, not as a rigid structure to be followed undeviatingly:

American Geography: Survey and Synthesis

I.   CONTEXT
  1. The Geographical Enterprise since the 1950s
  2. Geography, Science, and Public Policy
  3. Geography and American Culture
  4. Ideological, Conceptual, and Methodological Trends in Geography

II.   METHODS
  [Cross-Cutting Themes: Science and Technology]
  5. Information Systems
  6. Field Techniques
  7. Cartography
  8. Remote Sensing
  9. Quantitative Techniques

III.   PHYSICAL SYSTEMS
  10. Climatology
  11. Geomorphology
  12. Soils
  13. Biogeography
  14. Interrelationships among Physical Systems

IV.   HUMAN SYSTEMS
  [Cross-Cutting Themes: Regional Formation and Change;
  Historical Processes; Policy, Planning, and Practice]
  15. Economic-Political Systems
  16. Socio-Cultural Systems
  17. Demographic Systems
  18. Synthesis of Human Geographic Processes

V.   PHYSICAL-HUMAN INTERACTIONS
  19. Human Ecology
  20. Resource Management
  21. Environmental Modification
  22. Hazards

VI.   OUTLOOK

A question that engendered heated debate within the task force was whether the book should be organized around traditional divisions such as cultural geography, political geography, and geomorphology, as its predecessor had been, or whether a different structure should be created. The outline the task force produced was a hybrid. Task force members representing physical geography and methods were unwilling to abandon traditional rubrics. After considerable discussion, the human geographers agreed that the goal of highlighting common ideas would best be served by avoiding divisions along the lines of familiar specializations.

The task force recommended three editors, one each for human geography (John Borchert), physical geography (Melvin Marcus), and methods (Judy Olson). (John Borchert subsequently declined appointment as co-editor. In early 1987, Ronald Abler was appointed in his stead.) The task forced further recommended that there be no steering committee for the project or, if the AAG Council believed a steering committee was needed, that the task force should serve as a steering committee.

The task force concluded that Gaile and Willmott's proposed *Geography in America* did not fulfill the mandate outlined by the AAG Council, inasmuch as its focus was primarily on specialty groups and their respective interests rather than on cross-cutting and unifying concepts and methods. It recommended that AAG proceed with the *Survey and Synthesis* project.

The editors devoted three meetings in November 1987, March 1988, and February 1989 to reworking the *Survey and Synthesis* outline. The revised outline was based on advice received from a large number of scholars to whom various versions of the project outline and goals had been sent. In the editors' judgment, the task force outline focused too much on subdivisions within the discipline and too little on themes that are common to many or all specialties. Furthermore, the detailed portrait of specialties being prepared by Gary Gaile and Cort Willmott, who had decided to proceed with their *Geography in America* volume, would obviate the need for separate treatment of individual specializations in the manner of *American Geography: Inventory and Prospect*. The outline proposed by the editors for *Geography's Inner Worlds* was similar in most respects to the final version incorporated in this book.

The first, brief meeting of authors was held on 21 March 1989 during the Baltimore AAG meetings. Authors were asked to submit chapter outlines during the following summer. A combined outline was compiled from the individual submissions, and in September 1989 the editors sent the authors suggestions for further refinement. Authors were asked to prepare first drafts by February 1990, and an extended meeting of authors and editors was held near San Francisco 18–20 May 1990. Chapter drafts and outlines were given a final polishing, and the chapter order was also changed to that which appears in this volume.

At its fall 1989 meeting, the AAG Council designated *Geography's Inner Worlds* as an AAG contribution to the 1992 Washington, D.C., International Geographical Congress. In recent decades countries hosting the congresses have prepared volumes that summarize the current state of geographic research and teaching in their respective countries, and the council decided to use *Geography's Inner Worlds* for that purpose. Although the editors and authors had some forewarning of the decision, it did engender some changes in perspective and presentation that were introduced when the chapters were edited in late 1990 and early 1991.

Our goal throughout the long period of correspondence, discussions, and meetings that preceded publication was to produce a *readable* and reasonably comprehensive account of contemporary American geography, one that would remind American geographers of all persuasions of the heritage they share. While we hope that interested nongeographers find the book helpful as an introduction to the discipline, its major purpose remains to reintroduce geographers to each other, and to introduce overseas geographers—especially those from outside the English-speaking realm—to American geography.

We owe many people a great deal for their assistance in bringing the book to print. The task force members, who continued to serve as a steering committee through 1989, were most generous with their time and thoughts. Their names were noted above. In addition, the following individuals provided reactions to outlines and suggestions for improvements at various stages of the project's evolution. Alice Andrews, George Mason University; Roger Bolton, Williams College; Larry Brown, Ohio State University; John Borchert, University of Minnesota; Helen Couclelis, University of California, Santa Barbara; Peter Gould, Pennsylvania State University; Susan Hanson, Clark University; John Hudson, Northwestern University; John Hunter, Michigan State University; Donald Janelle, University of Western Ontario; Bob Kates, Brown University; Jeanne Kay, University of Utah; Martin Kenzer, Louisiana State University; Greg Knight, Pennsylvania State University; Alan MacEachren, Pennsylvania State University; Mark Monmonier, Syracuse University; Richard Morrill, University of Washington; Peter Muller, University of Miami; David Reynolds, University of Iowa; Richard Skaggs, University of Minnesota; Larry Sommers, Michigan State University; Tony de Souza, National Geographic Society; Waldo Tobler, University of California, Santa Barbara; Bill Turner, Clark University; Thomas Vale, University of Wisconsin; Michael Watts, University of California, Berkeley; Tony Williams, Pennsylvania State University; Harold Winters, Michigan State University; Joe Wood, George Mason University; and Wilbur Zelinsky, Pennsylvania State University. It is no accident that a number of people who commented on outlines subsequently became authors or coauthors.

We thank also the universities and agencies that employ the chapter authors. Among other contributions, they underwrote the travel costs for authors who attended the meetings in Baltimore in 1989 and in San Francisco in 1990. The book would have been much less cohesive in the absence of their support.

Elizabeth Beetschen and Melissa Posey of the AAG office performed much of the mechanical work involved in production of the book, including entering corrections, retyping drafts, and obtaining permissions. We appreciate the cheerful good-will with which they undertook those tasks. Our thanks also to Mary Ann Marcus, who provided room, superb board, and vivacious company during several extended editing sessions in Tempe.

We are pleased to have been asked to undertake this task and even more pleased that it has been completed—for the time being. As in any effort of this kind, we believe we have succeeded in some respects and fallen short of our own goals in others, as noted in the Afterword. All in all, however, we are grateful for the opportunity to work closely with an energetic and talented group of authors on a project that is dear to our hearts—trying to articulate the essence of an ancient intellectual tradition that continues to bewitch us.

RONALD F. ABLER
MELVIN G. MARCUS
JUDY M. OLSON

About the Authors and Editors

RONALD F. ABLER is executive director of the Association of American Geographers in Washington, D.C., and professor of geography at the Pennsylvania State University. His research and teaching have been devoted to the geography of telecommunications networks and their uses and to the history of geography.

ANTHONY J. BRAZEL is professor of geography at Arizona State University and the state climatologist of Arizona. His research and teaching have focused on alpine, urban, and arid land climatology with emphasis recently on global climatic issues.

DWIGHT A. BROWN is professor of geography at the University of Minnesota. He teaches water resources, geographic information systems, and Quaternary landscape evolution. His current research focuses on modeling spatial processes in biogeography.

HELEN COUCLELIS is associate professor of geography at the University of California, Santa Barbara. Her research and teaching interests include spatial cognition, urban and regional modeling and planning, and the philosophy of science.

GARY L. GAILE is associate professor and associate chair of geography at the University of Colorado, Boulder. His research specialties include urban and regional planning, Third World development, and spatial statistics and models.

PHILIP J. GERSMEHL is professor of geography at the University of Minnesota. His research and teaching specialties are soils, agricultural policy, land evaluation, and geographic information systems. He has also engaged in a number of projects to improve geographic education in Minnesota and the United States.

PATRICIA GOBER is professor and chair of the Department of Geography at Arizona State University. Her research interests are in population movements at both the interregional and intraurban scales and in housing demography.

MICHAEL F. GOODCHILD taught geography at the University of Western Ontario for nineteen years before taking his current position as professor of geography at the University of California, Santa Barbara, where he is also director of the National Center for Geographic Information and Analysis (NCGIA), funded by the National Science Foundation. His research focuses on generic issues of geographic information and methods of spatial analysis.

WILLIAM L. GRAF is professor of geography at Arizona State University. He is a fluvial geomorphologist specializing in river mechanics, human impacts on river systems, and public land policy.

DEREK GREGORY is professor of geography at the University of British Columbia. Born and educated in England, he taught at the University of Cambridge for sixteen years before taking up his current appointment in 1989. His teaching and research interests focus on the relations between human geography and social theory and on historical geographies of modernity.

JOHN C. HUDSON is professor of geography at Northwestern University. His research and teaching interests include the cultural, economic, and physical geography of North America and the role of maps and models in geographic research.

DONALD G. JANELLE is professor and chair of the Department of Geography at the University of Western Ontario. His research and teaching concern the time geography of urban environments, transportation and urban systems development, locational conflict analysis, and the social structure of geography.

C. GREGORY KNIGHT is vice-provost and dean for undergraduate education and professor of geography at the Pennsylvania State University. His research focuses on the intersection of physical and human geography, with specific attention to human use of environmental resources in Africa.

DAVID A. LANEGRAN is professor of geography at Macalester College and coordinator of the Minnesota Alliance for Geographic Education. His research focuses on the nature of urban communities and the impact of urban planning in several cultures. His teaching is devoted to the general themes of human and urban geography.

PATRICIA F. MCDOWELL is associate professor of geography at the University of Oregon. Her research interests include fluvial and eolian geomorphology, the influence of climatic change on landscapes, and Quaternary environmental history.

ALAN MACEACHREN is associate professor of geography at the Pennsylvania State University. His research and teaching focus on cartography and spatial cognition, and on their interaction in wayfinding and scientific visualization in geography.

MELVIN G. MARCUS is professor of geography at Arizona State University, Vice-President of the American Geographical Society, and chair of the U.S. National Committee for the International Geographical Union. He is a physical geographer whose research and teaching have focused on glacier environments, climate, and environmental geography.

WILLIAM B. MEYER is research associate in the George Perkins Marsh Institute and the Graduate School of Geography at Clark University. His interests include environmental change, political geography, and the history of geographic thought.

JUDY M. OLSON is professor of geography and chair of the geography department at Michigan State University. Her research has focused on cartography, especially map design and map use. She teaches cartography and quantitative methods.

RISA I. PALM is professor of geography and dean of the College of Arts and Sciences at the University of Oregon. Her research has focused on societal responses to environmental hazards, particularly earthquake hazards in California and hurricane hazards in Puerto Rico.

JOHN PICKLES is associate professor of geography and director of graduate studies at the University of Kentucky. His research focuses on regional development and industrial peripheral states and on the philosophy of science. He is currently engaged in research projects on regional restructuring in South Africa and on political and economic change in Bulgaria.

BONHAM C. RICHARDSON is professor of geography at Virginia Polytechnic Institute and State University. His research deals with historical geography and human migration in the Commonwealth Caribbean. His most recent book is *The Caribbean in the Wider World, 1492–1992: A Regional Geography*.

B. L. TURNER II is professor in the Graduate School of Geography and director of the George Perkins Marsh Institute at Clark University. His research interests include cultural ecology, agricultural change theory, and the human dimensions of global environmental change.

MICHAEL J. WATTS is professor of geography and coordinator of development studies at the University of California, Berkeley. His research and teaching have focused on the political economy of the Third World, agrarian transformations in peasant societies, food security and famine in Africa, and political ecology.

CORT J. WILLMOTT is professor of geography and marine studies and chair of the geography department at the University of Delaware. His research and teaching interests include the influence of land-surface processes on climate, global climatic change, and quantitative approaches to examining geographic systems.

DAVID WOODWARD is professor of geography and director of the history of cartography project at the University of Wisconsin, Madison. His research and teaching have focused on the history of cartography and on cartographic design.

GEOGRAPHY'S INNER WORLDS

Contemporary American

Geography

The Editors

American geography perplexes many people inside and outside the discipline. Both groups have difficulty identifying cohesion and core because of the ways geographers cluster in specialist groups and because of the great diversity of topics to which geographers apply their expertise. How does one cope with a discipline that seems to incorporate physical scientists, social scientists, and humanists, and whose members study everything under the sun?

Geography, like anthropology and psychology, embraces the physical and social sciences. A few countries have chosen to separate physical and human geography in their universities, but they remain unified—if vaguely suspicious of each other—in American universities. Physical and social science perspectives by no means exhaust geographers' intellectual repertoires. If one insists on putting every geographer in a category of intellectual activity, some would fit best in the compartment that holds behavioral scientists, and a few would be happiest if they were classified as humanists. Some geographers devote their careers to acquiring substantial knowledge of a major region of the earth. Others delve deeply into systematic specialties such as climatology or environmental perception, giving scarcely a thought to places and regions as such.

Specialization is especially strong in American geography. Scholarly societies throughout the world have established study or interest groups to accommodate the wishes of specialist communities to share information about their respective pursuits. Whereas the Canadian Association of Geographers has eight interest groups and the Institute of British Geographers embraces eighteen study groups, the Association of American Geographers (AAG) boasted forty-one specialty groups in 1991. Although the AAG's specialty groups range across the spectrum of geographic content, methods, and philosophy, most cater to systematic interests; only six focus their attentions on regions.

Specialization and specialty groups foster better communication in a multi-faceted discipline. They permit scholars and practitioners with common interests to achieve identity without forming independent associations. They play prominent roles in organizing national meetings. Specialization and specialty groups also foster intellectual isolation by retarding the cross-fertilization that occurs when geographers encounter unexpected ideas. Despite the fact that the fission evident within geography is common in the physical and social sciences, specialization may have gone about as far as it can or should go in American geography.

Geography's Inner Worlds was commissioned to highlight the common elements within a discipline whose practitioners are in danger of forgetting their shared heritage and ideals. In their preoccupation with the great diversity of geographic problems with which they deal on a day to day basis, geographers are prone to overlook how much they share emotionally and intellectually with other geographers that they do not share with colleagues in other disciplines.

The diagnostic experience is that most typically geographic exercise—a field trip. Regardless of specialty, nothing reminds geographers of how much they share—and how much geographers differ from colleagues in other disciplines—than a multidisciplinary transect through almost any landscape in the world. Historians, sociologists, and political scientists will cluster in the back of the bus where they will chat in a desultory manner or sleep. Geologists may be roused into observational action by road cuts but will see little between them. Meteorologists will be helpless without their computers and models. Only the geographers—again, regardless of specialty—will incessantly rubberneck, gawk, point, explain, speculate, and argue about what they are seeing, more or less without regard to whether it is urban or rural, physical or anthropogenic, beautiful or hideous. In real places, much of what seems to separate geographers evaporates, and what unites them becomes vividly obvious.

This book is intended to be something of a field trip through American geography. It focuses on the things geographers do regardless of specialty or empirical interest. Accordingly, it is divided into four major sections that portray what geography is about, what geographers do, how geographers think, and why geographers think the way they do.

What geography is about is the earth. Throughout the two thousand years it has existed as a distinct intellectual enterprise, geography has always focused on the earth as created by nature and as modified by human action as abiding and distinctive objects of study. Chapter 2 ("Geography's Worlds") explores recent American manifestations of the ancient and persistent interest in geography's two worlds, the physical world and the human world, and the interconnections between the two. As detailed in chapter 3, "Places and Regions" are critical and distinctive focuses for geography, keys to understanding the world that are as venerable as geographers' interest in natural and human processes on the earth's surface. Places and regions are pebbles and figures in a global mosaic. They are of interest individually and as parts of global patterns. Because geographers have always been interested in global patterns, maps have traditionally been close to the core of geography in three ways: they are products of geographic inquiry, they are analytic tools, and they are themselves objects of study as a means of conveying information. Chapter 4 ("Representations of the World") addresses those topics.

What geographers do differs from what other scholars do only in detail. Geographers apply methods that are common to all sciences—observation, visualization, analysis, and modeling—but as do their colleagues in other disciplines, in distinctive ways. Like other scholars, they communicate the results of those processes among themselves and to others. As suggested above, geographers are keen observers of the world. As described in chapter 5 ("Observation"), they observe and capture data across a spectrum ranging from field work on the ground, through laboratory experiments and archival work, to the use of information captured by aerial photography and satellites. The need to see broad patterns on continental and global scales and the masses of data with which geographers often deal have led, early and late, to the use of graphic methods for viewing both data and analytical results. Those methods have profoundly affected inference and understanding. As emphasized in chapter 6 ("Visualization"), maps and other images are central in geography and distinctive of the enterprise. "Analysis" is the focus of chapter 7. Observation, classification, and visualization are linked closely to inference in geography, which ranges from qualitative intuition to complex quantitative analysis. As in all disciplines, inference is significantly influenced by changing technologies and philosophies. Analysis forms the basis for "Modeling," the subject of chapter 8. Models of all kinds link concepts, substance, and methods into structures that serve as yardsticks for explanation and prediction. The point of what geographers do is to understand the world. Geographers share their understandings with each other, with their students, and with the public through publication and teaching, but especially through maps and atlases, as befits their subject matter and the importance of visualization to understanding phenomena on regional, continental, and global scales. Chapter 9 ("Communication") describes the ways geographers do so.

How geographers think differs from the ways other scholars think. Geographers use procedures that are common to science and the arts, as detailed in part 2 of this volume, but they do not think exactly the same way other scholars do; it is the set of concepts and theories that a discipline brings to its research and teaching that distinguishes it from others. A small set of basic concepts permeates geographic thinking and distinguishes geography from other disciplines. One such concept is a focus on knowing the world through the perspectives of "Location, Place, Region, and Space" (chapter 10). The spectrum of geographic perspectives, ranging from location at one end through abstract views of terrestrial space at the other, provides the conceptual underpinnings of geography as a discipline. As explicated in chapter 11, geographers are also sensitive to "Movements, Cycles, and Systems," to both static and dynamic structures, and to the movements, interactions, and balances that produce and alter those structures. Geographers examine those features at all scales along a "Local-Global Continuum" (chapter 12). Fruitful and valid geographic inquiry depends critically on understanding "Scale in Space and Time" (chapter 13), on choosing degrees of resolution for both time and space that are appropriate for specific analyses.

Geographers think the ways they do because of the ways they are taught to think about the world, which in turn arise from overarching views about how the world can be known, what the appropriate ends of knowledge are, and what kinds of people become geographers. Chapter 14 ("Paradigms for Inquiry") reviews the broad views that geographers have used to organize their inquiry and exposition in recent years and shows how different perspectives have enriched the discipline. Although scientific and humanistic components have long been present in geography, they have coexisted happily for much of the discipline's recent history. At the moment, however, tension between the two cultures postulated by C. P. Snow is rising, a debate covered in chapter 15 ("Humanism and Science in Geography"). Much of that conflict arises from differing beliefs regarding the appropriate uses of geographic knowledge. As noted in chapter 16 ("Applications of Geographic Concepts and Methods"), geography embraces the spectrum from basic research to active advocacy for the implementation of the results of geographic research in policy and programs. A major challenge for the future is to ensure that this ongoing debate yields constructive rather than destructive results. No survey of this kind would be complete without an analysis of *who* becomes a geographer in the United States. Therefore chapter 17 ("The Peopling of American Geography") examines the origins and education of American geographers and the networks of communication and influence they create.

The volume concludes with the editors' views on what the book has accomplished and where it has fallen short. Part of that critique addresses the book as a process as well as a product: the lessons that emerged in the course of organizing and editing the book, and what that sequence of events can reveal about American geography. The Afterword suggests that a basis for greater coherence and unity certainly exists within American geography, but that some of the difficulties encountered in trying to bring scholars with diverse interests to that common ground do not augur well for widespread acceptance of a unifying perspective.

The chapters in this volume were not intended to be comprehensive. Authors were asked to take a broad view of their topics and to include all relevant perspectives. They were also instructed not to attempt to be definitive but to base their presentations on selected representative examples. In the interest of producing a readable volume and avoiding pedantry, authors were also asked to restrict references to those that were essential and to focus on the recent work. The charge that guided the preparation of this volume was to capture the essence of American geography as of the early 1990s. Contemporary research and teaching have therefore been emphasized at the expense of a historical perspective. In few instances is there any consideration of work published before 1950. That omission is deliberate.

The absence of traditional rubrics as chapter titles or as subheadings is also deliberate. A number of geographers who have reviewed outlines or drafts have decried that structure, asking how it is possible to have a book that purports to describe contemporary American geography that does not address urban geography, historical geography, geomorphology, and the like. The answer is that a focus on such topics—which are admittedly major components of current American geography—would direct attention toward the discipline's peripheries rather than to its core.

The core of geography is the set of assumptions, concepts, models, and theories that geographers bring to their research and teaching. The phenomena to which they apply those tools are of interest and they are important, but they do not constitute the discipline. At one time geographers applied those tools to the world's seas and oceans. Very few now do, that conversation having largely ceased within geography after the distinct discipline of oceanography evolved. Yet the conversation among geographers persists, using many of the same intellectual tools, focused now on other phenomena.

A discipline *is* no more than an ongoing conversation among a group of scholars who share a common outlook that is defined by the phenomena they study, the methods they apply, and the theory they use. This book consists of fragments from that conversation in the United States in the late 1980s and early 1990s, along with considerable commentary and interpretation. It is far from a complete transcript, and the commentary often focuses on what is implicit in the conversation rather than on what was actually said. That implicit subtext reveals the persistent, inhering themes that unify what sometimes sounds like a babble of conflicting voices.

PART I

WHAT GEOGRAPHY IS ABOUT

Geography's Worlds

C. Gregory Knight

⋯ ———————————————————— ◯ ———————————————————— ⋯

> I readily believe that there are more invisible Natures in the universe
> than visible ones. Yet who shall explain to us this numerous com-
> pany, their grades, their relationships, their distinguishing features
> and the functions of each of them? What do they do? What places
> do they inhabit? (Burnet 1692:68)

Geography embraces both the physical and human worlds. Indeed, the first
geography department in the United States aimed to bridge natural and human
sciences; the University of Chicago's department (founded in 1903) was to lie
"intermediate between geology . . . and history, sociology, political economy
and biology" (Mikesell 1974:2). Both the environmental-ecological and the
cultural-social worlds of geography contain a seeming plethora of approaches,
philosophies, and foci. Some perspectives abide largely or even entirely in the
physical or in the human domain; others are more richly informed by interactions between the two realms and are expressed in approaches such as natural
resources, environmental perception, human ecology, and places, regions, and
landscapes.

For geographers, regardless of their respective specialties, the physical and
human worlds constitute warp and woof, the elementary fabric of spatial experience and understanding. Consider this fabric a map, a chart that incorporates and underlies the perspectives held by communities of geographers, past,
present, and future. And consider this chapter as a gazetteer to that map, with
vignettes that will provide a general itinerary for what will follow in this volume as warp and woof are examined in more detail.

Geography's worlds are many, yet one. Academically and intellectually, geographers see innumerable worlds that parallel their individual and collective viewpoints. Some of geography's worlds are visible and tangible; others remain hidden and elusive. Some worlds represent an ordered reality, as seen from the enduring but vulnerable consensus of science. Others are ephemeral, the creations of humanistic minds that seek understanding and meaning in geographical patterns and processes, and in transient ideas that explicate society's internal interactions and its relationships with its physical environment—interactions made manifest in the visible landscapes that always reveal their makers. Some of geography's worlds consist of naturally and socially derived objects and events, and some are existential—meanings in a world of human intentions (Tuan 1971:182). Yet there is one world amidst that multiplicity of perspectives, for all focus on society-and-earth as the singular geographic experience.

Some geographers represent their worlds with words, others make maps, and still others prefer graphs or mathematical models. Words, maps, and models all express geographic phenomena and forces as well as the understandings geographers bring to their classifications and explanations. In spite of the different viewpoints and representations they choose, geographers share a deep conviction that environmental and human processes create the reality in which humankind lives, and that those processes are keys to understanding the complexity of places.

Geographers' worlds also vary with their professional responsibilities. Academic pursuits of teaching and research, government service, private sector employment, citizen participation, and voluntary organizational activity produce different but complementary worlds. Indeed, because every person must navigate within built structures and traverse outdoor locations in the course of daily life, everyone is inherently a geographer. Place and places are part of every human life.

The wealth of stances taken by geographers seeking geographical understanding and explanation produces the richness of modern American geography. At the same time, the many worlds geographers profess sometimes cause the public and other scholars to misapprehend the field. The ambiguity apparent, and seemingly inherent, in geography is evident in Robinson's suggestion that "to be a geographer is to participate in an intellectual activity characterized by confusion and excitement that derives from the lack of a coherent organizing principle" (1982:177).

FIVE VIGNETTES

The rich diversity of geography's worlds can only be suggested by five examples that elucidate geographic concern with the physical environment and with interactions between human society and the environment. One focuses on the role of solid water in global physical processes and on what it can tell us about climate change. The second deals with liquid water, a universally important natural resource whose allocation is strongly conditioned by society through the modality of law. As shown in the third example, environmental conditions affect the food production that sustains society, as do wider socioeconomic forces. Societies face natural and technological hazards in the fourth example; geographers have given primacy to both environment and society in their research on calamities. Finally, the geography of the city is seen as a product of the socioeconomic environment, of forces that drive economic systems and social interaction.

From these examples, discussion will be expanded to suggest some major issues in contemporary geography that concern human interaction with environment, the human and physical worlds in geography, and, most importantly, the geographer's fundamental understanding of society-environment interactions.

Glaciers and Climatic Change

From the time of Alexander von Humboldt's mapping of the glacial snow line on mountains in relation to latitude, geographers have recognized that glaciers are related to wider environmental conditions. Indeed, hypotheses that allowed Louis Agassiz to conceptualize widespread continental glaciation derived from changes in the extent of mountain glaciers and from landform remnants found in valleys then free of glaciers. Today, geographers and other scientists see the varying world of ice and snow—some 10 percent of the earth's surface, depending on season, plus a like area of periglacial landscapes—as an indicator of regional and global environmental conditions and events.

As an undergraduate research assistant in 1948, Mel Marcus began a four-decade-long study of Lemon Creek Glacier in the Juneau ice field of southeastern Alaska, an investigation that spanned Marcus's dissertation during the International Geophysical Year (1957–58), work of other scientists, and his own field sessions in 1989 and 1990 (Marcus 1964; Marcus, Chambers, and Miller 1990). The basic strategy was to assess the glacier's extent and dynamics through studies of its mass balance in relation to local and mesoscale climatological conditions that affect snow input and ablation and melting losses (Yarnal 1984). Such studies are undertaken by field mapping and remote sensing, using complementary planimetric and volumetric strategies. For Lemon Creek Glacier, Marcus found that net mass losses over four decades were consistent for lower elevations but did not hold true for glaciers at higher elevations during the recent decade. These losses are then studied in relation to local and regional climatic patterns, including synoptic conditions, with widespread regional concordance of glacial dynamics potentially indicating changes of global climate.

Glacial studies are a traditional, fieldwork-intensive component of physical geography that also have social significance. Glaciers are a notable source of runoff, providing water resources for society, particularly critical in arid downstream lowlands. Glaciers and mountain snow have important recreational benefits and also present hazards such as interruptions to transportation, flooding, and avalanches. Most salient, glaciers are a herald of environmental change. As geographers extend their understanding of interannual changes in glacial mass and dynamics in relation to local atmospheric processes and energy fluxes, the foundation will have been laid for documenting the direction and magnitude of longer term climatic and synoptic changes signaled by glaciers.

Water and the Law

Water is fundamental to the existence and structure of the earth's biosphere because of its biological role and extraordinary thermodynamic properties. All life depends on water. So too is water acutely important to society as an environmental resource. Sites of human settlement depend upon the location of potable water. Communities located away from fresh water have had to build elaborate transfer systems (pipes, canals, *qanats*) or water purification or distillation facilities to ensure water availability. Water resources change with the seasons, but seasonal variability can be buffered by scheduling human activities to harmonize with natural calendars or by storing water. Seasonality of water resources governs the annual agricultural cycle for the majority of the world's peoples. Although water is treated as a free good in the American economy, its contributions to the economy are unfathomable.

The Colorado River is a critical resource in the southwestern United States, providing hydroelectricity, water for urban and agricultural use, and recreation. Many of the needs for Colorado River water have developed at locations far distant from the river and even from its basin. Substantial seasonality and interannual variability of flows create an unreliable natural resource that has been increasingly managed, in effect, to buffer periodicity and uncertainty (Graf 1985). The greatest instability will be social: international and interstate allocations of the Colorado's water, set during a period of unusually high flows, far exceed the longer term average of resource availability.

Throughout history, the power to control water has been the power to control society and ensure development (Walker and Williams 1982; Worster 1985). Control over water has usually been exercised through the law, and the intersection of water (the environment) and law (human society) in America has been explored by Matthews (1984), who documented how legislative, administrative, and judicial authorities affect the availability and allocation of water as a critical resource for human consumption, recreation, agricultural and industrial productivity, hydroelectric power, waste disposal, and navigation—

each of which is subject to a variety of legal controls. Water in similar hydro-geological circumstances in America may or may not be available as a resource to the landowner or others, depending on whether state water law is based on the riparian or the prior appropriation doctrine. The older riparian tradition links water rights to land ownership; the prior appropriation doctrine is common in the American West; it sequentially allocates water on the basis of temporal priority of first use. Matthews further considers whether these principles of surface water rights apply to ground-water, noting the evolution of additional legal principles affecting the geography of underground water resources.

Equally perplexing is the role of water in demarcating land units, whether political boundaries or property metes and bounds. In the natural processes of flooding and channel evolution, rivers change their positions. Matthews has researched the consequences of channel instability on boundary marking, noting differences in international, national, and land-tenure implications of river meandering. Where rivers separate states or nations, the geographical boundary is typically changed only by revision of a treaty, with landlocked areas of one jurisdiction being isolated across the river's new channel until boundaries are adjusted. In contrast, American legal precedents give newly formed land to the adjacent landowner and offer little but condolences to former holders of land now washed away. A map of water resources only has meaning for society when overlain by a map of the legal principles that govern water ownership and rights to its use—its human meaning.

The Cultural Ecology of Food

Food is as fundamental to human society as water. Materials for human sustenance are produced through annual fluxes of water, solar energy, and soil nutrients that are channeled to animals and people by plants via photosynthesis. Geographers have focused research on issues of agricultural production and food supply stability in many non-Western societies. By the 1960s, geography's close interaction with anthropology enhanced the evolution of cultural ecology, the study of human society as a system adapting to exigencies of the local environment. These disciplines shared a concern with adaptive mechanisms and with ethnoscientific knowledge of the environment. East Africa was a major locus for research, and Philip Porter was prominent among American geographers undertaking cultural-ecological research there. Porter collaborated with an anthropological team (Porter 1965) in studying several societies in differing ecological settings. He soon realized that for scientist and African alike, rainfall availability, seasonality, and variability are key environmental elements to understanding the food production of African societies (Porter 1976). Porter adapted C. Warren Thornthwaite's (1948) findings about soil water regimes to model soil water availability and variability for agricultural production and crop risk in East African localities. From these models, he deduced the functional utility of African cropping technologies and strategies. Porter's explanation of the human-environment system in Africa was seen as adaptation and the "functional coherence [of] a group's appraisal and management of its environment" (Porter 1979).

As Porter's work progressed through several lengthy research seasons in Africa through the 1970s, he became critical of contemporary development policies (Porter 1987) and development theory (Souza and Porter 1974). At the same time, he became increasingly dissatisfied with the scope of cultural ecology as a framework that viewed local, largely preindustrial, Third World societies as isolated from larger political and economic forces. Thus Porter contributed to the evolving concept of political ecology or ecopolitical economy (Yapa 1979; Bassett 1988). He viewed local society-environment interaction as conditioned, if not controlled, by exogenous political and economic forces, as well as by tenacious internal structures governing access to resources (Porter 1979). Food production is the most basic human-environment nexus. Without food, nothing else is possible. Porter and others argue that so fundamental a relationship, even in seemingly traditional societies, cannot be seen as simply a matter of people's understanding and use of the local environment. Models of human adaptation to environmental uncertainty can have little meaning if they fail to incorporate pervasive and powerful social forces.

Natural and Technological Hazards

Just as East Africans must cope with interannual variability in rainfall, all humans are vulnerable to other natural hazards that cause damage to life and property: floods, tornadoes, hurricanes, earthquakes, and the like. Based upon innovative research by Gilbert White at the University of Chicago, several generations of scholars have studied human response to environmental hazards. With important applications to American public policy originally focused on floodplain hazards, the research paradigm linked the natural process of flooding to human perception and resulting policy developments. It was later extended to international settings and to other environmental hazards (White 1974; Burton, Kates, and White 1978). Natural hazard research contributed to the emergence of environmental perception studies in geography. In the case of floods, a natural hazard syndrome was posited, in which perceptions of the risk of flooding underestimate the real risk and diverge increasingly from reality with the passage of time since a flood. Nevertheless, natural hazard researchers found evidence of rational human responses within the boundaries of limited knowledge and awareness.

Two important research directions emerged from natural hazards investigations. The first addressed technological hazards, ranging from chemical and oil spills, highway accidents, waste dumps, and nuclear power plant failures to threats of thermonuclear war. Here again, focus was on the nature of the event, assessment of the risk, and human perceptions and concerns (Cutter 1984). Zeigler, Johnson, and Brunn (1983) contributed to this research by specifying the geometry and temporal and spatial scales of technological hazards, including global hazards such as carbon dioxide buildup, ozone depletion, and ocean contamination. As in the case of natural hazards, attention centers on evacuation, emergency planning and relief, public policy, and dissonance between real and perceived risk. Technological and natural hazards research by geographers and interdisciplinary risk assessment and management studies have developed useful ideas regarding the hazardousness of places and risk mosaics (Zeigler, Johnson, and Brunn 1983:48–70). Related research in medical ecology and in medical geography focuses on risks to human health. Gould's analysis of the Chernobyl explosion consequences is an excellent example of linkages between studies of technological hazards and of human health (Gould 1990).

The second research direction that emerged from the natural hazards research school focused on concepts of societal vulnerability to hazards, including the social magnification of risk and social causes of vulnerability to hazards. Hewitt (1983a:24–25) suggested that hazards are "not explained by, nor uniquely dependent upon the geophysical processes." Perceptions and reactions to hazards depend on the structure and values of society, and the causes, nature, and results of disaster must be seen in the workings of society and its characteristic linkages to environment. Hewitt and his contributors viewed natural calamities as functions of the vulnerability of human society created by contemporary life (Hewitt 1983b). Later, Montz and Gruntfest (1986) documented this vulnerability in the form of increasing flood damage in America over recent decades despite widespread use of structural flood controls, because of the encroachment of development onto floodplains. As Watts (1983:56) wrote, "Hazard response is . . . contingent upon the social context of the responding units and upon their situation in the productive process." Even chronic natural hazards such as soil erosion (Blaut et al. 1959; Blaikie 1985; Blaikie and Brookfield 1987) and sudden catastrophes such as earthquakes must be viewed from the perspective of political ecology or political economy.

Cities

The city, like water, food, and hazards, is an inexorable part of modern life. American geographers have long been interested in the structures and functions of urban phenomena, both within and among cities. The last example of human-environment research focuses on Los Angeles, the archetypical contemporary city. In many cities, the physical environment is a human construction, distant from nature and largely social in character. City development is understood in terms of economic, political, and social forces that structure the productive system, and the visible landscapes of the urban area are "seen as transformations of social and political ideologies into physical form" (Duncan and Duncan 1988:125). The key to deciphering Los Angeles is an understanding of the dynamics of modern capitalism.

For Allen Scott, urban research has both empirical and theoretical dimensions. Scott traced the development of the urban economy and the geography of intraurban areas as products of the division of labor and the workings of local capital and labor markets. He sees urban structures as products of modern capitalism: "The comparative advantages of industrial regions are not always based simply on given natural endowments . . . they are also *socially produced* by the internal logic of regional growth and change" (1988:59). Industrialization and urbanization are related aspects of agglomeration based on linkage costs for the movement of goods, information, labor, and intangible interactions. Contemporary developments such as vertical disintegration of the firm, subcontracting, and linkage structures within an industry create "tightly convergent complexes of interrelated economic activity," which in turn create labor markets and are closely associated through the division of labor with social space (Scott 1988:115). Thus urban industrial, commercial, and residential patterns are interdependent and spatially related by accessibility of workers to employment, yet they are distinguished by occupational divisions of labor, social class, and ethnicity.

The modern metropolis, then, is an agglomeration of human labor and capital linked through "the logic of production, the formation of local labor markets, and the dynamics of community" (Scott 1988:231). Underlying all of these are capitalist rules of order played out through local production complexes and labor markets. Scott's analysis incorporates cartographic, statistical, and graphic models, detailed empirical analyses, and prose conceptualizations. He recognizes that the scheme he offers cannot fully account for urban residential and social structures, although he suggests they are closely connected to the system of production. The environment of a city for Scott is not an ecological phenomenon; it is the historical evolution of capitalism and technology in a particular urban place.

> Where processes of the social division of labor, industrial specialization, diversification, and so on come into play, development can frequently occur even when there is no underlying resource base whatsoever. Thus, regions lacking in natural wealth . . . may sometimes take off into rapid industrial growth if an initial impulse, however fortuitous its origins, pushes them to the threshold of complex formation. (Scott 1988:59)

The Orange County high technology complex, for example, emerged with the exurban decentralization of large plants in the aerospace and electronic industries. An internal momentum of agglomeration economies furthered that expansion and created a concomitant pattern of industrial activity and division of labor, "potent testimony to the genius of capitalism for constructing and deconstructing the social and geographical conditions of its own existence as each new regime of accumulation and mode of social regulation comes and goes" (Scott 1988:201).

Edward Soja addressed urban phenomena in Los Angeles from the viewpoint of critical social theory informed by an equally critical spatial perspective (Soja 1985). For Soja, "critical historical discourse . . . sets itself against abstract and transhistorical universalizations . . . ; against naturalisms, empiricisms, and positivisms which proclaim physical determinations of history . . . ; against religious and ideological fatalisms which project spiritual determinations and teleologies" (1985:14–15). Fundamental historical and geographical questions are interpretive. Contemporary Los Angeles is understood in "the historical geography of capitalism via analyses of the evolution of urban form in the capitalist city, the changing mosaics of uneven development within the capitalist state, and the various reconfigurations of an international spatial division of labor," all cast in a "historical imagination" that ties together history, biography, and a reasserted spatiality developed from Henri Lefebvre, Michel Foucault, and others. "Just as space, time, and matter delineate and encompass the essential qualities of the physical world, spatiality, temporality, and social being can be seen as the abstract dimensions which together comprise all facets of human existence" (Soja 1985:25).

These perspectives lead Soja to see in Los Angeles a "case study in the historical geography of urban and regional restructuring" (1985:191). His interpretive essay begins a historical, economic geography vignette of the city's restructuring, deindustrialization and reindustrialization, labor market segmentation, and reorganized social space: what Soja refers to as a "technopolitan industrial complex" (1985:204). He explores the emergence of greater social, economic, and spatial disparities noting ironically that the center of the city is now its periphery, a host to foreign ownership, immigrant populations, and enterprise zones. It is a "corporate citadel of multinational capital" (Soja 1985:217). Soja notes Scott's earlier work on Orange County and focuses on similar industrialization and deindustrialization processes and on spatial and social divisions of labor. But Soja turns from analysis to visualization when he embarks on a graphic description of the visible, material landscape in a narrative spiced with reflections on history and meaning and with interpretations from the viewpoint of critical theory. Los Angeles, he argues, defies the definitions of urban and suburban; it is "a confusing collage," beneath which lies "an economic order, an instrumental nodal structure, an essentially exploitative spatial division of labor" (1985:245, 246). Soja acknowledges that his description and interpretation of the city is "purposeful, eclectic, fragmentary, incomplete, and frequently contradictory" (1985:247), but underlying his social critique is a clearly spatial, geographical perspective. His and Scott's concern is the same: the urban imprint of capitalism. The real place is the same: Los Angeles. For Scott, understanding lies in political economy and the order inherent in the dynamics of capitalism, with humanistic overtones. For Soja, meaning is revealed through critical human discourse, buttressed with political economy. As in the four previous examples, different perspectives provide distinctive insights, each addressing the very same world and mutually enriching our comprehension of it.

THE GEOGRAPHIC PERSPECTIVE

What is common to the ways geographers have studied the rich array of topics represented by these five selected examples? All societies must meet basic biophysical and social needs and all societies experience space-time challenges. Resources are distributed over space, whereas humans and their settlements are, in comparison, points. Resource availability may be concentrated in time, such as seasons of harvest, whereas human needs, particularly for food, are continuous through time. These challenges require spatiotemporal organization to bring resources and society together in space and time. Spatial organization must, as we have seen, address cycles (seasonality) and perturbations or, at the extreme, calamities. Societies may be seen as complex adaptive systems in which new and variable behavior is tested through links to a potentially changing environment.

Human societies differ in:

- their uses of nonrenewable or capital resources, which may be tapped (with little impact compared to the resource base) or mined (with significant decrease in the resource base);
- the nature and intensity of their technologies;
- their uses of energy, particularly nonrenewable versus flow resources;
- the spatial scale of their resource dependency;
- the longevity of impacts of environmental use;
- the complexity or simplicity of their primary productivity;
- their interdependence with other regions and societies;
- the distance of individuals from sustaining resources; and
- the complexity of regulation of resource-exploitation decisions.

This list of commonalities and differences in society-environment interaction is far from exhaustive, but it gives a flavor of the relationships that humankind widely shares.

Myriad society-environment issues are important to contemporary geographers, including agricultural production and food security, population growth, human perception and behavior in both natural and built environments, resource assessment and management, environmental and regional planning, tourism and leisure studies, medical geography, landscapes as cultural and ideological symbols, the cultural and environmental causes of environmental degradation, and geographical analyses of the causes and consequences of global change. One could argue that virtually any element of the relationships that bind humans and society to environment that has geographical expression is of potential geographical interest.

Human Worlds

In an intriguing essay written for landscape architects, "The Beholding Eye," Donald Meinig (1976) suggests that a landscape can be seen from different perspectives. A landscape may represent nature, a human habitat, an artifact of human activity, a system, a problem, a reflection of wealth, an expression of ideology, a history, a place, or an aesthetic reality. Meinig's various viewpoints constitute cultural subsystems, each with its own meaning or goal, each with its specific decision methods (Farness 1966). For example, landscape as nature calls for objective, dispassionate scientific investigation in the search for empirical truth. Landscape as a system calls for conscious design and improvement in the quest for efficiency. Landscape as ideology conjures an image of politics, law, and jurisprudence with their goals of equity, order, and justice—all subjectively and historically derived. History is especially important to human geographers. Processes of innovation and diffusion occur over time and are strongly conditioned by dimensions of geographical space. The diversity of geographical viewpoints results from the profound complexity of the human world. It also suggests that the understanding of each world must be sought from a different but perhaps complementary direction, and that expressions of understanding will take many forms of discourse. Whether rooted in dominant positivism, existential humanism (including phenomenology or aesthetics), Marxist or radical thought, gender perspectives, or some other paradigm, each mode of thought suggests different approaches and questions at different scales, and each mode reveals different realities.

Physical Worlds

Physical geography has its intellectual roots in the concepts of an objective, positive, empirical science and has substantive roots in the antiquity of human inquiry into nature. Physical geography—with recent roots in eighteenth- and nineteenth-century natural science, uniformitarianism, Darwinian appreciation of change through time, and the apogee of European exploration and discovery—has sought to understand the earth's environment through positivist methods. The empirical study of the environment to develop and test hypotheses as a basis for theories or laws was paralleled in post-1950s human geography with the search for laws governing spatial processes. The boundaries of physical geography are permeable, both to individuals and to the diffusion of ideas across the earth and biological sciences. With a scientific core devoted to studying processes at the earth's surface, physical geography includes a focus on understanding the environment in relationship to society, on human modification of environmental processes and patterns, and on physical bases for environmental assessment and management. Emel and Peet suggest that "the unity of geography never comes closer to realization than in resource geography" (1989:49). Physical geographers play leadership roles in development of geographical methodology, including systems theory, remote sensing, geographic information systems, and application of quantitative models.

Society and Environment

In his monumental *Traces on the Rhodian Shore*, Clarence Glacken (1967) traced the development of three persistent questions. Two date from antiquity: Is the Earth purposefully designed? Have the features of the Earth influenced the moral and social nature of individuals and cultures? By the mid–nineteenth century, ideas of a designed earth had been supplanted by the growth of science. The question regarding environmental influence as moral and social behavior competes with a third question that became prominent in the eighteenth century: How has humankind changed the Earth?

The second question provides the origin of the competing conceptualizations of human-environment interaction. Environmental determinism was a logical consequence of the ideas of evolution and positive science that prevailed in the late nineteenth century. A form of environmental determinism persists today as a mild environmentalism that includes concern over environmental effects on humans, over environmental hazards, and over the impacts of global environmental change on society through, for example, changing crop yields or sea-level rise. In its contemporary sense, environment is seen as a contributing cause that may be buffered or accentuated by human action and that can be addressed through a variety of mitigation mechanisms. A change in perspective from the dominance of environment to the primacy of society is illustrated by the concept of possibilism. Possibilism suggests that choices are available to societies in a given environment, and that similar environments could host different human adaptations. A variant of possibilism, probabilism, posits that for a given society, one possibility among others is more likely or probable. Technological materialism designates technology as the dominant factor that determines both society's use of environment and its socioeconomic complexity. As we have seen, for many human geographers the world is one created by human perception, intention, and behavior.

Geographers today conceive of human-environment interactions from the viewpoint of systems in which components are mutually interactive and causal. The roots of this viewpoint can be found in ecology and in Sauer's (1925) concept of the cultural landscape, which recognized a mutual interaction between society and environment. Sauer's geography was closely related to contemporary cultural anthropology (Solot 1986). Whereas anthropology sought to understand the origin, diffusion, and distribution of human cultural traits in their environmental context, anthropogeography focused on the cultural landscape as an artifact of human interaction with environment, an environment that is both a cause affecting society and an effect of society's activities.

Dissatisfaction with environmental determinism is also expressed in the related concepts of environmental perception and cognitive behaviorism. In a simplistic sense, environment affects humans through the intermediary of the human mind; an environment may be perceived and used by different societies and individuals in divergent ways (Brookfield 1969). The related concept of cultural ecology has been utilized by many geographers. The mode of understanding in cultural ecology is largely functional explanation, as in Bishop's (1990) study of the complex spatial and temporal components of the cultural-ecological systems in the Nepal Himalayan region. Bishop modeled the upper Karnali River subsistence system, suggesting that land degradation is caused by stress in the arable agricultural and livestock subsystems—stress that is in turn related to pressures of population growth and sociopolitical and economic change since 1950. In political ecology, society and environment interact through the organization of labor, space, and resources in a system of production. Geographers have contributed to a Marxist or radical interpretation of society-environment issues, such as population growth, pollution, drought, conservation, land use, public policy, and natural resource management. For these geographers, nature and natural resources are human constructs that reflect political and social ideology as well as economic process.

Some conceptualizations of human environmental interactions mentioned are linear (figure 2.1), whereas other schema are systemic (figure 2.2). Determinism (2.1A) posits an active environment as cause and passive society as effect. Mild environmentalism, the persistent interest in the impact of environmental events on society, treats environment as less dominant (2.1B). Possibilism (2.1C) and probabilism (2.1D) offer choices of equal or enhanced opportunity. Finally, technological materialism (2.1E) focuses on technology as the intermediary in society-environment interaction. Among systemic conceptualizations, a mutual interaction between society and environment creates the cultural landscape (2.2F). In cognitive behaviorism, human perceptions mediate between environment and society (2.2G). The connections illustrated indicate both effects and feedback mechanisms. Cultural ecology is an interacting system embracing environment, society, and technology (2.2H). Society creates and perpetuates a sustaining technology that in turn defines resources and extracts them from the environment for human use, with a resulting bio-

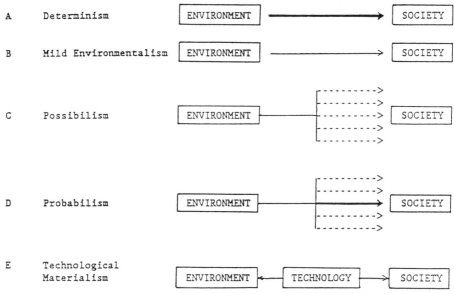

Fig. 2.1. Linear Conceptualizations of Human-Environment Interactions. Some aspects of this scheme are derived from Bennett 1976:165.

physical imprint on environment. Political ecology stresses the importance of technology and labor in creating wealth and power in societies which, in turn, define ownership and access to resources (2.2I). Finally, an adaptive systems framework includes, in addition to the familiar components, feedback and feedforward processes, and adaptive mechanisms and strategies created over time (2.2J).

THE GEOGRAPHER'S ENVIRONMENTS

For geographers, the environment is at once a resource, a hazard, and most importantly, a place. Concerning environment as a resource, geographers want to understand the natural, physical environment from the viewpoint of science. But geographers are also concerned with the preservation, conservation, and management of environment as a resource conditioned by social phenomena: custom, law, power, wealth, beauty, meaning. In ecology, everything is related to everything else. This equally applies to geography. Geographers are, by nature and by nurture, eclectic, inquisitive, and holistic. For geographers, finding order in the seeming chaos of spatial relations that tie humans and environment together requires a multiplicity of perspectives. Our queries into the challenges of society in a changing global milieu are enriched by the commonality of our purpose and the diversity of our viewpoints. In his role as president

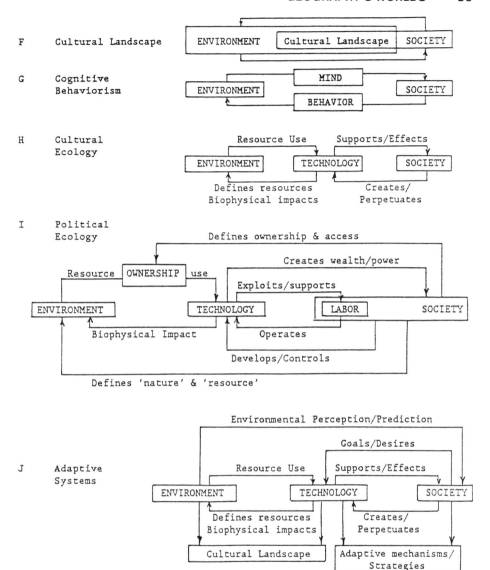

Fig. 2.2. Systems Conceptualizations of Human-Environment Interactions. Some aspects of this scheme are derived from Bennett 1976:165, Price 1977:72, and Bennett and Chorley 1978:100.

of the International Geographical Union, Akin Mabogunje (1984) argued that geography is a logical bridge between the social and natural sciences in addressing contemporary environmental challenges. He saw physical geographers as building that bridge with assistance from human geographers, who view landscape as an artifact of human-environment interaction. Mabogunje argued that geography is enhanced when its efforts are directed "towards a

much enhanced understanding of how to make the earth a better home for man"
(1984:6).

Most profoundly, environment is place—place with a content, physically
tangible and socially subtle, a locale palpably experienced by living, sentient
people in both material and cultural dimensions. Whether resource or place,
environment pervades everything geographers study, often explicitly but frequently implicitly. The human and physical environments are the geographer's
worlds. As Yi-Fu Tuan (1971:181) stated, "Knowledge of the earth elucidates
the world of man . . . to know the world is to know oneself."

REFERENCES

BASSETT, THOMAS J. 1988. The political ecology of peasant-herder conflicts in the
northern Ivory Coast. *Annals of the Association of American Geographers* 78:
453–472.

BENNETT, JOHN W. 1976. *The ecological transition: Cultural anthropology and human adaptation*. New York: Pergamon.

BENNETT, R. J., and R. J. CHORLEY. 1978. *Environmental systems: Philosophy, analysis, and control*. Princeton: Princeton University Press.

BISHOP, BARRY C. 1990. *Karnali under stress*. Research Papers 228–229. Chicago:
University of Chicago Department of Geography.

BLAIKIE, PIERS. 1985. *The political economy of soil erosion in developing countries*.
London: Longman.

BLAIKIE, PIERS, and HAROLD C. BROOKFIELD. 1987. *Land degradation and society*.
London: Methuen.

BLAUT, JAMES M., RUTH M. BLAUT, NAN HARMAN, and MICHAEL MOERMAN. 1959.
A study of cultural determinants of soil erosion and conservation in the Blue
Mountains of Jamaica. *Social and Economic Studies* 8:403–420.

BROOKFIELD, HAROLD C. 1969. On the environment as perceived. *Progress in Geography* 1:51–80.

BURNET, T. 1692. *Archaeologiae philosophicae sive doctrina antiqua de rerum ori-
ginibus*. London: Libri Duo.

BURTON, IAN, ROBERT W. KATES, and GILBERT F. WHITE. 1978. *The environment as
hazard*. New York: Oxford University Press.

CUTTER, SUSAN L. 1984. Risk cognition and the public: The case of Three Mile Island. *Environmental Management* 8:15–20.

DUNCAN, J., and N. DUNCAN. 1988. (Re)reading the landscape. *Environment and
Planning D: Society and Space* 6:117–126.

EMEL, JACQUE, and RICHARD PEET. 1989. Resource management and natural hazards.
In *New Models in Geography*, ed. Richard Peet and Nigel Thrift, 49–76. London: Unwin Hyman.

FARNESS, SANFORD. 1966. Resources planning versus regional planning. In *Future
environments of North America*, ed. F. F. Darling and J. P. Milton, 494–502.
Garden City, N.Y.: Natural History.

GLACKEN, CLARENCE J. 1967. *Traces on the Rhodian shore*. Berkeley: University of California Press.

GOULD, PETER. 1990. *Fire in the rain: The demographic consequences of Chernobyl*. Baltimore: Johns Hopkins University Press.

GRAF, WILLIAM L. 1985. *The Colorado basin: Instability and basin management*. Resource Publication 1984-2. Washington, D.C.: Association of American Geographers.

HEWITT, KENNETH. 1983a. The idea of calamity in a technocratic age. In *Interpretations of calamity*, ed. Kenneth Hewitt, 3–32. Boston: Allen & Unwin.

———, ed. 1983b. *Interpretations of calamity from the viewpoint of human ecology*. Boston: Allen & Unwin.

MABOGUNJE, AKIN L. 1984. Geography as a bridge between natural and social sciences. *Nature and Resources* 20, 2:2–6.

MARCUS, MELVIN G. 1964. *Climate-glacier studies in the Juneau ice field region, Alaska*. Research Paper 88. Chicago: University of Chicago Department of Geography.

MARCUS, MELVIN G., FRED B. CHAMBERS, and MAYNARD M. MILLER. 1990. *Glacier mapping and mass balance: Lemon Creek Glacier, Alaska*. Final report. National Geographic Society Grant 4048–89.

MATTHEWS, O. P. 1984. *Water resources: geography and law*. Resource Publication 1984–3. Washington, D.C.: Association of American Geographers.

MEINIG, DONALD W. 1976. The beholding eye. *Landscape Architecture* 66:47–54.

MIKESELL, MARVIN W. 1974. Geography as the study of environment: An assessment of some old and new commitments. In *Perspectives on environment*, ed. Ian R. Manners and Marvin W. Mikesell, 1–23. Commission on College Geography Report 13. Washington, D.C.: Association of American Geographers.

MONTZ, BURRELL, and EVE C. GRUNTFEST. 1986. Changes in American urban floodplain occupancy since 1958: The experience of nine cities. *Applied Geography* 6:325–338.

PORTER, PHILIP W. 1965. Environmental potentials and economic opportunities— A background for cultural adaptation. *American Anthropologist* 67:409–420.

———. 1976. Climate and Agriculture in East Africa. In *Contemporary Africa: Geography and change*, ed. C. Gregory Knight and James L. Newman, 112–139. Englewood Cliffs, N.J.: Prentice-Hall.

———. 1979. *Food and development in the semi-arid zone of East Africa*. Syracuse, N.Y.: Syracuse University, Maxwell School of Public Affairs.

———. 1987. Wholes and fragments: Reflections on the economy of affection, capitalism, and the human cost of development. *Geografiska Annaler* 69B:1–14.

———. 1991. Cultural ecology. In *Modern geography: An encyclopedic survey*, ed. Gary S. Dunbar, 38–39. New York: Garland.

PRYCE, REES. 1977. Approaches to the study of man and environment. Unit 2 of *Fundamentals of human geography, Section 1, Man and environment*, 45–90. Open University Media Development Group, 45–90. Milton Keynes: Open University Press.

ROBINSON, JOHN B. 1982. The quagmire of phenomena: An examination of the relationship between man and nature. In *Rethinking geographical inquiry*, ed. J. D. Wood, 177–265. Geographical Monograph 11. Downsview, Ont.: Department of Geography, Atkinson College, York University.

SAUER, CARL O. 1925. The morphology of landscape. *University of California Publications in Geography* 2, 2:19–53.

SCOTT, ALLEN J. 1988. *Metropolis: From the division of labor to urban form*. Berkeley: University of California Press.

SOJA, EDWARD W. 1985. *Postmodern geographies*. London: Verso.

SOLOT, MICHAEL. 1986. Carl Sauer and cultural evolution. *Annals of the Association of American Geographers* 76:508–520.

SOUZA, ANTHONY R. DE, and PHILIP W. PORTER. 1974. *The underdevelopment and modernization of the Third World*. Resource Paper 28. Washington, DC: Association of American Geographers.

THORNTHWAITE, C. WARREN. 1948. An approach toward a rational classification of climate. *Geographical Review* 38:55–94.

TUAN, YI-FU. 1971. Geography, phenomenology, and the study of human nature. *Canadian Geographer* 15:181–192.

WALKER, RICHARD A., and MATTHEW J. WILLIAMS. 1982. Water from power: Water supply and regional growth in the Santa Clara Valley. *Economic Geography* 58:95–119.

WATTS, MICHAEL. 1983. On the poverty of theory: Natural hazards research in context. In *Interpretations of calamity*, ed. Kenneth Hewitt, 231–262. Boston: Allen & Unwin.

WHITE, GILBERT F., ed. 1974. *Natural Hazards: Local, national, global*. New York: Oxford University Press.

WORSTER, DONALD. 1985. *Rivers of empire: Water, aridity, and the growth of the American West*. New York: Pantheon Books.

YAPA, LAKSHMAN S. 1979. Ecopolitical economy of the green revolution. *Professional Geographer* 31:371–376.

YARNAL, BRENTON M. 1984. Synoptic-scale atmospheric circulation over British Columbia in relation to the mass balance of Sentinel Glacier. *Annals of the Association of American Geographers* 74:375–392.

ZEIGLER, DONALD J., JAMES H. JOHNSON, and STANLEY D. BRUNN. 1983. *Technological hazards*. Resource Publication 1983–2. Washington, D.C.: Association of American Geographers.

Places and Regions

Bonham C. Richardson

───── ○ ─────

"If the possession of an especially comprehensive place knowledge, and the unquenchable desire to extend it, are the chief *raisons d'etre* of the geographer in the eyes of the man in the street, it is nevertheless true that we are often uncomfortable and defensive about this conception of our primary intellectual role. Unhappy that the general public should think of us in such unsophisticated terms, we are sometimes moved to a foolishly hasty denial of our basic heritage." (Clark 1962:232).

Accustomed to little publicity, most American geographers are pleased with the extraordinary media coverage given the discipline late in the 1980s. Many newspaper editorials and magazine articles revealing American geographic ignorance have suggested, naturally enough, educational remedies. Those involved in geographic education—a long-neglected subfield of geography—have suddenly found themselves in demand by school administrators, politicians, and funding agencies so that American schoolchildren might know more about the world. No one would have predicted a decade ago that an academic geographer would appear on national television for five mornings running to discuss geography's approaches to current events, as did Harm J. de Blij in late September 1989.

U.S. geographers cannot be faulted if their surprise at these events has soon been replaced by feelings of satisfaction. Many feel that they finally are receiving appropriate attention for what they knew all along were important methods and issues. It may not be long, if the flush of publicity about geography is sustained, before American geographers realize the level of respect accorded their colleagues throughout the British Commonwealth.

The attention given geography has been accompanied by a subtle reinforcement of the discipline's traditional definition in the eyes of most Americans and in the way that Andrew Clark expressed it a generation ago. Geography, to most members of the American public, is place knowledge: Alabama (the one on the right) is nearly symmetrical with Mississippi; the llanos and matto grosso are in Venezuela and Brazil, in that order; and the early trade links between Acapulco and Manila followed a clockwise direction because of the prevailing winds of the north-central Pacific. Academic geographers, as Clark suggested, may recoil at such oversimplification, but it is clear that a mandate has been delivered that professional geographers help educate the American public about places and regions of the world.

Pessimists may wonder—in more candid moments—if geographers truly have the collective expertise or inclination to fulfill these expectations. Every American geographer is aware of changes in the discipline (as also occurred in neighboring social sciences) during the 1960s and 1970s, a period during which academic geographers concentrated on spatial abstractions and quantitative data analysis. Less regional geography was done during this period when "our brightest and best . . . described us as a spatial science" (Kates 1987:526). Dissertations dealing with foreign areas were few, and a number of geography departments dropped their traditional regionally oriented "lands and peoples" courses. Foreign field research possibilities were reduced because of funding shortages, and access to Third World countries was increasingly difficult to obtain as local governments became less willing to subject their peoples to clinical investigations by overseas social scientists. More recently, the technological aspects of cartography and geographic information systems have captured the attention of many geographers and administrators, resulting in a demand for technically sophisticated geographers.

These internal and external pressures have transformed and modernized the discipline, but the American geographer's traditional focus on places and regions is alive and well. Many geographers continued to pursue studies of particular places, often while integrating new measurement techniques or geometrical perspectives into their work. A resurgence of interest in the study of place and region in geography is similar to recent changes within all the social sciences that have seen scholars turning away from physics or biology for inspiration to seek understanding from the humanities (Geertz 1980). At the beginning of the final decade of the twentieth century, the majority of American geographers would agree that places and regions, however defined, have an important role in the discipline.

FIVE EXAMPLES

This chapter examines five recent studies in American geography that exemplify major ideas and themes currently being explored within the context of places and regions. Brief comment is in order about what I do and do not attempt here. First, I present no exhaustive definition of either *place* or *region*. There has been no lack of discourse about the significance, scale, and definition of place and region in American geography, but for most of us, places and regions have been important by implication, not by our attempts to define them. Second, no discussion of *regional science* is included here. That term, associated especially with the work of Walter Isard and associates at the University of Pennsylvania, designates a specialized field of study at the interface between geography and economics. Third, I do not assess the important role that textbooks have played in maintaining the tradition of place and region in American geography. Place and region continued to provide the organization for leading high school and college geography textbooks and maintained these traditions in the classroom during the 1960s and 1970s.

Describing Atlantic America

The title of Donald Meinig's study, *Atlantic America, 1492–1800*, indicates the book's vast range and scope. It is the first of a projected trilogy, *The Shaping of America*, that eventually will cover the historical geography of North America to 1990. Meinig is now at work on the second volume, tentatively called *Continental America, 1800–1915*.

The first volume consists of four chronologically arranged parts whose titles suggest the book's flavor and substance: "Outreach: The Creation of an Atlantic World"; "Implantations: The Creation of American Diversity"; "Reorganizations: The Creation of an American Matrix"; and "Context: The United States circa 1800."

Meinig is North America's most prolific historical geographer, well-known among his colleagues for his influential books about the American West as well as a book about South Australia. He has been an unwavering devotee of traditional regional geography and one of the few attempting to define regions. Through his concepts of core, domain, and sphere he suggests that a region's cultural impulses flow outward from a core or nucleus, and he recognizes these ideas' ancestry in the work of earlier regionally oriented geographers and also of cultural anthropologists. In the preface to *Atlantic America*, Meinig explains that the volume is a study in geography that, although inseparable from history, provides "a special way of looking at the world." Geography differs from history in that the former pays "special attention to localities and regions, networks and circulations, (and) national and intercontinental systems." This geographical viewpoint, according to Meinig, is crucial for those Americans who want to understand how their present is informed by the past, an understanding that will remain incomplete without "this venerable but neglected field" of geography.

Unlike much research published by geographers, Meinig's book has been reviewed widely in the popular press and also by academics in neighboring disciplines. His comments about geography's viewpoints and the impressions his book creates, therefore, have weighty impact. For Yale historian William Cronon, reviewing *Atlantic America* in the *New York Times Book Review*, Meinig's study reinforces the public's image of geography: "Historical geography differs from ordinary history in its central fascination with place" (1986:12). In reviewing Meinig's book for the *Times Literary Supplement*, historian/essayist Garry Wills is struck by the way Meinig replaces static description with dynamic maps posted with "arrows, pointers, and traffic signs" (1987:405) indicating circulation, fermentation, and change. Wills feels that Meinig's characterization of the oppressive nature of the American frontier goes far in neutralizing the romance of individualism, egalitarianism, and opportunity that made up the frontier image fostered by the writings of the famous American historian Frederick Jackson Turner.

All reviewers of *Atlantic America* are impressed by the study's vast scope. By the term *Atlantic* Meinig does not mean simply the eastern seaboard of North America but, literally, the four continents connected by that ocean and how they have influenced America's evolution. The advantage of this mega-regional view is that it provides fresh and perhaps unique insight. For example, the early plantation settlements of South Carolina are interpreted as an African foothold on North American soil; Meinig sees these settlements not simply as the results of direct slave shipments from West Africa but as climaxes of a far-flung imperial plantation tradition involving African slaving stations, Carolina rice colonies, and intermediate experiments combining sugar cane and slavery in Brazil and the Caribbean. Meinig's grand scale also allows him to compare the more conventional transatlantic ties forged by Anglo-French farmers and fishermen in the north with those formed by contemporaneous Iberian soldiers, priests, and seekers of empire in the south.

Meinig recognizes the coexistence of several scales of investigation, regions within regions that became ever more important as Europeans gained footholds in varying locales and created distinctive landscapes based upon combinations of local environments and externally influenced commercial opportunities. The second section of *Atlantic America* consists principally of vignettes at this intermediate scale, from "Northern Coasts" to "Tropical Islands" and including the midlatitude regions in between. Although Meinig's focus rarely descends to the microscale, a source of concern for some reviewers since so much social history now deals with parish tax rolls and gender conflicts rather than kings and battles, he includes enough early prints of townscapes, city designs, and rural scenes in *Atlantic America* to people his places with flesh and blood. Meinig casts a watchful eye over the evolution of Atlantic America using a telescope that can be focused at several scales. It is, most importantly, a single telescope that can at once detect the transatlantic impulse created by a growing demand for fish in fifteenth-century Europe as well as the resultant travails of individual fishermen along the rocky coasts of Labrador and Newfoundland.

Next to breadth of coverage, the most compelling characteristic of *Atlantic America* is its description of continuous movements or circulation of people, commodities, foodstuffs, cultural traits, and the vehicles that transported all these entities. It is perhaps paradoxical to suggest that a study of regions is at the same time a study of human-induced movements, until one realizes that the early culture area or culture-circle anthropologists delineated regions and the traits therein precisely for the purpose of determining the change or diffusion that had taken place within and between culture areas. Meinig's maps are accordingly enlivened by arrows indicating continuous movement. Even his nomenclature indicates mobility. The early zones of colonization south of New England and north of Virginia—the so-called *middle colonies* described in textbooks—are labeled *entryways*. The principal direction of the arrows in Meinig's maps is, of course, from east to west because the thrust of his argument is that Atlantic America was created in the main by transoceanic expansions emanating in Old World cultural hearths in Europe (north and south) and Africa. The relentless east-to-west surge of peoples, commodities, and diseases occupies much of Meinig's narrative.

The circulation patterns depicted in Meinig's regions are not exclusively east to west, and he acknowledges that impulses from the Old World were often paralleled by countercurrents from the New. Most Europeans emigrating westward across the Atlantic stayed, but even the earliest transatlantic voyages depended upon return shipments of fish, lumber, tobacco, cotton, and raw sugar. In Europe, the New World staples represented much more than foodstuffs, stimulants, and building supplies. They inspired further plans for intensified production and resource exploitation in America. Economic planning was augmented by concerns about territorial expansion and control and geopolitical rivalries. Meinig's concentration on movement and circulation within Atlantic America suggests the ways in which regions on either side of the Atlantic were transformed by these interactions. All these movements were caricatures of change, and Meinig emphasizes geographic change as a major theme of *Atlantic America*. The evolution of America in particular was a centuries-long example "of the continuous reshaping of the geographic character of areas" (1986:xvii).

Meinig sees an underlying direction and motivation in the colonial transformation of America. That thrust was *imperialism*, another overarching theme of *Atlantic America*. Meinig realizes that the term may be controversial but sees no reasonable alternative to its use. Furthermore, territorial encroachment and subjugation of aboriginal and imported peoples was not an exclusively American trait: "Such events are apparently as old as human history" (xviii). The imperialism theme is represented in an exceptionally effective way by Meinig's regional approach. Europeans and Native Americans interacted immediately upon European arrival, and thereafter these interactions almost invariably were animated by conflict. Europeans pushed from east to west, and Indians pushed back as hard as they were able in the other direction. Meinig maps the patterns of these conflicts as *zones of encounter* (fig. 3.1). He re-

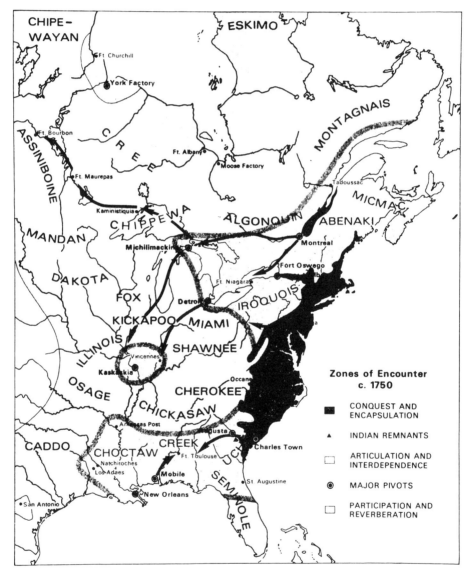

Fig. 3.1. Colonial Zones of Encounter, ca. 1750. Source: Meinig 1986:209. Copyright © 1986 by Yale University Press.

minds readers inclined to generalize to a global scale that similar encounters had taken place elsewhere earlier in history and would do so in the future. Throughout *Atlantic America* the importance of these disputed marginal areas or zones of conflict is evoked by Meinig's graphic prose. Early settlers rocked and swayed westward on their Conestoga wagons to "a ragged, bloody edge of empire" (212). And British designs of a New World empire ruptured along "geopolitical fracture lines" (384). By calling attention to these margins and edges of disputed territories, Meinig has effectively addressed the thorniest problem confronting academics dealing with regions—the near insoluble problem of boundary definition.

Within his regional and heavily empirical study, Meinig employs a number of spatial concepts that are useful for illustrating points. These are particularly evident in his maps and figures that deal with culture hearth, distance decay, and center-periphery ideas. Near the end of *Atlantic America*, Meinig presents idealized schematic maps showing linkages among Britain's transatlantic colonial empire and the *geographical morphology* of an emerging nation whose core lay along North America's Atlantic seaboard (fig. 3.2). Meinig's use of schematics exemplifies theoretical geography employed to clarify historical research.

The grace and clarity of Meinig's writing is a key strength of his work. His prose is bold, lucid, and disarmingly free of jargon. His nuanced *thick description* bridges rigorous description and explanation. Especially in part 2 of *Atlantic America*, Meinig's descriptions of the varied landscapes of colonial America sparkle with imagery that reminds many geographers of those first intellectual sensations that lured them into the discipline.

African Peasant-Herder Conflicts

Whereas Meinig's study is vast in time and scale, Thomas Bassett's (1988) is just the opposite. "The Political Ecology of Peasant-Herder Conflicts in the Northern Ivory Coast" is limited in area and confined to events of only the past two decades. Despite its narrow focus in time and space, Bassett's perspective is admirably broad because he extends the implications of his study to national and international levels.

Bassett's study region is the north central part of Ivory Coast, near the borders with Mali and Burkina Faso (erstwhile Upper Volta). It is in the savanna climatic zone of western Africa where southerly, moisture-laden winds bring heavy summer rains, conditions that contrast sharply with the low-sun season when the dry harmattan winds turn the landscape brown. Bassett assesses recent conflicts between the region's Senufo cultivating groups and the Fulani herding peoples who have moved into the area in the last few years. His analysis goes well beyond traditional explanations of conflict between West African cultivators and herders that have relied on population density issues and sim-

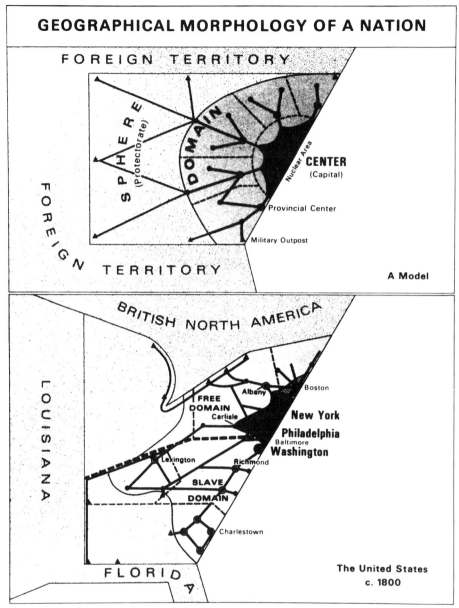

Fig. 3.2. Geographical Morphology of a Nation. SOURCE: Meinig 1986:402. Copyright © 1986 by Yale University press.

plified assumptions relating group political power to land tenure patterns. He shows that with the advent of the modern state, external power holders and decision makers have effected major changes in northern Ivory Coast. Government agents, inspired by financial debt crises to encourage local food production in order to reduce food imports, invited Fulani herders into the region. A complex series of misunderstood and mishandled range improvement and cultivation development schemes led to serious damage to Senufo crops. The resulting reprisals against the Fulani saw eighty of the pastoralists killed in the spring of 1986. Bassett suggests that a political ecology approach improves understanding of these events because it provides "detailed attention to the interrelationships between the political-economic and human ecological dimensions of production processes" (469).

Bassett never specifically mentions place or region, but his study is permeated by an intimate familiarity with the area where the peasant-herder conflicts occurred. He is unconcerned with bounding his study area; the large-scale maps in his article show his research area with a stippled overlay, suggesting that boundaries are permeable to movements of people, cattle, and ideas. Bassett can be classified as an Africanist or as a regional specialist on the basis of his work although he is not a theoretical regionalist. His deep knowledge of the study area is based on a total of seventeen months in the field.

Bassett and other field-oriented geographers have demonstrated that a researcher's familiarity with a region outside his or her culture involves lengthy observation and complete personal involvement. In a study region like Bassett's, it takes months simply to be accepted. Field research must be conducted in at least one foreign language. Interviewing is confounded by interviewees' understandable suspicions of outsiders and local (and changing) struggles for prestige, money, women, and cattle. When maps are available they are often guarded, at odd scales, badly dated, and inaccurate. Coping with unfamiliar diets and different climates continuously strains field researchers. More often than not they risk safety and health. And all these obstacles must be overcome without demonstrating fatigue or ill will, lest one's demeanor jeopardize rapport with local peoples. Even when a long field session is finished, researchers often feel that inadequate time has reduced their understanding. Further, because of such cyclical phenomena as drought, economic depression, or political strife, the field study may have been conducted during an unrepresentative period.

Field scholars like Bassett have maintained and improved a long standing tradition in geography through intensive data collection that is scientific in every respect. Because data about their regions of interest often are not available, research information has to be obtained from scratch. Bassett is not content simply to gather data concerning economic outputs and the land on which these outputs are produced. He also traces the eventual disposition of economic surpluses by assessing patterns of consumption and trade and how these patterns differ among groups and even among individuals. Terms such as *surplus transfers, extraction*, and *means of production* thereby accent his analysis, leading to notions about linkages: "While stressing the economic role of the state and its intervention in the agricultural sector, the intention has been to lay bare the mechanisms through which surpluses are systematically transferred from rural producers to different groups" (457).

Parts, wholes, and the connections between them are integrated in Bassett's work into systems, the overarching theoretical framework commonly used by cultural ecological geographers. Drawn from biological ecology and cybernetics, systems concepts have been exceptionally useful in regional geography. They encourage simultaneous scrutiny at differing scales and demonstrate that local issues are affected by, and in turn affect, more widespread events. Systemic thinking enriches Bassett's analysis in three specific ways: First, Fulani herders encroach on Senufo fields, owing to Ivory Coast government incentives and to the southward push of herding peoples by the great Sahelian Drought of the early 1970s. Second, externally influenced, money-producing agriculture tends to break up traditional lineage-based agricultural units; the resultant conjugal families therefore have less power to protest crop damage from Fulani herds. Third, local Senufo strategies to cultivate near Fulani cattle pens create impulses that move in the other direction and cause government planners to rethink the nature of their large-scale development schemes.

Bassett's system is significantly geographic because he places so much emphasis on his region of study. His work, and that of other geographers, fits neatly into the core-periphery or world-system thinking that has been influential among social scientists during the past two decades and that has been closely associated with the work of Immanuel Wallerstein. Knowledge of places is vital to world-system analysis; moreover, traditional regional geography may be said to have been infused with a modern theoretical vigor. That is the opinion of geographer Peter Taylor of the University of Newcastle upon Tyne, who feels sufficiently confident in the virtues of world-system thinking to proclaim that "a geography without knowledge of place is hardly a geography at all" (1987).

Taylor is one of several influential geographers who have called for a rebirth of a regional emphasis in the discipline, this time with an eye to ascertaining how particular places absorb and are affected by wider events. Doreen Massey pleads for a concentration on spatial processes with full acknowledgment "of the individual character and meaning of specific places and regions" because "geography . . . implies not only spatial distance but also physical differentiation" (1984:52). Locality studies recently have surfaced in the United Kingdom with incipient counterparts in the United States. They assess everyday life in small areas and how life is restructured by broad economic processes.

Bassett and other American cultural ecological geographers have helped bring regional specializations back to prominence in the discipline. Their commitment to local communities and local peoples—absolute necessities if detailed data are to be gathered—has earned them trust in foreign areas. Their interest in seeking answers to village-level problems often inspires pursuit of larger issues bearing on local problems. An important result has been a reconciliation of village-level events with international headlines that has helped bring region and place back to life in the classroom. The sorts of knowledge and expertise at the microscale possessed by Bassett and other cultural ecological geographers are immensely practical. A common conclusion or theme held by regional specialists is that a local ecosystem's particular characteristics must be thoroughly understood before externally conceived programs of development are applied. Given the important lessons their studies teach, political ecologists are likely to play greater roles in planning and development.

Rethinking the Great Plains

Planning and land-use control have long traditions in American geography. The third example of recent, region-oriented work in American geography lies squarely in the planning tradition. The work in question has been published in a series of brief articles by Frank J. Popper and Deborah Epstein Popper of Rutgers University. Their ideas are summarized in "The Great Plains: From Dust to Dust," which appeared in *Planning* in December 1987.

The region of concern to the Poppers is the short-grass plains country between the 98th meridian (running from San Antonio, Texas, in the south to Grand Forks, North Dakota, in the north) and the Rocky Mountains. The Poppers note the region's endemic economic malaise. Taking the long view, they attribute it to "the largest, longest-running agricultural and environmental miscalculation in the nation's history" (12). They cite fading energy prices, recurring drought that threatens to produce an endless series of dust bowls, and a cumulative waste of ground water that have left the vast grasslands region with a declining and aging human population that can hope for little improvement: "Agricultural market towns [in the Great Plains] get smaller, older, and poorer. Already modest downtowns become gap-toothed streets of increasingly marginal business. Entire counties lack a single doctor or a bank, and many more are about to lose them" (15).

The Poppers' solution to this regional dilemma is that the U.S. federal government—for decades directly active in the Great Plains through construction projects, irrigation works, and agricultural price supports—should begin restoring vast stretches of the Great Plains to a pre-European condition, creating a *buffalo commons* of native grass and livestock. A vast regional agency with a Plains-specific mandate, similar to the Tennessee Valley Authority, would buy back (deprivatize) large chunks of the region and place them under central control, or more accurately, a lack of control that would be more rewarding than anything "the unsuccessfully privatized Plains have ever offered" (18). The Poppers cite several studies suggesting that a restored short-grass Great Plains would support thousands of game animals that could yield tourist revenues not unlike those produced by U.S. national parks. Cities in the region, such as Lubbock and Cheyenne, would be locally self-supporting, and service centers along interstate highways would remain. But the region's small towns would wither, and the federal agency taking control of the region would provide "compassionate treatment for the Plains' refugees" (18).

The hostile reactions to the Poppers' ideas—which have been republished and summarized in several Great Plains newspapers—provide eloquent testimony to the importance of place and region in individual and collective human identities: What can transient researchers from New Jersey—a state that does not fulfill everyone's notion of a socially redeeming and well-planned environment—tell people from the Dakotas about the places where they always have lived? Newspaper editorials condemning the Poppers' ideas suggest that they are a pernicious blend of outsiders' (Easterners) meddling in local affairs and downright lunacy. Beyond the indignation over outside interference, local responses to suggestions that the Great Plains become a tourist center react to the demeaning prospect that today's ranchers, farmers, and oil-rig operators will be tomorrow's chambermaids, waiters, and tour-bus drivers. The Poppers' buffalo commons proposal has had impacts well beyond the Plains because their thesis has appeared in the *Washington Post* (which described it "a radical remedy") and has thus been syndicated in newspapers throughout the United States. The Poppers' undeniably intriguing proposal has hit a collective nerve within the American public at large. One wonders if their ideas will become a staple of in-flight magazines as was G. Etzel Pearcy's recasting of U.S. state boundaries in the early 1970s.

Much of the appeal of the Popper thesis derives from the Great Plains region itself, which, because of vastness and history, conjures imagery with universal attraction. The Plains' size and starkness made it a region to traverse rather than go to during the movement to the West, the compelling geographical trajectory in the American experience. Further, the despair and hopelessness of the depression decade of the 1930s is intimately bound up with the dust bowl and with the human movements across the Plains that touched everyone from the Midwest to California.

Academic geographers find the Popper thesis appealing for teaching purposes. One does not need to be a regional specialist to recognize the importance of the Great Plains region in the classroom; it is a laboratory ideal for exemplifying many of our concepts. Just as the Amazon Basin has become a region-as-metaphor for teaching students about wanton resource exploitation with devastating effects that extend far beyond northern Brazil, the Great Plains lends itself to classroom instruction better than any other large region of the United States. Both physical and human geographical concepts come to life on the Plains. It is the battleground between continental polar and maritime tropical air masses. When the former dominate, they produce the country's lowest temperatures, and when the latter prevail, temperatures are among the nation's hottest. Dramatic, even violent storms result from contacts between these air masses. Climatic variability, further, is no better exemplified than with runs of annual rainfall data derived from weather stations on the Great Plains. The Plains' relentless winds explain why soil erosion has been so common and so devastating there.

Classroom concepts from human geography are equally well exemplified by

case studies drawn from the Great Plains, and the Poppers' proposal is immediately appealing because the American experience in the region always has
been a combination of image and substance. The Great Plains were seen as a
desert and therefore became a zone of trails until technological innovations in
the late nineteenth century gave settlers the mechanical wherewithal to transform the region into a vast bonanza area of livestock grazing and grain production. Steel plows, railroads, barbed wire, livestock breeding techniques,
deep-water wells, and marketing telecommunication all came together in the
Plains to transform the region physically and to inspire confidence that initial
high yields always would be sustained. But this image of long-term success
was short-lived owing to a combination of local physical features and fickle
external markets.

The overarching appeal of the Poppers' proposal for the Great Plains may
be that it is ultimately a study of image or perception. It characterizes the
region for what it could be rather than what it is, in an area where actuality
always has been illusory and short-lived. Great Plains farmers have consistently underestimated drought, which in turn produced optimism, although
they also experienced a feeling of helplessness in the face of natural forces
(Saarinen 1966).

Newspaper articles usually classify the Poppers as geographers, although
only Deborah Epstein Popper is so fortunate. Frank Popper, whose advanced
degrees are in public administration and political science, chairs the Rutgers
University urban studies department. That his work is closely allied with geography demonstrates that geographers have no monopoly on studies about
place or region. Frank Popper's work, which is eminently geographic, also
shows how geography can be invigorated by insights from administration and
government.

The resource advocacy and planning strategies exemplified in the Poppers'
work follow naturally from geographers' concerns about places and regions.
Geographer-planners work actively with economists, political scientists, and
planners. As the Poppers' study demonstrates, geographer-planners also provoke controversy. Planning is synonymous with central or directed growth, and
plans that include control over regions and their peoples by outsiders are
viewed in parts of the United States as leftist, Washington-based enterprises.

Few geographical ideas have piqued the interest of as many Americans as
have the Poppers' suggestions for the disposition of the American grasslands
west of the 98th meridian. But many other geographic places and regions contain spatially defined conflicts that demand similar attention—inner-city slums
versus peripheral suburbs, Rustbelt versus Sunbelt, Third World versus the
developed world. Couple such issues with the environmental problems associated with crowding, rising pollution levels, recurring hazards or catastrophes,
and global atmospheric change, and it is easy to envisage important advising
and planning roles for American regional geographers in the future.

Wildland Fires

Policy recommendations also conclude the fourth example of recent regionally oriented work accomplished by American geographers, although policy matters are here not the heart of the research as they are for the Poppers. The research is that of Richard Minnich, a biogeographer at the University of California at Riverside. Minnich recently has written a series of articles, most notable among them a summary entitled "Fire Mosaics in Southern California and Northern Baja California" that appeared in *Science* early in 1983, that deal with the comparative frequency, extent, intensity, and histories of wildland fires in the western borderlands where the United States and Mexico come together.

Minnich identifies from Landsat imagery wildland fires that occurred in the region from 1972 to 1980. He further notes the locations and sizes of the fires, their seasons, and the affected vegetation types. Fires in Baja California during the period were numerous, relatively small, and often set by ranchers and farmers attempting to improve local grazing habitats. On the northern side of the international boundary fire suppression was much more common, yet severe wildfires raged on the California side, leading Minnich to analyze what appeared to be counterintuitive data. Fire suppression had little effect in the coastal sage scrub and grasslands of southern California, but it reduced the number of fires at intermediate elevations, where the sclerophyllous chaparral vegetation grows up to four meters high on rocky slopes of low fertility. Minnich suggests that chaparral vegetation—protected by fire suppression policies in southern California—ultimately becomes very flammable because after thirty years' growth diminished nutrient cycling leads to an accumulation of litter on the ground and dead leaves in plant canopies. He therefore concludes that further fire suppression policies in southern California, combined with the Santa Ana wind that blows seasonally off the Mojave Desert, almost guarantee large fire sizes in southern California's future. In subsequent studies Minnich has attempted to establish rough historical relationships between frequent large fires in southern California and the advent of local fire suppression practices at the turn of the century. He uses conventional history reports and Los Angeles newspapers' archives (1987).

Richard Minnich's policy recommendations—that controlled and frequent burning of the southern California wildlands may reduce the intensity and scale of the region's fires—appear to be of immediate practical value. They are consistent with a general U.S. Forest Service policy adopted in 1978 that called for controlled burning under certain circumstances. That policy is not universally agreed upon, however, and it came in for prolonged and intense critique after Yellowstone National Park's immense fire in the summer of 1988 (Pyne 1989). It follows that U.S. Forest Service policy at both the regional and national scales can profit from the kind of region-specific and historically sensitive perspective that Minnich's research exemplifies.

The regional tradition in academic geography initially was underpinned by physical differences on the surface of earth. Although American physical geographers in recent decades have focused on understanding physical processes at the expense of studying physiographic provinces or climatic regions, there remains a commitment to understanding field areas. The field tradition in geography, which in many ways involves a holistic microregional point of view, endures in current research among American geomorphologists, climatologists, biogeographers, and soils geographers.

Beyond the microscale or small region where so much physical geographic research is currently conducted, physical geographers continue to rely on a broader regional perspective that carries over into the classroom. Physical regions are delineated on the basis of processes and phenomena from the poles to the Equator, and there is no clearer global interrelationship between climatic processes and resultant regional climate than that found in the distinctive, summer-dry–winter-wet Mediterranean climate that is the setting for Minnich's research. The recommendations drawn from his observations and analysis of conditions in southern California are transferable to the Mediterranean borderlands of central Chile, the extreme southwestern tip of Africa, and southern Australia.

Minnich asserts that the regional perspective is central to his work and that it is what really distinguishes his biogeographical research from that done by botanists and ecologists. He argues that because much research in those disciplines centers on assessments of tiny areas, extrapolation of the homogeneity encountered in a ten-square-meter field plot to the larger universe can be inappropriate and misleading. For example, major differences in fire sizes between southern California and northern Baja California exist in part because of climatic subtleties within the larger region itself. The Santa Ana winds, which originate in the Mojave Desert and which are heated adiabatically as they descend into the Los Angeles Basin during the autumn months, almost never extend into northern Baja California. There, fire patterns are more closely associated with "onshore sea breezes and valley winds that prevail during summer" (1983:1292).

Minnich also relied heavily on Landsat images to determine the frequency and extent of wildland fire on either side of the border. Remote sensing is a valuable tool for the regional geographer because it provides broad coverage that "can be used to collect fundamental biophysical data" (Jensen et al. 1989:746). Minnich found that the Landsat imagery was exceptionally useful because it allowed him to determine reasonably precise boundaries between burned and unburned vegetation over a large study area. By combining Landsat imagery and conventional aerial photos, he also was able to differentiate broad physiognomic vegetation units. Using these techniques, Minnich replicated with remarkable precision the fire boundaries that had been mapped on the ground by the U.S. Forest Service.

Although Minnich's work as a biogeographer usually is pigeonholed as "physical" geography, its ultimate appeal may be that it is an example of physical geography relevant and thereby meaningful to the average, urban-dwelling American. When one considers that raging chaparral fires have been known to jump eight lanes of freeway in the Santa Monica Mountains in the northwestern part of the Los Angeles Basin, the significance of his work is obvious. Minnich's research augments the *hazards approach* in American geography (See Chapter 2). It reminds us that wildland fires also are examples of hazards affecting human communities, occasionally on a recurring basis. Importantly, it suggests that hazards in nature are not always altogether natural.

In a broader sense, Minnich's study also touches on one of the grand themes in traditional American geography: the impact of human activity on the landscape. It is hardly a revelation to suggest that the landscape of southern California has been affected greatly by human activity. But Minnich's study shows that what people may have thought was wildland is neither altogether wild nor entirely tamed or controlled by humans. Modernization has unleashed long-term processes that geographers neither understand nor fully control.

Shaping Urban Commercial and Residential Areas

The preceding examples of recent American geographic research deal with regions of varying sizes. The final example enlarges the scale to consider what most geographers would call *places*. Mona Domosh, a Florida Atlantic University geographer, recently published an article entitled "Shaping the Commercial City: Retail Districts in Nineteenth-Century New York and Boston" (1990). Her primary evidence comes from commercial registers and pamphlets published a century ago. Her secondary sources draw on all the neighboring social sciences, including architectural history, but her result is geographical because she focuses on material elements of the cultural landscape, particularly the urban built environment, to determine how social and cultural symbols originate, continue to express themselves, and provide sharp contrasts between the two cities.

Domosh contrasts the relative locations and extents of commercial retail districts and elite residential areas in Boston and New York in the late 1800s. In New York, the retail district expanded—both spatially and vertically—in a seemingly unlimited way, suggesting that "unbridled economic growth was the major goal of the economic and political systems" (296). Accordingly, the elite residential district that emerged along Fifth Avenue in New York was an extension of an earlier elite movement north from the central business district, a movement controlled mainly by land prices. Boston's smaller and more homogeneous nineteenth-century elite, in contrast, controlled the expansion of its retail district, in large part because unlimited expansion (New York–style) would have obliterated the fashionable Back Bay residential district they had built by filling the tidal flats near Boston Harbor after the Civil War. Boston's power holders eschewed the greater profits that might have come had limitless commercial expansion been allowed, and their decisions were reflected in their city's eventual urban form: "While Boston's elite were able to control the shape of the city, New York's were far too splintered for such a coordinated effort" (275). Domosh notes that national and global economic forces beyond the eastern seaboard of the United States were transforming the two cities as well, but she points out the important noneconomic ingredients in the shaping of Boston and suggests that perhaps "the inapplicability of urban economic theory . . . in late nineteenth-century Boston is far more common than theorists would allow" (280).

Cities and urban areas have received intense scrutiny by American geographers, especially in the latter part of the twentieth century. Geographers focused on a generic American city, however—with a central business district, slums, suburbs, a Manhattan metric, industrial zones, gentrification projects, racial turmoil, and political strife—an invaluable model with which particular urban places can be compared. Some of the recent reactions against such work in geography have stressed interpretative understanding. Mona Domosh's comparison of Boston and New York is an excellent example of a humanistic attempt to analyze urban patterns and forms because she proceeds beyond mere description or mere quantification to explain why particular urban forms were produced in particular places and how those forms continued to influence the inhabitants of the two cities.

Domosh's interpretation and reflection about the nature of particular places is buttressed by her historical approach. Theoretical norms developed by contemporary urban geographers have been applied and tested retrospectively. The results have underlined the importance of the human contexts of places. The historical trajectories of individual cities all followed particular paths that were neither illogical nor irrational but were based upon an uneven sequence of decisions related to particular sites, city hinterlands, and regional, national, and global markets. "History did not always happen as it should have" (Earle et al. 1989:159). It is therefore not surprising that urban places in the United States—and in the rest of the world for that matter—have distinctive characteristics.

Domosh's analysis of place lays bare the "inner workings of culture" that too many cultural geographers, focusing only on the patterns or caricatures of culture, have left to others (Brookfield 1964). Examples from her analysis of Boston help to clarify the point (fig. 3.3). That city's elite restricted and standardized the Back Bay area—strategically buffered from the commercial district by their preservation of Boston Common—to create in their residential zone "a spacious, homogeneous and ornamental environment" (275). Its architecture was planned to be conservative and anticommercial in character. Homes differed little from one another and often evoked the past. Public cultural and civic edifices such as schools, churches, and museums were enticed into the

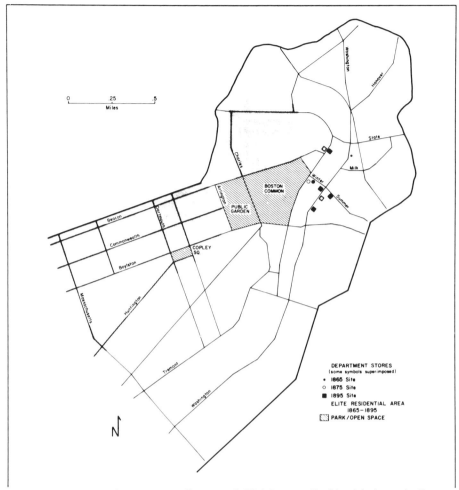

Fig. 3.3. Locations of Department Stores and High-Income Residential Areas in Boston, 1865–1895. SOURCE: Domosh 1990:273.

area to produce a tranquil, cultured atmosphere. In contrast, New York's lack of similar controls led to a variety of architectural residential forms that reflected both Boston-type conservatism and the exuberance of first-generation fortunes (fig. 3.4). "As a result, the Fifth Avenue streetscape was one of disorder, with buildings of all different sizes and shapes intermingled" (276).

One could bring up place-region distinctions by citing the obviously large size and dominance of both Boston and New York. Neither is a hamlet, and at the end of the twentieth century the two immense urban zones are intermingled into a megalopolitan colossus. But Domosh treats both cities as places; she maps the contrasts in commercial and elite residential zones within each city, and she includes late nineteenth-century photographs of contrasting individual residences. The photos make obvious the differences between the standardized

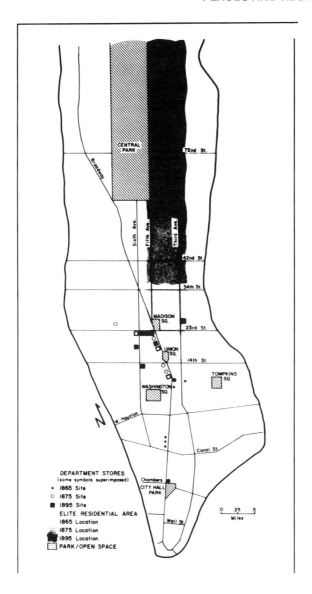

Fig. 3.4. Locations of Department Stores and High-Income Residential Areas in New York City, 1865–1895. Source: Domosh 1990:274.

conservatism of Back Bay Boston and heterogeneous Fifth Avenue. The look of the landscape is appreciated all the more because Domosh has explained how and why it came about.

Consistent with the magnified scale that her photographs provide, Domosh includes revealing contrasts between two major department stores in Boston and New York and vignettes about the different men who controlled them. Alexander T. Stewart emigrated from Ireland to New York in 1820. Before his death in 1876 he had become one of the city's richest men. His fortune was based upon his innovative siting of new department stores, notably in uptown Manhattan. His wealth derived in part from the locational risks he was willing to take. Stewart built one of the first mansions along Fifth Avenue, an opulent marble edifice with a private art gallery. Boston's leading department store during the same period was owned in part by Eben Jordan. Unlike Stewart, who located new retail outlets in different parts of New York, Jordan expanded his Boston store incrementally on the same site during the latter half of the nineteenth century; a new Jordan Marsh building was constructed only in 1907, and then on a lot adjacent to the original site. Eben Jordan, also unlike Alexander Stewart, expended a good part of his commercial wealth for public cultural endeavors, most notably as a founder of the New England Conservatory of Music. Such personal benevolence was consistent with the expectations of Boston's self-constrained elite.

Would Alexander Stewart have adopted a more constrained commercial strategy and lifestyle had he emigrated to Boston, not New York? Perhaps it would have been essential for success there. Then again, he might have moved on to New York where greater risks led to possibly greater rewards. The point of course is that Domosh, by contrasting two urban places, focuses on the people who created those places, and on their preferences, tastes, and decisions.

Domosh seems to have captured something of the contrasting personalities of Boston and New York by using conventional geographical insight and focusing on two different places. After reading her article, one has a better sense of the Boston Brahmin stereotype and of a city renowned for its cultural and educational opportunities. New York's variety, openness, and changing personality also come across. We understand better after reading her work why the latter city spawned robber barons and the former developed blue bloods. Domosh demonstrates, most importantly, that a geographical analysis of place can evoke genuine excitement and insight.

REGION AND PLACE AS COMMON THEMES

Every geographer reading this essay knows other examples and other themes that could illustrate recent work of American geographers on places and regions. Major book-length syntheses of large U.S. areas other than Donald Meinig's, such as John Borchert's *America's Northern Heartland* (1987) and Wilbur Zelinsky's *Nation into State* (1988), affirm the discipline's regional tradition for both geographers and nongeographers alike. The work of Lawrence Grossman (1984) concerning the effects of global economic penetration on the subsistence ecology of highland Papua New Guinea and Bernard Nietschmann's (1973) seminal research about similar effects on aquatic and terrestrial ecosystems of coastal Nicaragua are further examples of what Thomas Bassett and others now choose to label "political ecology." A number of geographers other than the Poppers have developed region-specific planning and policy studies, including Charles Good's (1987) research on the interrelationships between traditional and modern health care systems in East Africa and Michael Greenberg's (1987) work on human health and urban environments in U.S. cities. Peirce Lewis's (1976) eloquent study of New Orleans and Sallie Marston's (1988) insights about the development of ethnicity in early Lowell, Massachusetts, are only two among many analyses of urban places besides the one by Domosh.

Physical geographers other than Richard Minnich are pursuing new research directions with strong regional components, such as Waltraud Brinkmann's (1983) work on indicators of climatic change over the U.S. Great Lakes and Thomas Vale's (1987) investigations of vegetation changes at Yosemite National Park. Others delineate a focus on landscape type, as in Jack Ives and Bruno Messerli's (1989) analysis of Himalayan land use and ecosystems, Merlin Lawson and Charles Stockton's (1981) analysis of climate and attitudes toward it in on the Great Plains, and R. A. Sutherland's (1990) analysis of sediment storage in semiarid Kenya.

Although geographers may disagree about which studies best typify our concern with places and regions, such differences of opinion can only be healthy; they reflect the vast number of geographical studies that deal with telling issues wherein region and place are a fundamental common ground.

A promising development for future geographic research dealing with place and region would be the use of commonsense terminology. Such an unusual strategy would remind geographers, as Harold Brookfield wrote years ago, that we are on the same footing as everyone else. Geographers have much to learn from colleagues in other disciplines and from people who inhabit places and regions—and much to teach them. Geographers will do better jobs of both by speaking to people in terms they can easily comprehend.

Geographers in the past quarter-century have taken places and regions apart to see what makes them tick. Now it is time to reassemble them again; analysis must finally serve synthesis. The geographer Yi-Fu Tuan is fond of explaining to his undergraduates at the University of Wisconsin at Madison that *world* has its roots in *wer*, which means *man*, and that an expression such as *man in the world* is therefore redundant. Tuan's thinking, which often finds eloquent expression in his sensitive evocations of place and regions, is at once inspiration and metaphor for the future of American geography.

NOTE

I gratefully acknowledge the advice and assistance of Thomas Bassett, James Bohland, Susan Brooker-Gross, James Campbell, Mona Domosh, John Jakle, Peirce Lewis, Donald Meinig, Richard Minnich, Deborah Popper, Frank Popper, Vanessa Scott, James Shortridge, and Thomas Vale.

REFERENCES

BASSETT, T. J. 1988. The political ecology of peasant-herder conflicts in the northern Ivory Coast. *Annals of the Association of American Geographers* 78:453–472.

BORCHERT, JOHN R. 1987. *America's northern heartland: An economic and historical geography of the upper Midwest.* Minneapolis: University of Minnesota Press.

BRINKMANN, WALTRAUD A. R. 1983. Secular variations of surface temperature and precipitation patterns over the Great Lakes region. *Journal of Climatology* 3:167–177.

BROOKFIELD, HAROLD C. 1964. Questions on the human frontiers of geography. *Economic Geography* 40:283–303.

CLARK, ANDREW H. 1962. Praemia geographiae. The incidental rewards of a professional career. *Annals of the Association of American Geographers* 52:229–241.

CRONON, WILLIAM. 1986. The landscape of our past. *New York Times Book Review*, 17 August, 12–13.

DOMOSH, MONA. 1990. Shaping the commercial city: Retail districts in nineteenth-century New York and Boston. *Annals of the Association of American Geographers* 80:268–284.

EARLE, CARVILLE, and others. 1989. Historical geography. In *Geography in America*, ed. Gary L. Gaile and Cort J. Willmott, 156–191. Columbus, Ohio: Merrill.

GEERTZ, CLIFFORD. 1980. Blurred genres: The refiguration of social thought. *American Scholar* 49:165–179.

GOOD, CHARLES M. 1987. *Ethnomedical systems in Africa.* New York: Guilford.

GREENBERG, MICHAEL R. 1987. *Public health and the environment.* New York: Guilford.

GROSSMAN, LAWRENCE S. 1984. *Peasants, subsistence ecology, and development in the Highlands of Papua, New Guinea.* Princeton: Princeton University Press.

IVES, JACK D., and B. MESSERLI. 1989. *The Himalayan dilemma: Reconciling development and conservation*, London and New York: Routledge and United Nations University.

JENSEN, JOHN, and others. 1989. Remote sensing. In *Geography in America*, ed. Gary L. Gaile and Cort J. Willmott, 746–775. Columbus, Ohio: Merrill.

KATES, ROBERT W. 1987. The human environment: The road not taken, the road still beckoning. *Annals of the Association of American Geographers* 77:525–534.

LAWSON, MERLIN P., and C. W. STOCKTON. 1981. Desert myth and climatic reality. *Annals of the Association of American Geographers* 71:527–535.

LEWIS, PEIRCE F. 1976. *New Orleans: The making of an urban landscape*. Cambridge Mass.: Ballinger.

MARSTON, SALLIE A. 1988. Neighborhood and politics: Irish ethnicity in 19th century Lowell, Massachusetts. *Annals of the Association of American Geographers* 78:414–432.

MASSEY, DOREEN. 1984. *Spatial divisions of labour*. London: Macmillan.

MEINIG, DONALD W. 1986. Atlantic America, 1492–1800. Vol. 1 of *The shaping of America*. New Haven: Yale University Press.

MINNICH, RICHARD A. 1983. Fire mosaics in Southern California and northern Baja California. *Science* 219:1287–1294.

———. 1987. Fire behavior in Southern California chaparral before fire control: The Mount Wilson burns at the turn of the century. *Annals of the Association of American Geographers* 77:599–618.

NIETSCHMANN, BERNARD Q. 1973. *Between land and water: The subsistence ecology of the Miskito Indians, eastern Nicaragua*. New York: Seminar.

POPPER, DEBORAH E., and FRANK J. POPPER. 1987. The great plains: From dust to dust. *Planning* 53, 12:12–18.

PYNE, STEPHEN J. 1989. The summer we let wild fire loose. *Natural History* Aug., 45–49.

SAARINEN, THOMAS F. 1966. *Perception of drought hazard on the Great Plains*. Research Paper 106. Chicago: University of Chicago Department of Geography.

SUTHERLAND, R. A. 1990. Variability of in-channel sediment storage within a small semi-arid drainage basin. *Physical Geography* 11:75–93.

TAYLOR, PETER J. 1987. Editorial preface. In *The United States in the world-economy: A regional geography*, by J. Agnew, Cambridge: Cambridge University Press.

VALE, THOMAS R. 1987. Vegetation change and park purposes in the high elevations of Yosemite National Park, California. *Annals of the Association of American Geographers* 77:1–18.

WILLS, GARRY. 1987. Views from the tower. *Times Literary Supplement*. 17 Apr., 405–406.

ZELINSKY, WILBUR. 1988. *Nation into state: The shifting symbolic foundations of American nationalism*. Chapel Hill: University of North Carolina Press.

Representations of the World

David Woodward

⋯ ──────────────── ◯ ──────────────── ⋯

MAPS IN GEOGRAPHY

Of the many forms of representation geographers use to understand the world—numerical, statistical, verbal, and pictorial—maps often spring to mind as hallmarks of geographical inquiry. It is usually through them, for better or worse, that the public forms its first images of geography. Although their centrality, or even their very value to the discipline, has been challenged on several occasions, the essential links between cartography and geography in North America are stronger now than ever. This liaison continues to strengthen despite enormous technological changes and the explosion of map-making of all kinds outside the boundaries of academic geography. But whereas scholars outside geography have usually regarded maps only as tools—as means to other ends—many geographers endow maps with intrinsic artifactual importance. They assert that maps can be studied in their own right; that they have a territory of their own. Hence the inclusion of this chapter in the section on the substance of geography.

Although recognizing the value of other representations of the world that geographers employ, discussion here will center on cartographic representation. It will consider the relationship between representation and the places, regions, and things represented, the crucial roles of generalization and transformations of all kinds, the constraints that old and new technologies have placed on representations of the world, and some emerging issues that are currently being addressed.

An initial clarification: for the purposes of this discussion, the term *carto-graphic representation* designates a graphic, visual image of some sort that is manifest as an artifact. The definition of *map* adopted for *The History of Cartography* insists on this point: "Maps are graphic representations that facilitate a spatial understanding of things, concepts, conditions, processes, or events in the human world" (Harley and Woodward 1987, xvi). The word *graphic* here does not exclude tactile or three-dimensional representations of the earth. Rather, I use the term to mean a conventional graphic spatial analog: "this line means that road," or "these shaded contours mean that mountain." That the essential constituent elements of graphic maps can be coded within a data base—as a *digital map* to use the usual term—does not negate the importance of the holistic graphic image that can be derived from it. The graphic image of the data so stored provides additional meaning and insight; it is more than the sum of its parts. The point that must be remembered is that a representation of data can alter their content and meaning; it is therefore misleading to say that a map and the data base from which it is derived have identical content unless *content* is defined independent of the meaning attached to it by the user.

Cartographic representation is essentially humanistic. Despite laudable goals of objectivity, cartographic representation remains a human activity and therefore sensitive to the selective, generalizing nature of human minds. When the pioneer statistical cartographer John K. Wright wrote in 1942 that "mapmakers are human," he had no inkling of the digital revolution to come. Yet his statement is as valid today as it was then. All representation requires selection and abstraction. Representation is thus driven by human agents, and studies of representation belong to the humanities. The thoughts, feelings, and knowledge of mapmakers cannot be factored out of the cartographic equation. Choices of which maps to make and which not to make, which data to show or ignore, and which one among many methods of representation to use are, and will always be, subjective decisions. Much of the effort in digital cartography over the last twenty years has been devoted to mimicking human operations. Expert systems will succeed in cartography to the degree that they incorporate the accumulated subjective experience of cartographers.

REPRESENTATION AND REALITY

The purpose of cartographic representation is to inform someone or oneself about a geographic reality. That task is complex because the reality to be represented—the human world—is not one static, physical, and universal reality, but a multitude of dynamic, conceptual, and culture-specific and even individual realities or worlds. Cartographers have noted many times that a map or picture is not a representation of reality but a representation of ideas, usually highly conventionalized, about that reality.

The measured representation to scale that has been traditionally expected of maps reflects only one of a number of different ideas about reality. Cartographers can measure and map a material object with multiple observations, from which a mean and standard deviation can be calculated, specifying the range in which other measurements are likely to fall. But mapmakers also recognize that there are intangible or immaterial qualities of reality that should be mapped cognitively or affectively. Indeed, Wood (1973, 1977, 1978) has argued that the most appropriate measures and structures of the *cartography of reality* reside in the cognitive and affective realms of individual experience, and he shows how to construct cartographies and geometries that represent that reality. A considerable literature has developed on cognitive maps (often called *mental maps*) covering an array of concepts from preference lists for places to live through children's representations of their immediate environments (Downs and Stea 1977; Gould and White 1986).

In the 1970s, in an attempt to explain the mapping process, American cartographers embraced a model of cartographic communication that, in simplified form, involved the encoding and ordering of data about reality by the mapmaker, transmission of the resulting information (subject to technical constraints and noise) to map users, and its decoding by those users. That model diverted attention away from the data gatherer and the mapmaker and toward the transmission process; it thereby downplayed problems of data accuracy and precision and those of representation. Indeed, employment of the term *map user* connoted a barren and passive process rather than the rich, active operations of map interpretation. The communication model became a prominent paradigm for research. But few studies provided empirical data to fuel the model, and as it was progressively fine-tuned to incorporate factors such as feedback, it became unwieldy in comparison to the original schema. In the absence of substantive studies, the majority of papers on the topic summarized methodological work in a seemingly endless rehearsal of modifications to the model accompanied by appropriate flow diagrams. The model's main value was didactic.

The primary challenge to the communication model espoused the view that the important question is whether information communicated by the sender is meaningful to recipients (Guelke 1976). The meaning of a place, region, or pattern cannot be communicated without knowing context. It was therefore unlikely that a model borrowed from electrical engineering could accommodate the complexities of a human and physical landscape, even though it might promote sensitivity to the process of information transfer. In short, there was confusion among the definitions of data, information, and knowledge. *Data* are raw quantitative or qualitative facts used as a basis to create information. *Information* is data ordered and contextualized in ways that give them meaning. *Knowledge* is the cumulative understanding of information. Daniel Boorstin stated the issue clearly for the world of library and information science:

> The last two decades have seen the spectacular growth of the information industry. We are exhilarated by this example of American ingenuity and enterprise—the frontier spirit in the late twentieth century. The information industry enjoys the accelerating momentum of technology and the full vitality of the marketplace. Meanwhile, what has become of our knowledge-institutions? They anxiously solicit, and gratefully acknowledge, the crumbs . . . while knowledge is orderly and cumulative, information is random and miscellaneous. (1980)

A geographical information system contains data about the world, but real danger lies in assuming that a geographic information system is synonymous with geographical knowledge.

By focusing on the process of information transmission and ignoring the knowledge that could be interpreted from maps, the communication model reinforced the view that by simplifying space, maps inhibited human understanding of the environments they were supposed to represent. Theodore Roszak put it thus:

> But worst of all, to mistake maps for landscapes is to degrade every other way of knowing the world's terrain into some sort of illusion, and in this way to close off the richest sources of joy and enlightenment in the personality. Then we forget that to map a forest means less than to write poems about it; to map a village means less than to visit among its people; to map a sacred grove means less than to worship there. Making maps may be absorbing and useful; it may take enormous intellectual talent and great training; but it is the most marginal way of knowing the landscape. (1972:410)

Roszak's view highlights two crucial points. First, and paradoxically, it underlines the power of maps, a power that attracts hyperbolic invective and compares maps with traditionally revered forms of humanistic expression, such as poetry, fellowship, and worship. Cartographers should be pleased that their craft is spoken of in the same breath with such other noble pursuits. We could also point out that all ways of knowing a landscape require a veil of representation, so that it is not always possible to separate the representation from the represented. Second, there is another, equally valid, more positive approach to the map. Cartographers rely constantly on metaphors "to bring the unknown within the sphere of the comprehensible by an analogical extension" (Livingstone and Harrison 1980:128). It is often only when they empower maps with such metaphorical qualities that they can project experiences of place and landscape onto its face and thereby use the artifact to trigger deeper truths about reality. Then a map can become its own territory, not just a bare recording of logical geometrical relationships, but a narrative to be browsed (Wood 1987) or a repository of autobiographical memories:

> Every square inch of its paper landscape remains so familiar that it can be read at random, and almost sensed in sleep. Its place names are not just a roll call of neighborhoods, but of people, some now dead, others still crossing and recrossing the town's pavements and squares and the fields of the countryside. In such a way, the map has become a graphic autobiography; it restores time to memory and it recreates for the inner eye the fabric and seasons of a former life. . . . The map encompasses not so much a topography as an autobiography. No price can be put on this image of a familiar landscape. (Harley 1987:20)

The internal world of maps derives much of its power from the complex web of signs from which it is constituted, and some scholars have sought to probe the nature of map representation from a semiological viewpoint. The late development of this approach in the United States may be ascribed to the fact that the major works on the subject, those of Jacques Bertin and his colleagues at the Laboratoire de Graphique in Paris, were not translated into English until recently (Bertin 1978, 1983). But Bertin's qualitative approach, and his insistence on the superiority of a designer's expert intuition and logical thinking over controlled experimental studies, have continued to rest uneasily in American academic cartography. Bertin's ideas regarding visual variables—theoretically independent modifiable attributes of signs such as hue, shape, value (gray level), and size—were introduced in university cartographic teaching in the late 1970s, but his wider theory of graphics remained virtually unknown (Muller 1981). Had American academic cartography incorporated his approach, research might have progressed more in a semiotic and less in a psychophysical direction (Petchenik 1985).

The elements of semiology for cartography have been usefully summarized by Schlichtmann (1985). Map signs, like other signs, consist of content (a church) and form (a black square surmounted by a cross) linked by a code ("in this map, any black square surmounted by a cross always means a church"). Without the code, the sign is meaningless and does not exist. Map signs, unlike other kinds of signs, also include location as parts of their content (church at 43°N 93°W [globe position]) and form (a black square surmounted by a cross at x, y [projected position]).

In the context of cartographic representation, the semiology approach aims to make the form and content of a sign homologous; that is, the sign's structure should bear an identifiable relationship to its referent. The sign need not be pictorially mimetic, but it should be spatially analogous to the referent in some way. For example, it is unusual to represent a river with a single dot or a compact settlement with a line because the spatial analogy rings false. Even signs toward the abstract end of the abstract-mimetic continuum are mimetic to some degree. An apparently abstract sign such as the small black dot that represents a city is mimetic because it represents the compactness and density of its referent (Pitkänen 1981).

When visual variables have been applied to qualitative (nominal) and quantitative (ordinal, interval-ratio) map data, there has been an attempt to match them with the levels of data. Hue and shape have traditionally connoted quality, whereas value and size have connoted quantity. When a map reader's abilities and proclivities are borne in mind, however, these seemingly logical associations may not conform. Training seems to be a prerequisite for making associations, suggesting that what is logical to a cartographer may not be as logical to a user, although natural logical associations clearly help in training (Dobson 1975). In certain choropleth map reading tasks, such as matching values in the legend to counties on the map, hue expresses quantity more effectively than does value (Mersey 1984). Any analysis of the role of association in representation must, therefore, take account of the task to be performed.

A map sign assumes tacit agreement between maker and user. Within the confines of a given map, if a red line 0.012 inches wide means a state highway in one place, it means the same thing everywhere within the map border. In this narrow semantic sense, there is no place for ambiguity in map design, even if the context might make a given case implausible. For example, if the 0.012 red line were to be found running down the middle of a river, context would tell a reader that a compilation error had occurred, *not* that the red line did not represent a state highway. The power of this tacit code, and the resulting surprise when it is violated, renders mistakes on maps particularly amusing.

This semantic code approach has raised hopes for applying expert systems to cartography. If rules of syntax could be developed for making maps, an impediment to understanding their internal logic might be overcome. There have been several forays into the subject, but with the exception of an analysis of a State of Washington highway map, in which the structural properties of the map are related to its functional properties, no map has been so analyzed (Child 1984). As research on automatic feature recognition progresses, and as cartographers refine definitions of initial entities (bridge, for example), attributes of entities (steel bridge), and attribute values (14-foot steel bridge), systems that describe the internal logic of maps will be in demand (Moellering et al. 1988). As Zelinsky put it:

> In speaking of the map as alpha and omega, as the last, as well as the earliest, frontier of communication, I wish to convey the notion

that the grammar peculiar to that language known as maps has yet
to be deciphered, and that when this feat has been accomplished
there may be some quantum leaps upward in the communicative
skills of the mapmaker. (1973:4)

It is by no means clear that such developments will ever take place, or that we
would recognize them if they occurred, given the complexity, richness, and
multidimensionality of maps of all kinds. But cartographers' desires to reduce
these maps to digitizable units and to analyze them semantically will keep them
on the quest for that Holy Grail.

GENERALIZATION: IS LESS MORE OR LESS?

Calvin: "Did you finish your map of our neighborhood?"
Hobbes: "Not yet. How many bricks does the front walk have?"

It is impossible, of course, to capture all the details of geographical reality
in any representation, be it numerical, verbal, or visual. But are such representations impoverished by selectivity?

The languages we choose to employ for description, both our traditional cartographic languages, as well as our symbolic mathematical languages, seem terribly constrained and limiting. Often they
are damaging in that they not only limit, but actually crush out of
existence that which might be open to our paying heed. If we project, as we so frequently do, the multi-dimensional nature that characterizes the complexity of contemporary life on to the traditional
space of the geographic map, . . . we may well crush information
out of existence in the name of simplification. (Gould 1981:174)

There is much to be said for this view. It is possible that the dominance of the
thematic map, with its insistence on clarity and simplification, has been responsible for the view that maps are impoverished sources of information.
Claiming that "detail is often the staunchest ally of clarity," Edward Tufte
argues that humans can process enormous amounts of data visually: "The
[1749 Bretez-Turgot] Plan de Paris, and even standard road maps, illustrate
the poverty of information that can be displayed on our computer monitors,
which show only a few thousand characters or, on a very good day, a million
bits. Yet a single United States Geological Survey quadrangle sheet contains
over 150 million bits" (1988:110). Detail clearly attracts interest, as the philosopher W. V. Quine's review of *The Times Atlas* attests. Quine is fascinated

by the scale of the maps and the detail that is possible even in remote areas such as north central Siberia. "A scale of five million is thus a feast when you get that far out. . . . The maps are an inexhaustible store of lore and an unflagging delight to the eye" (1981:200, 202).

Whereas there are pragmatic needs to generalize and sample data about the earth, maps can also reflect the complexities of reality. The purpose of incorporating detail is not to obfuscate, but to stimulate interest and enable the map user to browse among the data and draw out far more connections than might be apparent. Use of space imagery to capture the delicate character of a coastline is a case in point. When the first photographs of the earth were received from space, and coastlines made familiar through generations of maps could be seen in the blink of an eye, we were struck by the intrinsic beauty of the coastal shapes. Each stretch of coast had a different character—here jutting headlands, there wispy sand bars. But in the thirty years since those first delicate filigree images were seen, what have cartographers learned about the representation of coasts? Maps—even those derived from such images—continue to exhibit a stiffness and crudeness inherent in representations constrained by the traditional line tools of the cartographer. The wispy shoreline of Cape Cod in the Landsat image is a lifeless imitation when rendered with a line tool (figs. 4.1, 4.2).

William Bunge addressed the issue of simplification as linked to the relationship between geometry and geography in his seminal book *Theoretical Geography*. Bunge concluded that maps were a subset of mathematics (1966: 38–71). It followed that if maps could be so classified, traditional maps were vulnerable. If mathematics does everything that maps do, and indeed demonstrably more, are not maps obsolete? Furthermore, if maps are mere geometry, should they not be criticized as being rigidly constrained, unrealistic, oversimplified, trivial, and generally a barrier to understanding the complexities of the human world? For Bunge, geography became the analysis of points and lines without regard to what these symbols represented (Sack 1972). Geometry is certainly one way of representing the earth in its static form, but explanation clearly requires understanding of processes and the nature of the phenomena being studied. Again, it is a question of understanding the distinction between information and knowledge.

The attraction of representing the environment with points, lines, and polygons is that such lines can be mathematically manipulated and generalized. To some extent, the adoption of the communication model had also encouraged the view of maps as transmitters of spatial data independent of their content. Thus reduced, it was claimed that geographic information systems (GIS) could map

the human brain, the way in which the AIDS virus travels through the human body or the natural environment of bacteria at a scale of

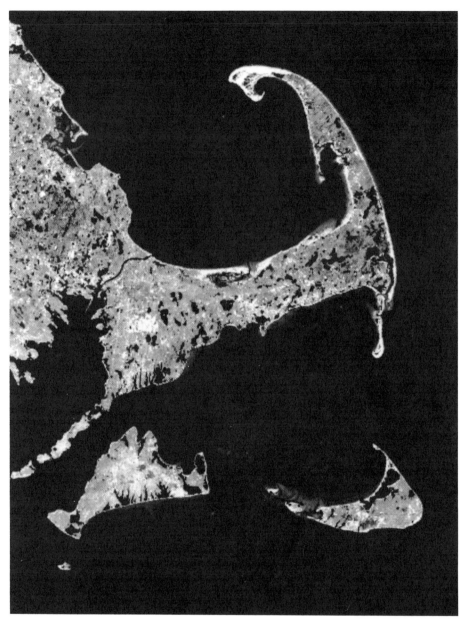

Fig. 4.1. Landsat Digital Image of Cape Cod and Environs. SOURCE: National Geographic Society 1985:45.

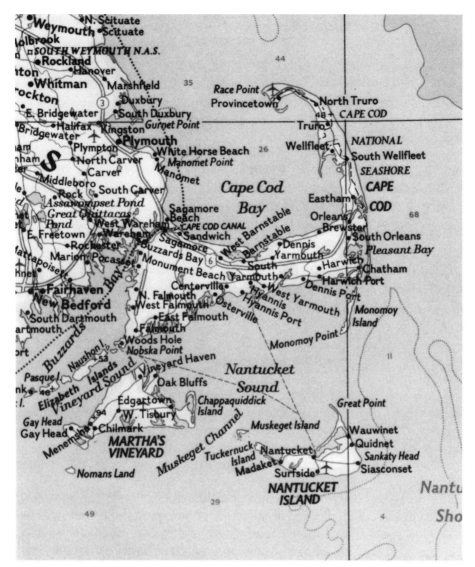

Fig. 4.2. Conventional Map of Cape Cod and Environs. SOURCE: National Geographic Society 1985:49.

perhaps 10,000,000:1. Using this technology, users may begin to map the seemingly spatial worlds of human decision making or crowd phenomena, or more effectively map the multiple dimensions of time, space and self-perception in which every human being exists. (Dangermond and Smith 1988:309)

Presumably, close familiarity with the subject matter would be necessary for such sweeping applications beyond the role of geographic mapping, and a serious question remains whether such generic representations could effectively work.

Most research in cartographic generalization in the 1970s assumed that the lines and points were generic, largely unrelated to content. In a study of how points are selected for retention in a generalized line, Marino (1979) concluded that cartographic training does not appear to influence the hierarchical selection. She supported the opinion that geometrical views of generalization are effective and that inputs by experts might not be necessary. The study, however, did not isolate subjects who had personal knowledge of any of the features selected, nor did it grapple with the problem of displacement of lines in relation to other elements. If the generalized representation of the Mississippi River, for example, does not pass between Minneapolis and Saint Paul, topological relationships between river and towns are violated.

Jenks identified two types of characteristic points for cartographic lines. The first type includes those that reflect the content of the sign—that occupy significant economic, political, or cultural locations. Examples are points at the intersections of boundaries of countries, the locations of rivers in relation to cities, and such economically important locations as San Francisco Bay and the Oslo fjord (Jenks 1981). The second type includes points that are perceived as giving the line its individual or distinctive formal structure, such as the geometrics Marino tested. Peuquet also draws a distinction between location-based information and object-based representation (analogous to the geometry/geography debate), both of which are needed for knowledge-based scene interpretation. She points out that attempts to provide a functional classification of object attributes have proved frustrating, noting that there is "a bewildering number of potential spatial relationships with seemingly infinite variations and . . . a resulting complex of often unpredictable side effects" (Peuquet 1988:390).

If geographical context were removed from consideration in line generalization, maps could be produced at any scale on any projection from the same data. One would simply apply one of the several generalization algorithms to the data. But, as experience in using large geographical data bases accumulated, it seemed that generalizing operations varied from scale to scale. One could not simply take a huge data base suitable for a series of town plans at 1:1,200 and shrink it to a reference map for an atlas at 1:5,000,000. Scale thresholds may exist at which different sign systems become necessary (Carstensen 1989).

A problem related to the generic approach involves the widespread claims being made for the use of fractal enhancement of cartographic line detail in generalization. Dutton justified the use of fractals in cartographic generalization:

But when mapmakers or map readers are presented with algorithms that add or displace coordinates, eyebrows rise. Somehow the notion of creating detail seems arbitrary, inappropriate, or untruthful. In careless hands, it may be argued, such algorithms can yield maps conveying a false sense of reality. But any map is an abstraction in which phenomena are selected, generalized, stylized, and empha- sized by the mapmaker. A map is not even an abstraction from re- ality, it is an abstraction of ideas about reality. (1981:34)

This argument makes an important point: cartographic representation depends on a set of largely conventional procedures. But the problem is more serious than Dutton believes, for the abstraction a cartographer does when reducing a line is different and less arbitrary than the procedure of using fractals to en- hance a line. The former is done to enhance the meaning; the latter either adds nothing to the meaning of the map or adds false meaning. Indeed, those who project twentieth-century values on the past ridicule the sixteenth-century car- tographer for gratuitously adding a coastline with rivers and inlets where no geographical knowledge existed (fig. 4.3).

There are justifiable uses of fractals. They are valuable, for example, as measures of the geographic accuracy or complexity of cartographic lines and planes. They are also valuable in simulating scenes that allow multiple solu- tions in planning, landscape architecture, and interior design to be visualized.

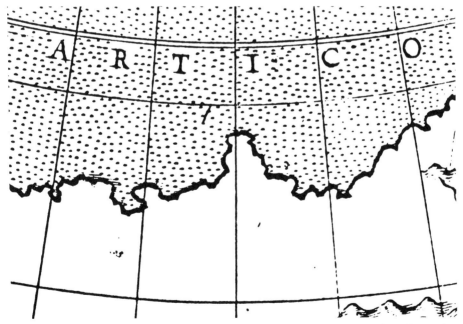

Fig. 4.3. Sixteenth-Century "Fractals." Detail of the unexplored Arctic Coast of North America on the Globe Gores. attributed to Giulio Sanuto, compiled in Venice ca. 1565. SOURCE: Private collection of Arthur Holzheimer.

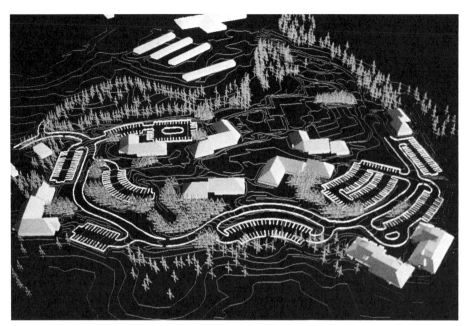

Fig. 4.4. Three-Dimensional Simulation of Buildings and Coniferous Trees Used Extensively for the Master Plan for Canyon Village at Yellowstone National Park. (Courtesy of Design Workshop, Inc., Denver, Colo.)

Their appropriateness for cartography has been argued to be simply an extension of such simulations. But it may still be objected that the simulation of aspen trees in a scene simulating a new national park facility is very different from the symbolization of trees on a conventional topographic map (figs. 4.4, 4.5). For the map, there is no claim that the tree signs represent individual trees unless it has been made explicitly clear, as in the designation of *lone tree* in figure 4.5. If the distribution of both kinds of tree signs regularly alternates over an area, the meaning conveyed is that the mix of trees is homogeneous. In the simulated scene, however, the photographic quality of the individual trees and the tacit code between maker and user implies that the trees are or eventually might be in the positions shown. If this is not true, understanding is violated and meaning is lost. In the map, it has been traditionally desirable to work back to the sources to check the authority of the work. The use of fractal enhancement, in creating detail ex nihilo, takes us farther from this goal.

The role of subjectivity in cartographic generalization is closely related to the geometry-geography issue. Familiarity with the landscape, for example, is regarded as essential to identifying important features that would otherwise be lost. An example is the inclusion of drumlin fields whose low relative relief would normally be filtered out by objective digitizing on a small-scale shaded

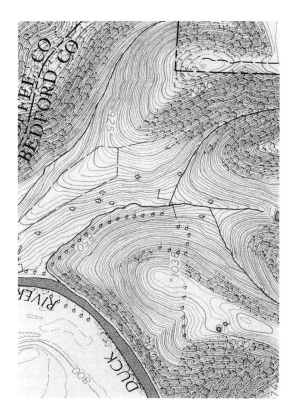

Fig. 4.5. Conventional Representation of Lone Trees, a Row of Trees, and Deciduous Forest Using Conventional Signs. SOURCE: Tennessee Valley Authority 1971.

relief map of Wisconsin (fig. 4.6). When it is possible to generate large and detailed shaded relief representations from digital terrain models, many such objections can be overcome. Furthermore, it is more likely that hidden structural trends in the landscape will emerge from the data in such detailed and less subjective representations. Details from a conventional, manually shaded relief map and one derived from a digital terrain model are compared in figures 4.7 and 4.8.

TRANSFORMATIONS

The need to distinguish the properties of projections is now more important than ever because the choice is vast and the flexibility alluring. Of the 179 main articles in the *American Cartographer* from its inception in 1974 to its change of title to *Cartography and Geographic Information Systems* in 1989, some 27 (15 percent) dealt with projection. Why this apparent importance? The iconograph power of projections is strong. The media attention on the comparative worth of the Peters and Robinson projections has stirred interest

LANDFORMS OF WISCONSIN

Geological and Natural History Survey

George F. Hanson, Director and State Geologist

UNIVERSITY EXTENSION, UNIVERSITY OF WISCONSIN

1971

0   30   60

SCALE OF MILES

Relief by David A. Woodward   Copyright © 1971 Regents, University of Wisconsin

Fig. 4.6. Landforms of Wisconsin with Relief Rendered in Charcoal Shading. Reproduced by permission of the Wisconsin Geological and Natural History Survey and the Regents of the University of Wisconsin.

in a subject that would normally leave the public quite unmoved. Whatever the flaws in Peters's arguments—and there are many—the reaction to them was extraordinarily intense (for example, Robinson 1985). Seven North American professional organizations in geography and cartography signed a resolution urging publishers, the media, and government agencies "to cease using rectangular maps of the world for general purposes or artistic displays" (Robinson 1990, Snyder 1989). The desire to fight back at the same emotional and political level that Peters initiated is understandable, but this is one of the rare occasions that a group of professional organizations has recommended censoring "artistic displays." Indignation over the correctness of projections and the layout of maps has recently extended to the Alaska Legislature, which passed a resolution "that all major United States magazines, newspapers, textbook pub-

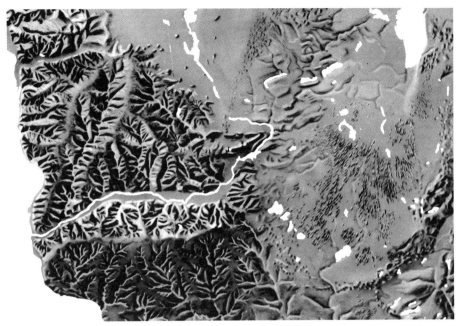

Fig. 4.7. Detail of the Devil's Lake Area. SOURCE: Wisconsin Geological and Natural History Survey 1969.

Fig. 4.8. Detail of the Devil's Lake Area. SOURCE: U.S. Geological Survey 1991.

lishers, and map publishers . . . [place] Alaska in its correct geographical position on maps of the United States, in the northwest corner, and that all geographical attributes of Alaska be correct in size and location" (Alaska 1990). These events underline the belief in the importance of physical geographical area in the status of political entities. They also illustrate that the *map* is a territory worth fighting over. Can it be long before the City Council of New York demands that its city be shown on official maps in correct proportion to its population?

Other kinds of transformations have been the focus of research in cartography. A generation of research on psychophysical perception of statistical maps sought to establish rules for perceptual scaling of cartographic signs. A circle representing 10,000 people would not be drawn to a size twice that of a circle representing 5,000 people, but to a size that would *look* to be twice the area (Flannery 1956). These studies provided students with a wealth of practice in hypothesis formulation and test design, but with important exceptions, such as studies on perceived equal steps of hue and value, they contributed only modestly to our understanding of how to design maps. The main problem was that much of the psychophysical testing treated subjects as if their responses to given stimuli were homogeneous, which was highly unlikely (Petchenik 1983). They were also tightly designed studies that held constant some of the most important variables affecting the decisions involved in making real maps.

CONSTRAINTS OLD AND NEW

The advent of automation heightened the urgency of many questions about cartographic representation. Traditional cartographic representation has so far been used as an analogue for a geographic information system. *Layers* of map overlays, for example, appear frequently in definition of the GIS. Terms emerged such as *digital map*, which implied that the visual concept of *map* was somehow embedded in a spatial data base. Furthermore, the primitives of vector cartographic representation—point, line, and area—were used from the very beginning of computer mapping (Tobler 1959) and have been the basis for many successful GIS software packages.

Several authors have argued for replacing the point-line-area approach. The technological constraints of traditional cartography have severely hampered the development of fluid and versatile tools of representation: "A technology based on lines of constant width cannot readily portray fuzziness or certainty or continuous change. . . . It is intuitively unreasonable that a technology optimized under the narrow constraints of pen and paper would turn out to be indistinguishable from one optimized under the much broader constraints of digital technology" (Goodchild 1988:311–312). There are grounds for disagreement. Traditional cartography was by no means limited to pen and paper and was far more versatile than is implied here. For example, continuous tone and color were often successfully used to represent uncertainty or continuous change, as in late eighteenth-century relief shading. It is unwise to throw out a rich tradition of cartographic representation that ranges from simple topological sketches to the complexity of multicolored representation until cartographers fully understand the variety and richness of the accumulated store of maps and the wisdom they represent. On the other hand, the unique abilities of the computer to provide a fast interactive environment are impressive. It is unlikely that an optimal state of mapping with digital technology will be identical to that of traditional cartography, but the ideas embedded in traditional cartographic representation will continue to trigger breakthroughs in digital representation.

Alongside the popular technology that encodes spatial reality in digital form for analytical purposes, there exists a parallel *digital cartography* in the prepress graphic arts industry, where some of the more exciting aspects of cartographic representation are being developed. In this technology, traditional maps are converted into digital form by high resolution raster scanners, as is common among commercial map and atlas publishers. The maps thus produced are indistinguishable from their traditional counterparts, but the analyses that can be performed on the data affect only the form or presentation of the map. These operations include manipulating hue, value, and intensity, resizing elements of the graphic, and adding special smoothing effects. Many university cartography laboratories are now converting wholesale from pen-and-ink and scribing technologies to computer aided design and drafting, illustration, and paint software. Their graphic primitives are similar to traditional cartographic primitives (Mattson 1989). The technical quality of the maps produced using these drafting programs with scanned input as a guide is often higher than that of their traditional counterparts. Their advantages lie in flexibility of representation, a capability that allows interactive experimentation, and in ease of editing. Effective use of this technology will continue to require traditional cartographic knowledge.

As increasing demands are being made on cartographic representation by a burgeoning but reductionist computer technology primarily concerned with data and information, cartographers also have a continuing responsibility to demonstrate the value of holistic thinking and to show how these data can be transformed into knowledge. The use of maps as graphic models to predict different outcomes from scenarios will continue to serve an important function in geographic research and planning. Far from limiting or constraining action, such images will reveal abundant possibilities.

Today the use of maps is rapidly becoming computerized and automated. Much storage and analysis of locations and patterns can be done within a computer and without any graphics. Meanwhile, however, computers have increased the capacity for graphic output; and the geographic method will continue to require visible maps and mental maps. Thus, as a result of this same revolution, the use of graphics for geographical communication, hypothesizing, and theorizing may well increase. (Borchert 1987:388)

The increasing interest in visual thinking in the history of science and art history and visualization in mathematics and the physical sciences, and now in geography and geographic information systems (see chapter 6), lends support to Borchert's view.

A major constraint of traditional maps has been their cumbersome representation of change. Traditional maps have been assumed to capture a slice of geographic space at a given time. This assumption overlooks the capability to show data from widely differing dates plotted on the same map and ignores the fact that even a remote sensing image relies on the sweep of a mirror or of a visual broom. At the extreme, a map of the heavens represents stars that exploded millions of years ago and millions of years apart. The impossibility of truly capturing a moment in time simply reinforces the difficulty of showing change in a systematic and logical way.

A related constraint of traditional maps has been the representation of movement. Movement can be perceived in a static map by the visually sensitive, as the experience of Harold Shapinsky, an abstract expressionist, shows: "As a small boy he spent hours poring over weather maps in the newspaper, his imagination stirred by what he calls 'the compression of distances and scale, the layering, the sense of pressures building and spilling into winds, the gracefully sweeping front lines, the three-dimensional density expressed through the simplest of abstract graphic means'" (Weschler 1985:59). But for general map readers, movement needs to be made more explicit by animation. In addition, cognitive problems of understanding maps of change, as with those showing historical development, need to be better understood. They may be different from animated multiple views of a single model (the now common fly-throughs in military and architectural simulations). Expert visual interpretation may still be the most effective approach to time-series data, as is true for a space satellite flyby. Representing change and movement is one of the most exciting research frontiers in cartography.

POSTSCRIPT

The enormous demand for geographic information systems at all levels of government, and indeed in all spheres of life that need rapid access to spatial information, will doubtless drive much of the world's mapping effort as the twentieth century closes. Cartographic representation, once thought to be a peripheral specialty, seems to be coming into its own. As the amount of raw geographical data and processed information continues to burgeon, ways to make sense of them—to turn them into spatial knowledge—become ever more critical.

As we have seen, an important part of this process is to recognize the subjective nature of representation and the need to take cognitive realities into account. The prevailing paradigm argues that mapmaking has followed a steady progression toward planimetric accuracy. Although that is broadly true, important parallel functions of maps have little to do with measured location and much to do with symbolic cosmography and iconography. When the definition of maps is enlarged to include representations of nonmetric worlds, the history of cartography shows that this dual function of maps—the measured and the symbolic—has existed at every period from prehistory to the present day, and in almost all cultures (Harley and Woodward 1987). Even the iconographic value of a GIS is emerging as worthy of note. In debates about where to locate controversial facilities in the environment, for example, a cartographic representation from a GIS can often be as powerful a tool politically as the data base is analytically.

A deconstructionist view of the history of cartography offers additional insights into current cartographic representation. It "allows us to challenge the epistemological myth (created by cartographers) of the cumulative progress of an objective science always producing better delineations of reality" (Harley 1989). Taking its ideas from postmodernist thought, deconstruction proposes an alternative epistemology based not on scientific positivism but on social theory, a "search for the social forces that have structured cartography and to locate the presence of power—and its effects—in all map knowledge" (Harley 1989:2). As developed by Harley, deconstructionist applications to cartography can be traced back to the analysis of the layers of meaning contained in art, wherein the deepest layer consists of latent iconographic meanings that reveal the basic social, religious, and philosophical attitudes of the artist (Blakemore and Harley 1980; Harley 1985). Such external codes illustrate the role maps play in creating myths by expressing subtle chauvinisms, auras of science, social consciousness, modernity, sophistication, and utilitarian messages of wealth and power (Wood and Fels 1986).

Reading "between the lines of the map . . . to discover the silences and contradictions that challenge the apparent honesty of the image" is worthwhile (Harley 1989:3). Indeed, the subject of cartographic ethics has been revived as it has become clear that messages can be blatantly manipulated to serve the compiler's ends. But while our eyes should always be open to human foibles and latent social messages hidden beneath the scientific veneer of maps, we must surely admit that being able to make and use maps to show where things are physically located has great practical value. The two positions are not ir-reconcilable; a map can be considered a product of both scientific objectifica-tion and of the social and cultural circumstances in which it was made.

Understanding the relationships among representations and multiple reali-ties, the nature of map generalization and transformation, the dangers of ge-neric reductionism, and the subtleties of cartographic ethics will enhance everyone's sensitivity to every map's own territory and its own intrinsic worth:

> A map is not normally thought of as an aesthetic object; . . . the mind is filled either with imaginations of the landscape the country would really offer, or with thoughts about its history and inhabi-tants. These circumstances prevent the ready objectification of our pleasure in the map itself. And yet, let the tints of it be a little subtle, let the lines be a little delicate, and the masses of land and sea somewhat balanced, and we really have a beautiful thing; a thing the charm of which consists almost entirely in its meaning, but which nevertheless pleases us in the same way as a picture or a graphic symbol might please. Give the symbol a little intrinsic worth of form, line, and color, and it attracts like a magnet all the values of the things it is known to symbolize. (Santayana 1936:158)

NOTE

I gratefully acknowledge the helpful advice of Nicholas Chrisman, Matthew Edney, Brian Harley, Mark Monmonier, Phillip Muehrcke, Barbara Petchenik, and Yi-Fu Tuan in the preparation of this chapter.

REFERENCES

Alaska. Legislature. 1990. *Joint resolution of the legislature No. 64 (State Affairs).*

BERTIN, JACQUES. 1977. *La graphique et le traitement graphique de l'information.* Paris: Flammarion.

———. 1978. Theory and communication of "the graphic." *International Yearbook of Cartography* 18:119.

———. 1983. *Semiology of graphics.* Madison: University of Wisconsin Press.

BLAKEMORE, M. J., and J. B. HARLEY. 1980. *Concepts in the history of cartography: A review and perspective.* Cartographica Monograph 26. Toronto: University of Toronto Press.

BOORSTIN, DANIEL J. 1980. *Gresham's law, knowledge or information? Remarks at the White House Conference on Library and Information Services, Washington, November 19, 1979.* Washington, D.C.: Library of Congress.

BORCHERT, JOHN R. 1987. Maps, geography, and geographers. *Professional Geog-rapher* 39:387–389.

BUNGE, WILLIAM. 1966. *Theoretical geography.* 2d ed. Edited by S. Nordbeck. Lund

Studies in Geography, Series C. General and Mathematical Geography, no. 1. Lund, Sweden: C. W. K. Gleerup.

CARSTENSEN, LAURENCE W., JR. 1989. A fractal analysis of cartographic generalization. *American Cartographer* 16:181–189.

CHILD, JOIS CATHERINE. 1984. *Creating a world: The poetics of cartography.* Ph.D. diss., University of Washington.

DANGERMOND, JACK, and LOWELL KENT SMITH. 1988. Geographic information systems and the revolution in cartography: The nature of the role played by a commercial organization. *American Cartographer* 15:301–310.

DOBSON, MICHAEL W. 1975. Symbol-subject matter relationships in thematic cartography. *Canadian Cartographer* 12:52–67.

DOWNS, ROGER, and DAVID STEA. 1977. *Maps in minds: Reflections on cognitive mapping.* New York: Harper & Row.

DUTTON, GEOFFREY H. 1981. Fractal enhancement of cartographic line detail. *American Cartographer* 8:23–40.

FLANNERY, JAMES J. 1956. *The graduated circle: A description, analysis and evaluation of a quantitative map symbol.* Ph.D. diss., University of Wisconsin, Madison.

GOODCHILD, MICHAEL F. 1988. Stepping over the line: Technological constraints and the new cartography. *American Cartographer* 15:311–319.

GOULD, PETER. 1981. Letting the data speak for themselves. *Annals of the Association of American Geographers* 71:166–176.

GOULD, PETER R., and RODNEY WHITE. 1986. *Mental maps.* 2d ed. Boston: Allen & Unwin.

GUELKE, LEONARD. 1976. Cartographic communication and geographic understanding. *Cartographica* 13, 2:107–122.

HARLEY, J.B. 1985. *The iconology of early maps.* In *Imago et mensura mundi: Atti del IX Congresso Internazionale di Storia della Cartografia,* ed. C. C. Marzoli, 29–38. Rome: Istituto della Enciclopedia Italiana.

———. 1987. The map as biography: Thoughts on ordnance survey map, six-inch sheet Devonshire CIX, SE, Newton Abbot. *Map Collector* 41:18–20.

———. 1989. Deconstructing the map. *Cartographica* 26, 2:1–20.

HARLEY, J. B., and DAVID WOODWARD. 1987. *The history of cartography: Cartography in prehistoric, ancient, and medieval Europe and the Mediterranean.* Chicago: University of Chicago Press.

JENKS, GEORGE F. 1981. Lines, computers, and human frailties. *Annals of the Association of American Geographers* 71:1–10.

LIVINGSTONE, DAVID N., and R. T. HARRISON. 1980. The frontier: Metaphor, myth, and model. *Professional Geographer* 32:127–132.

MARINO, JILL S. 1979. Identification of characteristic points along naturally occurring lines: An empirical study. *Canadian Cartographer* 16:70–80.

MATTSON, MARK. 1989. Desktop mapping at Temple University. *Cartographic Perspectives* 3:3–13.

MERSEY, JANET ELIZABETH. 1984. The effect of color scheme and number of classes on choropleth map communication. Ph.D. diss., University of Wisconsin, Madison.

MOELLERING, H., L. FRITZ, T. NYERGES, B. LILES, N. CHRISMAN, C. POEPPELMEIER, W. SCHMIDT, and R. RUGG. 1988. The proposed national standard for digital cartographic data. *The American Cartographer* 15:9–142.

MULLER, JEAN-CLAUDE. 1981. Bertin's theory of graphics: A challenge to North American thematic cartography. *Cartographica* 18, 3:1–8.

NATIONAL GEOGRAPHIC SOCIETY. 1985. *Atlas of North America: Space age portrait of a continent.* Washington, D.C.: National Geographic Society.

PETCHENIK, BARBARA BARTZ. 1983. *A map maker's perspective on map design research.* In *Graphic communication and design in contemporary cartography,* ed. D.R.F. Taylor, 37–68. New York: John Wiley and Sons.

———. 1985. Review of Jacques Bertin, *Semiology of graphics. American Cartographer* 12:71–72.

PEUQUET, DONNA J. 1988. Representations of geographic space: Toward a conceptual synthesis. *Annals of the Association of American Geographers* 78:375–394.

PITKÄNEN, RISTO. 1981. *On the analysis of pictorial representation.* Amsterdam: North Holland.

QUINE, W. V. 1981. *The Times Atlas.* In *Theories and things,* 199–202. Cambridge: Belknap.

ROBINSON, ARTHUR H. 1985. Arno Peters and his new cartography. *American Cartographer* 12:103–111.

———. 1990. Rectangular world maps—no! *Professional Geographer* 42:101–104.

ROSZAK, THEODORE. 1972. *Where the wasteland ends: Politics and transcendence in postindustrial society.* Garden City, N.Y.: Doubleday.

SACK, ROBERT DAVID. 1972. Geography, geometry, and explanation. *Annals of the Association of American Geographers* 62:61–78.

SANTAYANA, GEORGE. 1936. *The sense of beauty.* New York: Scribner's.

SCHLICHTMANN, HANSGEORG. 1985. Characteristic traits of the semiotic system 'map symbolism.' *Cartographic Journal* 22:23–30.

SNYDER, JOHN P. 1989. Geographers and cartographers urge end to popular use of rectangular maps. *American Cartographer* 16:222–221.

TENNESSEE VALLEY AUTHORITY, 1971. *Normandy Dam 1:12,000.* Sheet 86 NW-2.

TOBLER, WALDO R. 1959. Automation and cartography. *Geographical Review* 49:526–534.

TUFTE, EDWARD R. 1988. Attention to detail, or less is a bore. *PC/Computing,* Nov. 110–115.

U.S. GEOLOGICAL SURVEY. 1991. Shaded relief from 1:250,000 Digital Elevation Model.

WESCHLER, LAWRENCE. 1985. A strange destiny. *New Yorker,* 16 Dec. 42–84.

WOOD, DENIS. 1973. The cartography of reality: First thoughts on the cartography of reality. North Carolina State University School of Design, Charlotte. Typescript.

———. 1977. The geometry of ecstasy: More on the cartography of reality. North Carolina State University School of Design, Charlotte. Typescript.

————. 1978. What color is the sky? An introduction to the cartography of reality. North Carolina State University School of Design, Charlotte. Typescript.

————. 1987. Pleasure in the idea: The atlas as narrative form. *Cartographica* 24, 1:24–45.

WOOD, DENIS, and JOHN FELS. 1986. Designs on signs: Myth and meaning in maps. *Cartographica* 23, 3:54–103.

ZELINSKY, WILBUR. 1973. The first and last frontier of communication: The map as mystery. *Bulletin of the Geography and Map Division, Special Libraries Association* 94:2–8,29.

PART II

WHAT GEOGRAPHERS DO

Observation

Philip J. Gersmehl
Dwight A. Brown

─── ○ ───

It was six men of Indostan
To learning much inclined,
Who went to see the Elephant
(Though all of them were blind),
That each by observation,
Might satisfy his mind.

Geographers avidly observe their worlds. They ask for window seats on airplanes; they choose less-traveled roads and prefer not to drive at night; they arrange their professional meetings in exotic places; they strike up conversations with strangers about what they see around them; and they continually compare what they observe with what they remember from other places or have seen on maps.

Careful observation of the landscape is not unique to geography, but it is certainly a distinctive geographic trait. An adequate record of the location of something observed is all that is needed to make observation geographic; the phenomena being observed can include giant rodent fossils (McPhee, Ford, and McFarlane 1989), modern plants (N. Brown 1989), habitat-plant associations (Dodge 1989), epidemic diseases (Cliff and Haggett 1989), historic bank notes (Black 1989), modern use of police weapons (Fridell 1989), private education (Bradford 1989), innovative retail methods (Martin and Cain 1989), and the amount of foreign investment in U.S. cities (Bagchi-Sen 1989).

Observations that include locational information provide the raw materials for visualizing and analyzing spatial patterns and for modeling the processes that may be responsible for those patterns (see chapters 7 and 8). Despite its pervasiveness among geographers and the apparent simplicity of observing, attempts to describe geographic observation almost immediately get embroiled in questions of prior knowledge and theory and how they affect what geographers see in the world. The purpose of geographic observation is simply to record the traits and locations of features with enough precision for valid pattern visualization and analysis. To see this as a linear process, however, is to overlook two fundamental truths of observational science: (1) the need to understand something about what is being observed in order to know what and how to observe; and (2) the need to observe in order to acquire that understanding. In the real world, observation and theory are entwined in a tight spiral that sometimes seems hopelessly circular: theory and predisposition affect what we observe, and vice versa (Brush, 1989).

Consider an innocent-sounding question: "How do you know that's a loblolly pine?" Questions of that kind are often asked by people who are just learning to observe trees, but the answer to this particular question was hardly what the student expected: "I know it's a loblolly pine because it looks like a loblolly pine. And, I suppose, because I know loblollies are likely to grow on that kind of site." That response shows how easily a seasoned observer can gloss over knotty problems of taxonomy, inference, context, and extrapolation—topics that should be part of any discussion of geographic observation. The purpose of this chapter is to look into those issues and to see how they apply to geographic observation. The chapter is organized as a sequence of comments on seven goals of geographic observation. To evade semantic problems, the following definitions have been adopted for phrases that often have fuzzy meanings:

Study area	Region of observation (e.g., Detroit, corn belt)
Observational unit	Individual observation (e.g., household, field)
Reporting unit	Aggregation of sites (e.g., census tract, county)
Phenomenon	What we observe (e.g., income, corn)
Measurement unit	Used in observing (e.g., dollars/year, bushels/acre)
Category	Range of measurements that characterize a given reporting unit in terms of the observational units within it (e.g., 16–20% below poverty line, medium corn yield)

With those definitions in mind (at least for the duration of this chapter), let us examine the goals of geographic observation. Relevant articles published in 1989 will illustrate key points, provide a cross section of the kinds of geographic observations currently being made, and illustrate the kinds of publications in which they appear.

RIGOROUS OBSERVATION

The First *approached the Elephant,*
 And happening to fall
Against his broad and sturdy side,
 At once began to bawl:
"God bless me! but the Elephant
 Is very like a wall!"

Most geographic observation relies on a small number of widely accepted ways to measure or classifying things. These methods were not handed down to a group of awestruck protogeographers on a mountaintop. In most cases they were forged over a long period of time, often in heated debates involving several disciplines. Recent examples include Dorn on rock varnish (1989) and Potter on squatter settlements (1989). Ironically, observation is often most acute during times of uncertainty and debate, whereas accepted taxonomies and observational units are often the hallmarks of a period of lazy observation, as Sauder (1989) shows for early pioneers and as Skidmore (1989) demonstrates for modern digital terrain analysis. In effect, people probably saw loblolly pines more clearly when they were trying to decide what qualifies as a loblolly pine; now that there is a widely accepted definition of the species and some experience in seeing it in various settings, people no longer observe loblolly pines as carefully as they once did.

In most cases, such complacency is justified; after all, one purpose of science is to make statements that need not be reconsidered every generation. But one of the mandates of this chapter is to make sure that we don't forget how we got this far. To do that, we must ask what comprises an accepted taxonomy of observational units and, in some cases, what is an appropriate observational unit. Loblolly pines, for example, have the advantage of existing in discrete units that are easy to distinguish, whereas an individual soil, climate, linguistic region, or ethnic neighborhood is much more difficult to describe.

By definition, classifications are human inventions, sets of words devised to impose order on observations and thus to make observing more efficient. First, observational units must be defined and then observed in all of their complexity; once categories are defined, one can get by simply by noting membership in a particular class. It is much easier to look at a house and write down its category name—dog-trot log cabin, postwar bungalow, Georgian row house, craftsman gothic—than to try to describe all of the features of something like a New Orleans shotgun bungalow, let alone a gentrified Victorian mansion. How accurately those observations communicate, however, depends greatly on the degree to which readers are familiar with the images implied by terms such as *Victorian* or *gentrified*. Familiarity, in turn, requires that observers go through at least two distinct phases of consciously matching landscape features with suggested names—once as a discipline and thereafter as individuals.

Some scholarly groups have gone quite far in the direction of establishing rigid standards for observation. For example, a Class A weather station results from conscientiously following a list of detailed specifications for instrumentation, enclosure, site, and time and method of data recording. As a result, one can examine data from a Class A station anywhere in the world and make reasonable inferences about the comfort of the air in a nearby town. Hidden inside that sentence is an important point: the data reported by a Class A weather station do not tell us the actual temperature even five feet away from the thermometer. Air is a spatially continuous phenomenon, which means that it does not come packaged in easy-to-separate units that can be named and described as individuals (like pine trees). What *is* known is that the thermometer in a Class A station was designed, built, exposed, and read in a standard way, and, therefore, a given measurement (for example fourteen degrees Celsius) is likely to bear a predictable relationship to the temperatures at other sites near the station (Lahiri 1989).

Much less rigorous are supposedly standard taxonomies such as the SIC (Standard Industrial Classification). It is widely used by economic geographers to record observations about industrial buildings, and it is decidedly nonstandard. An observer notes the primary economic activity within a building by recording a two-, three-, or four-digit number, depending on the detail desired in the study. For example, all enterprises that are engaged primarily in processing food fall into SIC category 20. Those that handle dairy products go into subcategory 202 and makers of ice cream fit into group 2024.

At first glance, the SIC appears to serve geographers well. It allows easy comparison of the mix of industries in various regions; it affords a way of tracking industrial change within an area through time; and it provides a framework for evaluating linkages among firms within an area (Pollard 1989; Scott and Mattingly 1989). However, closer examination reveals some serious problems that are clearly illustrated in the dialogue begun by Chang (1989) and continued by Reid and O'hUallachain (1990) and Chang (1990). Many difficulties arise because the Census of Manufactures ultimately relies on a firm's definition of itself. That definition may be ambiguous in an age when a single highly leveraged conglomerate may produce automobile mufflers, potato chips, and pantyhose (an actual example). Asking that kind of firm to classify itself invites all kinds of error. Trying to get it to subdivide itself into smaller, homogeneous entities only begs an even messier question: any given category of well-known household products includes items produced by different industrial processes. Take, for example, a wristwatch. Is it primarily a precision mechanical instrument, an integrated electronic chip, a product of a jeweler's imagination, or a fashion accessory? At least some watches would fit into (or be excluded by) any one of those categories, and the categories themselves fall into different SIC classes.

At a different scale, consider a world-renowned manufacturer of super computers. The enterprise may be big enough to affect the industrial profile of an entire metropolitan region. It appears at first glance to fit solidly in SIC 3573 (computing equipment), but certainty crumbles with the realization that most of its patents fall into the realm of chemical engineering (refrigerant technology, SIC 3585) because of the need to maintain safe temperatures in its tightly packed box of purchased electronic chips. Moreover, most employees are engaged in writing software for a machine with standard parts but a radically new structure. Those employees should therefore be counted in SIC 7372, which differs from 3573 even at the first-digit level. In another corner of the same industry, a small manufacturer of sink strainers made a minor adjustment to its stamping machine and started turning out cabinets for computer disk drives. Is it still a metalworking firm or has it become part of the electronics industry?

Pervasive ambiguities in the Standard Industrial Classification are especially troublesome because of the nondisclosure requirements placed on the census. To avoid revealing information about individual firms, the census reports only the taxonomic categories and general size statistics of firms within a census tract or an even larger reporting unit. The individual firms (observational units) remain hidden, and the dimensions of the reporting units are beyond the control of the investigator.

The Standard Industrial Classification is a creaky old boat, riddled with wormholes caused by technological changes, corporate takeovers, and census tract redefinitions. Yet its advantages are precisely the same as those of the Class A weather station. The observer of the industrial scene may not record the actual industrial process, but the observation is made in a standard way, which a knowledgeable observer can correlate with the process that is taking place within the factory walls. In short, the advantages of assumed comparability seem to outweigh the problems of frequent but usually predictable misclassification.

Geographers trying to observe human predispositions to particular kinds of behavior have an even less standardized taxonomy. Consider those who study human perception of (and response to) natural hazards. They often use an open-ended questionnaire in which the respondent completes a sentence that begins with a phrase such as "During a hurricane warning, I . . ." or "Getting ahead in the world is a result of. . . ." Over the years, researchers have noted that people who make particular choices or use particular words when answering these questions often exhibit predictable degrees of fatalism or aggressiveness when they confront environmental problems. Those conclusions must be tempered with the observation that subtle differences in the meanings of words from one place to another can cast doubt on conclusions about attitudes. Despite such problems, semantic-differential or word-association tests provide useful windows into the reasons for differing human behavior in dissimilar places.

Survey research that requests responses to multiple choices or to a standard interview form (Clark 1989) may appear, at first glance, to be less vulnerable to these ambiguities. One should consider, however, how much knowledge of probable responses is needed to design a suitable set of questions or a reasonably exhaustive list of alternative responses. For this reason, survey research often includes two phases: a pilot phase or exploratory study that determines the range of probable responses, and a final study that establishes the statistical validity of the results (Valentine 1989).

The final example in this section is the U.S. Department of Agriculture Soil Taxonomy, a set of mutually agreed criteria that govern soil resource surveys. Soil scientists have had to wrestle with the fact that a soil profile results from a host of independent influences ranging from climate and local topography to the mechanical aptitude of the person who tilled the soil fifty years ago and the psychological propensity for risk avoidance of the present landowner. Because of this complexity, soil specialists have spent more time than most other scientists working out the details of their classifications. The result is a taxonomy that does better than most in making its presumptions and criteria explicit. Geographers could learn much by watching soil surveyors work; one soil scientist can pick up the results of another's observation and draw reasonably accurate conclusions about the capability and limitations of a particular tract of ground. In the grand scheme of things, that is a commendable accomplishment, a tribute to what a mutually agreed-upon set of categories and procedures for observation can produce.

The moral: Have a standard ruler in hand before trying to describe the dimensions of an elephant's side. A good observation should be rigorous enough that another person can reconstruct the essential nature of the phenomenon from a recorded observation. Such rigor usually implies using accepted measuring instruments, observational units, and taxonomic categories. When interpreting an observation, remember that a taxonomy inevitably reflects current theoretical understanding of the world, which makes the proper definition and interpretation of taxonomic categories a never-ending task. The more that is learned about the world, the more there is need to change categories to reflect new knowledge.

REPRESENTATIVE OBSERVATION

The Second, *feeling of the tusk,*
Cried, "Ho! what have we here

So very round and smooth and sharp?
To me 'tis mighty clear
This wonder of an Elephant
Is very like a spear!"

The deepest intellectual wellspring of geography is our acute awareness that truth can depend on location: a theoretical statement may be valid only under specific conditions that exist in a particular place. In the real world, this means that geographers usually begin by assuming that at least some of whatever they think they know in one place is likely to be wrong elsewhere. That awareness is part of the reason geographers have done so much theoretical and empirical research on questions of scale and sampling. This research is a quest for theoretical guidance in choosing the appropriate spatial scale at which to observe a phenomenon that may exhibit different kinds of variation at different scales.

Selecting an appropriate observation strategy must begin with a clear statement of the goal of the observation. A few good photographs may be enough to demonstrate a difference in the degree of architectural homogeneity in different neighborhoods. As Larry Ford (1989) has argued, architectural uniformity is often associated with demographic homogeneity, which, in turn, implies boom-and-bust cycles and resulting inefficiencies in the use of public facilities such as schools, parks, and medical centers. Identifying differences in the appearance of housing, however, is much easier than trying to delimit areas in which a specified degree of architectural diversity is the rule. The difficulty of this task is one reason there is a large and rapidly growing body of literature dealing with problems of regional delimitation, problems that are now seen as common in a variety of topics ranging from urban morphology (Ladanyi 1989; Slater 1989) and voter behavior (Shelley and Archer 1989) to landform analysis (Kemmerly 1989) and soil resource appraisal (Gersmehl, Baker, and Brown 1989).

A sizeable fraction of that literature directly addresses the topic of spatial autocorrelation—the tendency for neighboring phenomena or data values to be similar to each other. Statisticians often view spatial autocorrelation as a major obstacle to making valid inferences about a population (Odland 1988). Their research shows that random sampling of a spatially autocorrelated phenomenon is likely to produce errors in estimating population means or in correlating one phenomenon with others. For geographers, this particular bogeyman is seldom frightening because spatial autocorrelation is precisely what they hope to see on their maps. For example, a statistically valid estimate of the average density of population in San Bernardino County, California, is far less interesting to a geographer than the simple mappable fact that almost all of the people live in one corner of the county (fig. 5.1). Awareness of this kind of strong autocorrelation within various counties can improve the interpretation of county-based national maps of other phenomena, such as occupational structure, average income, or ethnicity.

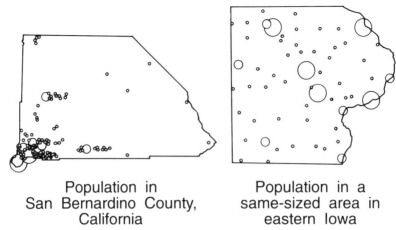

**Population in
San Bernardino County,
California**

**Population in a
same-sized area in
eastern Iowa**

Fig. 5.1. Population Patterns in Iowa and California. Mutually reinforcing patterns of uneven accessibility and the restricted availability of water have produced a strongly autocorrelated pattern of population in San Bernardino County, California. By contrast, the essential homogeneity of the resource base and historic patterns of transportation have combined to produce a poorly autocorrelated distribution in a similar-sized area in eastern Iowa.

A pretest of observational methods can help minimize scale and autocorrelation problems, although designing such a test is far from a trivial matter. Alternatively, geographers can draw on a rich tradition of regional analysis, which has as one of its goals the delimitation of reasonably homogeneous areas, within which a given observational strategy will deliver similar results (Ebbe 1989; Gallant et al. 1989). For a geographer, the goal is to know how much a new environment must differ from a known one before another pretest of observational methods is needed.

A broadly trained geographer can draw on a wide range of clues that raise flags when an investigator enters unfamiliar territory and needs to be warned to pay more attention to the frequency and method of observation. For example, the presence of a number of deer-hunting clubs may indicate a larger than average proportion of land owned by industrial corporations, with a host of implications concerning access, local tax structure, rural housing patterns, school issues, and infrastructural needs (Mason 1989). Differences such as these may be apparent in tabulated census reports or in plat books, but noting the locations of a particular form of recreational service can provide the impetus to look at tabular data. That coin also has a flip side: a computer search for unusual ratios among different kinds of census data can suggest places where field observation is warranted. For example, comparing housing tenure patterns can help isolate areas where particular kinds of protest movements are likely to win popular support (Drakakis-Smith 1989).

The moral: Scan the whole elephant before devoting the notebook to a description of one tusk. A good geographic observation should truly represent the phenomenon under study. The inherent variability of most phenomena and the rich nuances of spatial association with other phenomena make the selection of representative observations a challenging task.

SIMULTANEOUS OBSERVATION

The Third *approached the animal,*
    And happening to take
The squirming trunk within his hands,
    Thus boldly up and spake:
"I see," quoth he, "the Elephant
    Is very like a snake."

The mobility of the trunk through time might have been the decisive clue that helped the third blind man discriminate between what he thought was a snake and what might otherwise have been deemed a fire hose. Knowing the time of observation is one key to valid interpretation of data on a wide variety of topics, from plankton productivity (Baird and Ulanowicz 1989) to governmental investment (Grier 1989). Timely observation is especially important when the purpose is to compare places; one cannot draw valid conclusions by comparing traffic flow on one freeway during the afternoon rush with similar observations of vehicles on another freeway after midnight.

Atmospheric scientists were among the first to require simultaneity for their basic observations. *Zulu Time* (formerly Greenwich Mean Time) is the internationally accepted clock that governs meteorological observations and their forwarding to forecasting centers, airports, television and radio stations, and private consultants around the world. Having to translate Zulu Time to local time is a minor inconvenience when weighed against the advantage of having simultaneous observations to use in the models that produce weather forecasts. The need for simultaneity of observation has long been a major justification for expensive remote-sensing platforms, whether they were balloons in 1906 or satellites and high-altitude airplanes almost a century later (Baker 1989; Ilbery and Evans 1989).

Weather satellites have geosynchronous orbits that keep them over the same place at all times. The data they gather are therefore simultaneous for the areas they observe. LANDSAT and SPOT satellites fly at lower altitudes in order to record smaller areas in greater detail. Even so, their areas of observation are large by human standards, and the thirty seconds that LANDSAT needs to record its observation of a single 185-by-185–kilometer scene is, for most practical purposes, instantaneous. This illusion of simultaneity begins to break down when we try to compare data for places that are not adjacent to each other. For example, the image for an area just north or south of a given LANDSAT scene is recorded a few seconds earlier or later, but the images for areas immediately to the east or west are captured on a completely different day or even in a different week, depending on orbit geometry of the number of satellites in the network.

The *sun-synchronous* orbits of these satellites are designed so they pass over the ground at almost exactly the same time of the day, which ensures that the east-west angle of the sun will be roughly identical in all scenes. The latitudinal position of the sun, however, is inevitably different, which can pose problems in trying to compare reflectivity from supposedly identical surfaces. The calendar of a phenomenon can add other sources of confusion: a 1 July scene may catch farmers out cutting hay in one part of a scene, but similar fields will mature two weeks later at the poleward end of the same scene. The spectral signatures of hayfields in these two locations will differ because a mown field looks different from an uncut one. In short, data within a single scene may be chronologically simultaneous, but that does not automatically mean that they are operationally comparable.

The Census Bureau is like LANDSAT in one important respect; both try to make standard observations within a time window that is narrow enough to permit direct comparison of places. Census observations require enormous amounts of human effort and high levels of cooperation at all scales, from individual to national or even international (Wise 1989). The resulting cost has an unfortunate consequence: a full-blown census usually occurs only at rather long intervals of time. Some of the resulting problems are easy to cite:

1. Criteria between categories can change. Should the definition of *poverty* be indexed to an inflating dollar or kept numerically comparable from decade to decade?
2. Places can change. How can one account for the fact that the census shows a place as an area of dilapidated housing, even though our eyes tell us it has been razed by a condominium developer?
3. People can move. Should census tracts be combined or subdivided in order to keep them comparable in size? And if so, can comparability with previous census information be maintained?
4. Topics may become irrelevant. Now that most homes have radios, a formerly important question about that consumer product has become almost meaningless. Should the census now ask about home robots or some other technological innovation?

Clearly, analysts and policy makers need reliable methods of updating census data and interpolating between census years (King and Killingbeck

1989; Kloos and Adugna 1989). But even when definitions stay nominally constant and reasonably valid, timing can affect the validity of census results. Cattle, for example, may be born on an Arizona range in spring, spend summers grazing in Wyoming, and eat alfalfa hay in California or sorghum in a Kansas feedlot the next winter. The month of an agricultural census, therefore, can affect perceptions of the importance of cattle in various regions. Likewise,the timing of a population census can misrepresent the residential patterns of migrant workers, homeless people, or the elderly, and as a result, the services that cater to motor homes, time-share condos, and "Winter Texans."

The moral: Recognize the futility of trying to picture a whole elephant from a baby elephant's tooth, a bull elephant's ear, and the vulture-picked remnants of a dead elephant's leg. Observations should be truly simultaneous if their purpose is to compare places, and the unfortunate truth is that simultaneity is more than a simple chronological fact.

TIMELY OBSERVATION

The Fourth *reached out his eager hand,*
And felt about the knee.
"What most this woundrous beast is like
Is mighty plain," quoth he;
"'Tis clear enough the Elephant
Is very like a tree!"

Observers must also take spatial and temporal cycles into account in trying to decide how, where, and when to make observations. One of the clearest examples of this principle is the idea of mean sea level, the datum against which many features, from tides to land elevations, are measured. The elevation of the surface of the sea varies with a host of nested cycles:

1. Ripples, which change sea level every fraction of a second
2. Waves, which change sea level every few seconds
3. Wave groups, which change sea level every few minutes
4. Tides, which change sea level several times a day
5. Moon phases, which change sea level from week to week
6. Seasonal cycles, which change sea level from month to month
7. Oscillations such as El Niño, which change sea level from year to year

8. Weather cycles, which change sea level from decade to decade
9. Changes in solar income, atmospheric dynamics, erosional processes, glacier size, and continental position, which can change
   sea level over hundreds, thousands, or millions of years

Despite all these variations, mean sea level serves as a datum for topographic
mapping because geographers and geodesists have a mutually agreed-upon
convention that ignores them. But is it any wonder that there is little consensus about whether the greenhouse effect is observable in the level of the
sea (Pirazzoli 1989)? In that light, the presidential and congressional proposals
for a crash two-year program of research on climatic change seem implausible
at best. To most climatologists, a valid observation of temporal trends requires comparable measurements throughout several relevant cycles, and the
mean duration of some of those cycles is orders of magnitude longer than two
years.

What is needed is a simple and accurate *deflator*, a factor that can adjust for
cycles and trends and thus make it possible to compare with confidence observations made at different times. Take an example from an industry notoriously
inaccurate in its perception of trends: adjustments for the growth of the market
and the changing value of the dollar should be made when the gross revenues
of recent films such as *Batman* or *Star Wars* are compared with the box office
returns of an older classic such as *Gone With the Wind*.

More and more observers are becoming aware of the need for deflators when
comparing places (Rosine and Walraven 1989; Soule and Meentemeyer 1989).
The quest after comparability does not end there, however; there may be
geographic differences in the validity of the trend equations and probability
functions that are used in trying to interpret data through time. Chipanshi
(1989) has noted this need for drought frequency, and Sherman, Griffin, and
Buerger (1989) see a similar need for crime indices. Consider, for example,
the phenomenon often described as the *baby boom*, a persistent postwar demographic wave with a wide variety of consequences for housing, savings rates,
and governmental programs (Amel and Jacowski 1989). The gross shape of a
demographic wave should not blind us to the fact that a given neighborhood
can have a unique demographic cycle that is driven primarily by the average
size of its houses and the dates of their construction; this kind of deviation
from national or global patterns can affect the daily, weekly, and seasonal
cycles of use of neighborhood facilities such as medical facilities or parks (Jim
1989).

*The moral: Watch an elephant walk a few steps before deciding that a leg
always touches the ground.* Geographic observations should be either lengthy
enough to capture the normal range of variations in a phenomenon, or carefully
adjusted to compensate for such variations. Unfortunately, knowledge of this
variation requires prior observation, another of the inevitable circularities that
help make observation such a challenging task.

SPATIALLY SIGNIFICANT OBSERVATION

The Fifth, *who chanced to touch the ear*
 Said: "E'en the blindest man
Can tell what this resembles most;
 Deny the fact who can,
This marvel of an Elephant
 Is very like a fan!"

Geographers who gather information for computerized geographic information systems (GIS) have had to wrestle with a kind of uncertainty principle.
Heisenberg's original principle dealt with subatomic particles, for which, he
asserted, a measurement of position automatically precludes an accurate measurement of velocity, and vice versa. The geographic equivalent of this dilemma involves an inherent contradiction between data-gathering procedures
that seek to minimize error in either (1) describing traits or conditions at a
specific location (the *tag approach*), or (2) assessing the quantity or spatial
extent of something within a specified area (the *count approach*).

Accurate description of individual sites is prerequisite to site-specific activities such as tax appraisal or zoning. The tag method of data recording is site
specific and usually requires some form of areal measurement to determine
what category of a phenomenon, such as vegetation cover or ethnic group,
occupies the largest area within an observational unit. In creating a tag-based
GIS, observers are obliged to ignore small or irregularly shaped features. Such
features do not dominate an observational unit and cannot fairly be said to be
valid descriptors of the unit as a whole. At the same time, they are part of
the landscape, and there are situations in which one might want to know
about them.

The alternative is a count approach, in which the observer uses a point-
sampling strategy to make a statistically valid estimate of the total quantity of
some phenomenon within a large reporting unit. Foresters have been among
the most creative users of this approach, partly because trees rarely complain
when sampling does what it inevitably must do occasionally, which is to place
a sample point in the middle of a tiny pond in a large pasture and then to record
the entire data cell as water. With budget-imposed limits on the number of
sample points, such misclassification of the dominant traits of a few individual
data cells (observational units) is the price that must be paid for a valid estimate
of the areal extent of particular phenomena within an entire region (fig. 5.2).

Problems arise when a tag-based GIS is used to answer a count kind of
question, or vice versa. In the 1970s, the state of Minnesota designed a statewide GIS to answer a number of tag questions. It chose to observe land cover
by recording whether forest, pasture, cropland, wetland, or urban features oc-
cupied the largest fraction of each forty-acre cell in the statewide survey grid.

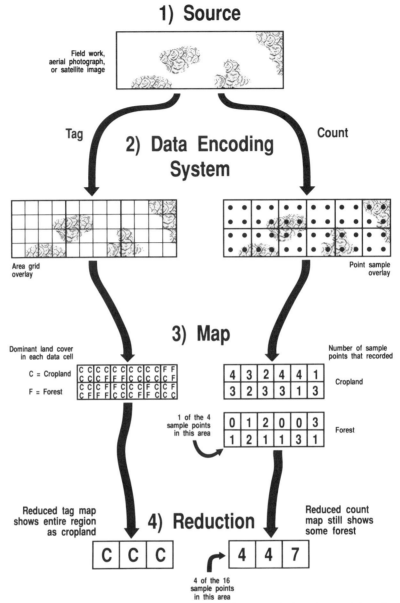

Fig. 5.2. The Process of Encoding Data into a Medium-Resolution Geographic Information System. The method of data encoding depends on initial assumptions about the kinds of questions the GIS should answer. At virtually any resolution above the microscopic, choosing the tag method of data encoding will yield major errors in answering count questions, and vice versa. SOURCE: C. Gersmehl et al. 1987.

This system produced some spectacular maps of land cover for the entire state, but a tag-based system of this kind fails miserably when faced with a count kind of question such as the number of acres of duck habitat in a given county. In many parts of Minnesota, the undulating glacial moraine is pockmarked with thousands of potholes. These small wetlands are prime waterfowl habitat, but a typical pothole is seldom more than a few acres in size and is therefore unlikely to be the dominant land cover in a forty-acre grid cell. The inevitable undertabulation of the extent of that particular resource in the Minnesota GIS can easily approach an order of magnitude (table 5.1).

The problem is less easy to see, but potentially even more serious, when tag maps of different phenomena are combined electronically to answer a count question as, for example, when a planner overlays maps of soil type and ownership patterns in order to find out how much Class IV land is in the hands of a particular group of owners. Unfortunately, vendors and users of GIS software seldom make systematic field observations to test the reliability of their overlays. Most are content to devote the largest share of their evaluation budget to assessing the accuracy of mechanical procedures such as data entry or digitizing. In most modern systems, these procedures have a potential error of a few percent, while an order-of-magnitude greater error can lie buried in original assumptions about the end purpose of the data base (Brown and Gersmehl 1989; Heuvelink, Burrough, and Stein 1989).

The moral: Avoid the trap of measuring ear-sized things if you're trying to count an elephant's eyes. The appropriateness of observational methods often

Table 5.1. Tag Maps and Count Questions

Minnesota County	Soil category[a]	Area depicted by Tag maps (Km²)	
		Atlas[b]	Survey[c]
Cottonwood	Poorly drained	140	670
Freeborn	Organic	70	130
	Poorly drained	620	920
Lyon	Poorly drained	310	440
Stevens	Poorly drained	190	790
Wright	Organic	30	220
	Poorly drained	30	310

SOURCE: *Gersmehl and Brown*
[a]Count question concerns the extent of poorly drained soils within a political unit
[b]*Minnesota Soil Atlas,* 1:250,000 (medium-coarse resolution)
[c]County soil surveys, 1:15,380 or 1:20,000 (medium-fine resolution)

depends on whether the goal is to describe the features of individual reporting units (the tag approach) or to make an inventory of the total amount of some phenomenon within a study area (the count approach). Unfortunately, in a world of limited budgets that decision often must be an either-or choice. Even if some kinds of data are obtained at a resolution detailed enough to answer both kinds of questions with acceptable accuracy, different kinds of data with different resolutions must often be combined to answer a question.

FULLY RELATIONAL OBSERVATION

The Sixth *no sooner had begun*
About the beast to grope,
Than, seizing on the swinging tail
That fell within his scope,
"I see," quoth he, "the Elephant,
Is very like a rope!"

Most geographic observations are made for purposes of comparison. For example, weather is observed to compare it with plantings and thus derive an estimate of potential yield. Or, in a classic map reproduced as fig. 7.2, the spatial pattern of water pumps was observed to compare it to the distribution of cholera deaths. Ibiebele and Sokari (1989) provide a modern answer to the same question. Such comparisons often leave us vulnerable to spatial versions of the ecological fallacy, which is a shorthand term for what happens when analysts assume that each individual within a reporting unit has all the traits described for that unit. For example, a census might show that a given tract has a high average income and a large number of domestic servants; a naïve interpretation might conclude that domestic service must be a high-income occupation.

The ultimate escape from the ecological fallacy is *punctual relationality*: gather all relevant information from precisely the same observational unit. Achieving punctual relationality is elusive, even in physical geography where the observational units often stay in one place and do not object to repeated measurements.

Comparing data that may not come from the same observational units is much more problematic in human geography (especially for someone trying to use off-the-shelf data), because a statutory nondisclosure rule does not allow the Census Bureau to release data that can be traced to identifiable individuals. A census without such a rule could not be tolerated in a free society, but because the rule exists, all statistical comparisons of census data must be viewed with a good deal of caution. For this reason, the Public Use Samples (and the

Manuscript Censuses, which are released after a hundred years) have become important tools for research in social and economic geography (Pudup 1989). These samples are strictly relational; all data come from individuals, even though their locations are deliberately left ambiguous.

For those dealing with questions that are more recent than the 1890s or more topical than the questions in the Public Use Samples, the path out of the ecological trap will involve either an expensive longitudinal study of individuals (Liaw and Wort 1989) or a sound theory that informs observation. (A good hunch will also do.) If one suspects, for example, that exposure to an environmental feature causes a disease, one may delimit reporting-unit boundaries to make it possible to compare samples taken at different distances from the feature. To make the case even stronger, samples and measurements should come from the same houses in which the presence or absence of the disease is recorded. (Philips 1989 begins an intriguing trio of articles that address the ethics as well as the mechanics of medical observation.) At the very least, there should be an effort to make observational units congruent with the areas in which the phenomena of interest are situated. As the scholars examining the results of the first free election in the Soviet Union concluded: "It is urgently necessary to quickly begin gathering socio-cultural and economic statistics by electoral district" (Berezkin et al. 1989:633).

The moral: Verify that the tail just pulled is indeed connected to the elephant that just trumpeted so angrily. Good geographic observation should strive to obtain all data from the same observational units, so that the results can be correlated without fear of nonrelationality; in the real world, however, justifiable concerns about confidentiality make that kind of fully relational observation difficult.

UPDATABLE AND EXTENDABLE OBSERVATIONS

> *And so these men of Indostan,*
> *Disputed loud and long,*
> *Each in his own opinion*
> *Exceeding stiff and strong,*
> *Though each was partly in the right,*
> *And all were in the wrong!*

Geographic observations, by their very nature, are vulnerable to at least four qualitatively different kinds of obsolescence:

1. Locational. Things can change position. People may migrate to the Sunbelt.

2. Taxonomic. Things can change their traits. A factory can start making a new product.
3. Quantitative. Things can change in degree. Maize yields may decline in an area.
4. Inferential. Consequences can change as human ideas change. New technology has made a waste rock called taconite a prized mineral resource.

The National Resources Inventory is a good example of an institutionalized method of providing longitudinal data that can answer questions about change through time. This program of the Soil Conservation Service has identified specific tracts of land for reexamination at intervals in order to assess long-term changes in the pattern of land use in the United States. Plat books or the Sanford fire insurance maps are a similar resource for urban areas; their detailed records of individual landholdings or buildings at a specified time are easy to compare with modern observations.

The moral: Mark the elephant so that it can be reexamined when inadequacies in earlier observations become apparent.

OBSERVING CAREFULLY

A good map is one of the best ways of passing the substance of observations on to the next generation of geographers—if the map or the accompanying text is clear about the procedures that were used in making the observations. Here, we end on a somber note: good examples are rare because geography's record of documenting the procedures as well as the results of its observations is not as good as it should be. The recently established policy of the National Science Foundation requiring that all data be archived in electronic form as a condition of grant approval is one step in the right direction. But taxonomically sound, timely, relational, and well-documented observation will remain rare until it receives the same stature accorded a well-conceived and persuasively argued theory.

REFERENCES

AMEL, D. F., and M. J. JACOWSKI. 1989. Trends in banking structure since the mid-1970s. *Federal Reserve Bulletin* 75, 3:120–133.

BAGCHI-SEN, SHARMISTA. 1989. Foreign direct investment in U.S. metropolitan areas, 1979–1983. *Urban Geography* 10:121–137.

BAIRD, D., and R. E. ULANOWICZ. 1989. The seasonal dynamics of the Chesapeake Bay ecosystem. *Ecological Monographs* 59:329–364.

BAKER, S. 1989. San Francisco in ruins: The 1906 aerial photographs of George R. Lawrence. *Landscape* 30:9–14.

BALL, R. M. 1989. Some aspects of tourism, seasonality, and local labour markets. *Area* 21:35–45.

BEREZKIN, A. V., V. A. KOLOSOV, M. E. PAVLOVSKAYA, N. V. PETROV, and L. V. SMIRNYAGIN. 1989. The geography of the 1989 elections of people's deputies of the USSR (preliminary results). *Soviet Geography* 30:607–634.

BLACK, I. 1989. Geography, political economy, and the circulation of finance capital in early industrial England. *Journal of Historical Geography* 15:366–384.

BRADFORD, M. 1989. Spatial polarisation of private education in England. *Area* 21:47–57.

BROWN, DWIGHT A., and PHILIP J. GERSMEHL. 1989. Geographic information systems, data, and water resources. Special issue, *Journal of the Minnesota Academy of Science* 55:14–17.

BROWN, N. 1989. ITE and digital mapping since 1973. *Cartographic Journal* 26, 2:91–95.

BRUSH, S. G. 1989. Prediction and theory evaluation: The case of light bending. *Science* 246:1124–1129.

CHANG, K-T. 1989. Japan's direct manufacturing investment in the United States. *Professional Geographer* 41:314–328.

———. 1990. Reply to Reid and O'hUallachain. *Professional Geographer* 42:226–227.

CHIPANSHI, A. C. 1989. Analysis of rainfall probabilities to determine maize species suitability: An agrometeorological study of Zambia. *Singapore Journal of Tropical Geography* 10, 2:110–118.

CLARK, WILLIAM A. V. 1989. Revealed preferences and neighborhood transitions in a multi-ethnic setting. *Urban Geography* 10:434–448.

CLIFF, A. D., and PETER HAGGETT. 1989. Spatial aspects of epidemic control. *Progress in Human Geography* 13:315–347.

DODGE, S. L. 1989. Forest transitions and buried glacial outwash within the beech-maple region of Michigan. *Geografiska Annaler* 71A, 3–4:137–144.

DONLEY, M. W., S. ALLAN, P. CARO, and C. P. PATTON. 1979. *Atlas of California.* Culver City, Calif.: Pacific Book Center.

DORN, RONALD I. 1989. Cation-ratio dating of rock varnish: A geographic assessment. *Progress in Physical Geography* 13:559–596.

DRAKAKIS-SMITH, D. 1989. Urban social movements and the built environment: An analysis of housing provision in North Australia. *Antipode* 21:207–231.

DRUCKER, E., M. P. WEBBER, P. A. McMASTER, and S. H. VERMUND. Increasing rate of pneumonia hospitalizations in the Bronx: A sentinel indicator for human immunodeficiency virus. *International Journal of Epidemiology* 18, 4:926–933.

EBBE, O. N. I. 1989. Crime and delinquency in metropolitan Lagos: A study of "crime and delinquency area" theory. *Social Forces* 67:751–765.

FORD, LARRY. 1989. Personal communication.

FRANKLAND, P. and S. AIROLA. 1978. *Atlas of selected Iowa services.* Iowa City: University of Iowa Press.

FRIDELL, L. 1989. Justifiable use of measures in research on deadly force. *Journal of Criminal Justice* 17, 3:157–165.

GALLANT, A. L., T. R. WHITTIER, D. P. LARSEN, J. M. OMERNIK, and R. M. HUGHES. 1989. *Regionalizations as a tool for managing environmental resources.* Corvallis, Ore.: U.S. Environmental Protection Agency, Environmental Research Laboratory.

GATRELL, A. C. 1989. On the spatial representation and accuracy of address-based data in the United Kingdom. *International Journal of Geographical Information Systems* 3:335–348.

GERSMEHL, CAROL A., DWIGHT A. BROWN, and PHILIP J. GERSMEHL. 1987. "Point sampling vs. area overlay in a water-resources GIS." Poster session at the 1987 Portland meeting of the Association of American Geographers.

GERSMEHL, PHILIP J., and DWIGHT A. BROWN. 1987. Maintaining relational accuracy of geocoded data in environmental modeling. In *GIS '87, the Second Annual International Conference* 1:266–275 San Francisco.

GERSMEHL, PHILIP J., B. BAKER, and DWIGHT BROWN. 1989. Effects of land management on "innate" soil erodibility: A potential complication for compliance planning. *Journal of Soil and Water Conservation* 44:417–420.

GRIER, K. B. 1989. On the existence of a political monetary cycle. *American Journal of Political Science* 33:376–389.

HEUVELINK, G.B.M., P. A. BURROUGH, and A. STEIN. 1989. Propagation of errors in spatial modelling with GIS. *International Journal of Geographical Information Systems* 3:303–322.

IBIEBELE, D. D., and T. G. SOKARI. 1989. Occurrence of drug-resistant bacteria in communal well water around Port Harcourt, Nigeria. *Epidemiology and Infection* 103:193–202.

ILBERY, B. W., and N. J. EVANS. 1989. Estimating land loss on the urban fringe: A comparison of the agricultural census and aerial photograph/map evidence. *Geography* 74:214–220.

JIM, C. Y. 1989. Changing patterns of country-park recreation in Hong Kong. *Geographical Journal* 155:167–178.

KEMMERLY, P. R. 1989. The karst contagion model: synopsis and environmental implications. *Environmental Geology and Water Sciences* 13, 2:137–143.

KING, RUSSELL, and J. KILLINGBECK. 1989. Carlo Levi, the mezzogiorno and emigration: Fifty years of demographic change at Aliano. *Geography* 74, 2:128–143.

KLOOS, H., and A. ADUGNA. 1989. The Ethiopian population: Growth and distribution. *Geographical Journal* 155:33–51.

LADANYI, J. 1989. Changing patterns of residential segregation in Budapest. *International Journal of Urban and Regional Research* 13:4555–4572.

LAHIRI, M. 1989. Physioclimate of Indian cities. *Singapore Journal of Tropical Geography* 10:27–42.

LIAW, K-L, and S. A. WORT. 1989. Intraurban mortality variation and income disparity: A case study of Hamilton-Wentworth region. *Canadian Geographer* 33:131–145.

LOWEL, K. E., and J. H. ASTROTH, JR. 1989. Vegetative succession and controlled fire in a glades ecosystem: A geographical information system approach. *International Journal of Geographical Information Systems*, 3:69–81.

MARTIN, D., and S. CAIN. 1989. Factory outlet retailing. *Urban Land* 48, 4:14–18.

MCPHEE, R. D. E., D. C. FORD, and D. A. MCFARLANE. 1989. Pre-Wisconsinian mammals from Jamaica and models of late Quaternary extinction in the Greater Antilles. *Quaternary Research* 31:94–106.

MASON, D. S. 1989. Private deer hunting on the coastal plain of North Carolina. *Southeastern Geographer* 29:1–16.

ODLAND, J. 1988. *Spatial autocorrelation*. Newbury Park, Calif.: Sage.

O'MAHONY, M., A. LAKHANI, A. STEPHENS, J. G. WALLACE, E. R. YOUNGS, and D. HARPER. 1989. Legionnaires' disease and the sick-building syndrome. *Epidemiology and Infection* 103:285–292.

PIRAZZOLI, P. A. 1989. Present and near-future global sea-level changes. *Paleogeography, Palaeoclimatology, Paleoecology* (global and planetary change section) 75:241–258.

POLLARD, J. S. 1989. Gender and manufacturing employment: The case of Hamilton. *Area* 21:377–384.

POTTER, R. B. 1989. Urban housing in Barbados, West Indies. *Geographical Journal* 155:81–93.

PUDUP, M. B. 1989. The boundaries of class in preindustrial Appalachia. *Journal of Historical Geography* 15:139–162.

REID, N., and B. O'HUALLACHAIN. 1990. Comments on "Japan's direct manufacturing investment." *Professional Geographer* 42:223–226.

ROSINE, J., and N. WALRAVEN. 1989. Drought, agriculture, and the economy. *Federal Reserve Bulletin* 75, 1:1–12.

SAUDER, R. A. 1989. Sod land versus sagebrush: Early land appraisal and pioneer settlement in an arid intermountain frontier. *Journal of Historical Geography* 15:402–419.

SAXE J. G. 1882. The blind men and the elephant. In *The poetical works of John Godfrey Saxe*, 111–112. Boston: Houghton Mifflin.

SCOTT, ALLEN J., and D. J. MATTINGLY. 1989. The aircraft and parts industry in southern California: Continuity and change from the inter-war years to the 1990s. *Economic Geography* 65:48–71.

SHELLEY, F. C., and J. C. ARCHER. 1989. Sectionalism and presidential politics: Voting patterns in Illinois, Indiana, and Ohio. *Journal of Interdisciplinary History* 20:227–255.

SHERMAN, L. W., P. R. GARTIN, and M. E. BUERGER. 1989. Hot spots of predatory crime: Routine activities and the criminology of place. *Criminology* 27:27–55.

SKIDMORE, A. K. 1989. A comparison of techniques for calculating gradient and aspect from a gridded digital elevation model. *International Journal of Geographical Information Systems* 3:323–334.

SLATER, T. R. 1989. Medieval and Renaissance urban morphogenesis in eastern Poland. *Journal of Historical Geography* 15:239–259.

SOULE, P. T., and V. MEENTEMEYER. 1989. The drought of 1988: Historical rank and recurrence interval. *Southeastern Geographer* 29:17–25.

VALENTINE, G. 1989. The geography of women's fear. *Area* 21:385–390.

WISE, S. 1989. Planning for the 1991 census. *Area* 21:313–315.

Visualization

Alan M. MacEachren
In collaboration with Barbara P. Buttenfield, James B. Campbell, David W. DiBiase, and Mark Monmonier

Cartography is central to geography's worlds. In a mapping of geography's disciplinary space, Goodchild and Janelle (1988) found cartography, along with historical and applied geography, to be at the center of geography's intellectual world (fig. 6.1). Their diagram depicts the result of multidimensional scaling (MDS) applied to a matrix of cross-memberships in special interest groups that focus on the indicated subfields of geography. MDS is an analysis technique that "provides a convenient means for *visualizing* structure in the matrix of cross-memberships" (Goodchild and Janelle 1988:12). Even when dealing with nonspatial relationships, geographers are most comfortable with a depiction that allows them to visualize relationships and connections that in turn lead to hypotheses about underlying causes for the patterns that become apparent when data are presented in a spatial format.

In chapter 4, "Representations of the World," David Woodward discussed maps as substance, as a subject of study within geography. In this chapter, maps are examined as visualization devices that are used across the spectrum of geography. Maps were important visualization tools well before the advent of spatial statistics or computer graphics, even before geography was a recognized academic discipline. Development of thematic mapping methods near the close of the eighteenth century might be considered the origin of *cartographic visualization*. Thematic maps focus on the location and distribution of one class of feature, the map theme. Early biogeographical examples are Humboldt's map of world vegetation, and his subsequent theories of vegetation change with latitude, altitude, and direction of slope. Humboldt's effort has an earlier parallel in ancient Greek depictions of longitudinal zones (*climata*) on their maps of the world.

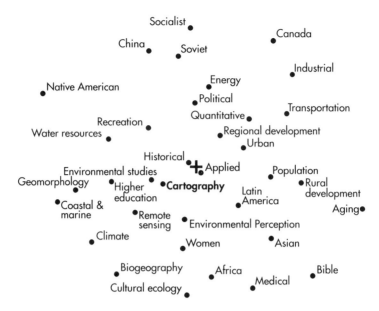

Fig. 6.1. An MDS Procedure. MDS procedures iteratively search any matrix of measures indicating relative similarity between pairs of objects for locations such that distances between objects are in the same rank order as the similarities. If the space for which locations are determined is two dimensional, the resulting solution resembles a map. SOURCE: Modified from Goodchild and Janelle 1988:13.

In what would now be called medical geography, use of visualization as a hypothesis-generating and problem-solving tool can be traced at least to John Snow's 1854 map of cholera cases in London (see chapter 7). Cartographic portrayal of his data allowed Snow to identify the link between cholera incidence and a specific water pump.

The key relevance of Snow's problem to modern geographic research and applications is that although a geographic information system (GIS) could allow us to reach the same conclusion without using a map, the analysis would be impossible without knowing what questions to ask. The portrayal of data in map form allowed Snow and would allow us today to see the relationship and generate a hypothesis without having a priori knowledge that cholera could be related to water sources.

Geographers use many visualization tools in addition to maps. Remotely sensed images play an increasing role in geographic visualization and, in various problem contexts, graphs, sketches, diagrams, photographs, and other two-dimensional representations are an integral part of geographic analysis. Although this chapter emphasizes maps as the geographical visualization tool, it also considers geographic applications of the full range of tools.

GEOGRAPHIC VISUALIZATION DEFINED

Visualization, as considered here, is not restricted to a method of computing, the key phrase from the 1987 report on visualization in scientific computing submitted to the National Science Foundation by the panel on Graphics, Image Processing, and Workstations (McCormick, DeFanti, and Brown 1987). Visualization is foremost an act of cognition, a human ability to develop mental representations that allow geographers to identify patterns and to create or impose order. The mental representations formed and the patterns people see are closely linked to expectations they bring to a given situation.

To illustrate the point, close your eyes for about ten seconds and imagine first an empty stage, then the entrance of a single actor. Now, on which side of the stage did your actor enter? If you had an answer, you have an understanding of what is meant by visualization. Although visualization can be prompted by verbal instructions or cues, science has recently rediscovered the power of more direct visualization tools in stimulating visual thinking. These tools influence the expectations geographers bring to visualization and help them examine problems and events from new perspectives. Recent interest in scientific visualization has emphasized development of new and more powerful visualization tools such as computer animation and three dimensional modeling. Geographers have much to contribute to scientific visualization because of a long tradition of visualizing the world through maps.

Geographic visualization will be defined here as the use of concrete visual representations—whether on paper or through computer displays or other media—to make spatial contexts and problems visible, so as to engage the most powerful human information-processing abilities, those associated with vision. Of concern in this chapter, then, are both representations and their ability to help define questions, develop hypotheses, and find answers. This chapter only touches upon the use of maps and graphics to communicate what we already know, that is, presentation. Emphasis instead is on the use of visualization tools to investigate what we do not know.

THE GEOGRAPHER'S VISUALIZATION TOOLS

The geographer's visualization tools vary in degree of abstraction from images that mimic what an observer sees from a particular vantage point, such as a photograph of vegetation on a slope or a remotely sensed image of settlement patterns in the ridge and valley section of Pennsylvania, to graphics that represent relationships that may or may not be visible and may not even have spatial extent, such as stream velocity over time or distance from farm to market for agricultural products (fig. 6.2, top). The crucial difference between

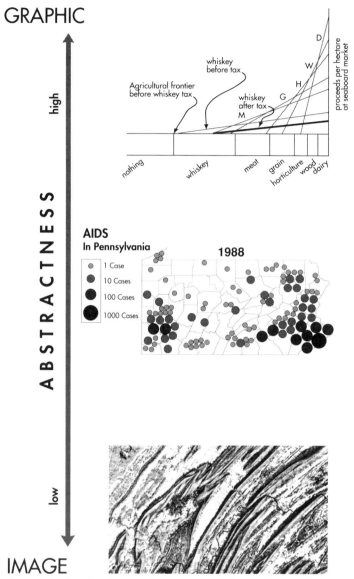

GRAPHIC

high

A B S T R A C T N E S S

low

IMAGE

Fig. 6.2. An Image-Graphic Continuum. *Top*, a distance decay diagram showing price per hectare of production in 1794 for various agricultural products in relation to the distance from the market at which they are sold. Sᴏᴜʀᴄᴇ: Modified from Abler, Adams, and Gould 1971:352); *middle*, incidence of AIDS in Pennsylvania for 1988 based on county-level statistics (Sᴏᴜʀᴄᴇ Portion of a map of "Aids in Pennsylvania" [Hoover 1989]; *bottom*: a portion of a Landsat image of central Pennsylvania.

images and graphics is that images are composed of ambiguous marks with no independently defined meaning (that is, lines, tones, and shades), whereas graphics consist of relatively unambiguous symbols precisely defined by convention or a key. The result is a difference in emphasis from a question of What is that? on the part of the viewer to What are the relations, in an abstract sense, among the things symbolized?

In the case of an image, arbitrariness is quite low: the thought process used by an individual is perceptual and will be likely to engage many diverse patterns depending upon the viewer's background and experience. Remote sensing occupies a place near this end of the image-graphic continuum (fig. 6.2, bottom). Whereas maps have been specifically tailored to present information with visual and logical clarity, remotely sensed images present information much more directly, with much less processing, abstraction, or selection. Remote sensing analysts must, therefore, interpret and select required information from the many different kinds that are available on any given image. This characteristic is an advantage in that remotely sensed data present a rich opportunity for learning about the earth, but is a disadvantage in that much more skill and effort are required to derive this knowledge. Patterns are not likely to be obvious unless the analyst has some preconceptions about what patterns to expect.

Moving to the midpoint of the continuum, arbitrariness increases. The viewer must learn the linkages between referents (the things represented) and the signs (map symbols) used to represent them. A map of AIDS incidence exemplifies this position on the continuum (fig. 6.2, middle). Once the linkage between referent and sign is understood, rapid and accurate recognition of complex relationships becomes possible. The act of symbolizing introduces a conscious recasting of information, however, that can both aid and hinder interpretation by focusing the representation on selected features, total numbers versus density.

At the graphic end of the continuum a viewer's spatial abilities are required to interpret interrelationships between phenomena in two or three visual dimensions representing two or more data dimensions. The distance decay diagram, which relates interaction to distance, (fig. 6.2, top), is a typical graphic used by geographers.

FUNCTIONS OF VISUALIZATION TOOLS IN THE RESEARCH SEQUENCE

DiBiase (1990), building from Tukey's (1977) concept of exploratory data analysis, suggests that visualization tools may play different roles at different stages of scientific research. DiBiase identifies four stages: exploration, confirmation, synthesis, and presentation. The first two represent largely private vi-

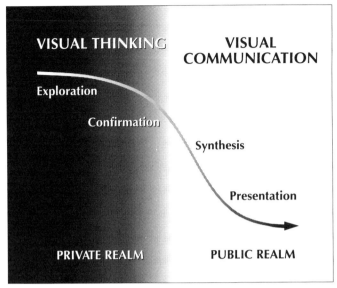

Fig. 6.3. The Functions of Graphics in the Research Sequence. From the investigator's point of view, graphics change from reasoning tools to communication tools as an investigation expands from a private to a public endeavor. The downward slope of the curve suggests the hypothesis that visual thinking involves higher-order cognitive tasks (for the investigator) than does visual communication. SOURCE: DiBiase 1990.

sual thinking while the latter are visual communication processes whereby a scientist shares ideas with a peer community or the general public (fig. 6.3).

Geographers typically use visualization tools for exploring worlds that are too distant in time or space for direct human observation. Satellite images and maps also give us access to the nonvisible. Reflectance in nonvisible portions of the electromagnetic spectrum can be displayed using remote sensing imagery. Maps also portray topics that can not be directly viewed or sensed, such as soil permeability, educational attainment, and the incidence of AIDS.

Exploration

At the exploratory stages, investigators often are not even sure of the questions to pursue. Visualization tools provide an advantage in such early stages of research because they allow investigators to sift massive amounts of data quickly and efficiently, using human vision to identify patterns and relationships worthy of pursuit. In many instances, an insight results from creating a representation to think through a problem with no definite hypothesis in mind. At this stage, visualization often remains undocumented because it occurs in the mind of an individual researcher or in the private realm of a small group of researchers. The representations are often schematic and generally not considered publishable.

A fascinating example of the process of exploratory visualization at work was provided by a group of colleagues. As part of an evolving research effort dealing with issues of global climate, Daniel Leathers, Brent Yarnal, and Michael Palecki produced a choropleth map, originally hand drawn with colored pencils, depicting correlations between U.S. winter temperatures and the Pacific North American (PNA) teleconnection, a zonal index of the midtropospheric circulation over North America (fig. 6.4). The map presented an intriguing and unexpected pattern. A high PNA index indicates the presence of a "deep, strong Aleutian low extending from the surface to the upper troposphere in the north-eastern Pacific, an upper level anticyclone centered over western Canada and a deeper than normal upper level trough located over the south-central United States" (Yarnal and Diaz 1986:197–198). Negative values on the index indicate absence of this pattern, that is, a jet stream with no dominant troughs or ridges. The choropleth map demonstrated a strong corre lation between the PNA index and temperature for both the northwestern and the southeastern U.S., with the Northwest being warmer than normal and the Southeast cooler.

In trying to explain this pattern, the researchers considered several possible

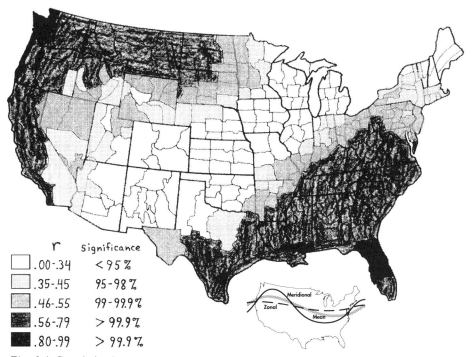

Fig. 6.4. Correlation between the Pacific North American Teleconnection Index and Mean Winter U.S. Temperature. This choropleth map represents values for each of the 344 divisions of the conterminous U.S. The inset illustrates the mean position of tropospheric flow in comparison to meridional and zonal flows (represented by positive and negative PNA indices, respectively).

explanations and connections to global circulation patterns. Going against the conventional wisdom that strong PNA patterns are a winter phenomenon and have little relationship to weather during other seasons, they found a similar relationship between the PNA index and temperature on an annual basis. A comparable relationship was identified for precipitation. Once discovered, the relationship was determined to have a strong physical basis, with topography exerting a blocking influence that reduces the association for the diagonal band from the Great Lakes to the Southwest. Working from the initial discovery, Leathers, Yarnal and Palecki produced a series of maps, did some numerical analysis, and demonstrated links among strong PNA patterns, El Niño events (warmer than normal ocean temperatures in the eastern Pacific), and polar circulation patterns. Their findings have significant implications for long-term weather prediction as well as for the regional impact of changes in global climate.

Confirmation

As a research initiative becomes defined, assumptions are made, questions posed, and hypotheses tested. Visualization tools at this stage of investigation play a confirmatory role. Graphic depictions of statistical analysis, maps of the outcome of a modeling effort, and images generated by combining or transforming the original observations all help confirm or counter hypotheses that have been posed. Frequently, visual display of anomalies or residuals (the cases that remain unexplained when a dependent variable is regressed against one or more independent variables) lead to the greatest insights.

An exciting recent application of confirmatory visualization in human geography involves the mapping of gentrification frontiers (Smith, Duncan, and Reid 1989). Citing use by developers and realtors of frontier images to depict the process of urban gentrification, the authors argue that gentrification involves a frontier of profitability. This frontier represents a dividing line between disinvestment ahead of the line and reinvestment behind it. Both graphs and a map are used as confirmatory tools to test this contention for New York City's East Village. Temporal disinvestment-reinvestment turning points were determined for twenty-seven census tracts by plotting number of units in arrears for each tract (fig. 6.5). The year determined for each tract was treated as a point on a continuous gentrification surface depicted on an isarithmic map (fig. 6.6). The map permits visualization of the gentrification surface. Advance of the gentrification frontier is readily apparent and confirms the researchers' suspicion that gentrification is far from a random process. The map shows a steady onslaught of the frontier from west to east with only a few pockets of resistance. In the East Village, the frontier apparently closed much earlier than generally thought. Using evidence from the gentrification map, the authors concluded that "reinvestment in the East Village began not in the area of deepest disinvestment and abandonment but on the borders where a killing

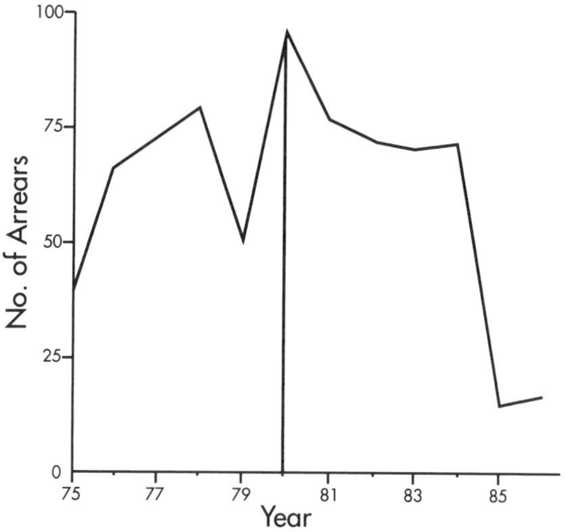

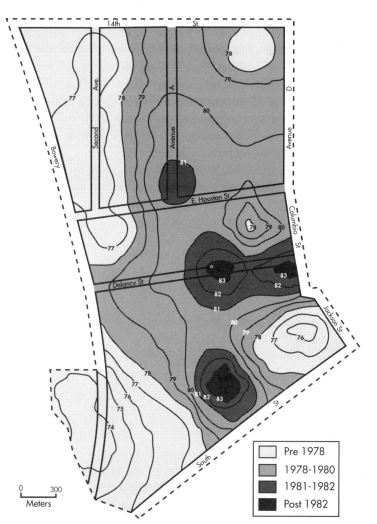

Fig. 6.5. Example of a Graph of Number of Housing Units in Arrears Plotted against Time. The turning point before disinvestment is the peak before the major decline, which occurred in 1980 in this census tract. SOURCE: Modified from Smith, Duncan, and Reid 1989:247.

Fig. 6.6. Isarithmic Map Showing Advancing Gentrification Frontier. Isolines depict the date of the onset of gentrification. SOURCE: Modified from Smith, Duncan, and Reid 1989:248.

could be made with little risk of being scalped, so to speak" (Smith, Duncan, and Reid 1989:250).

Synthesis

As investigators gain confidence in a theory or perspective on a problem, they use visualization tools to synthesize ideas and formulate a coherent abstract statement of what at that stage might be only loosely connected threads. Synthesis is potentially as creative an act as analysis. It is during the synthesis stage that investigators step back from data and specifics to put the problem into a larger theoretical context. The synthetic stage of visualization often leads to presentational graphics. Researchers seldom have an opportunity to peer into an investigator's mind to observe the process of synthesis; however, David DiBiase recently had that experience when he helped Robert Sack develop a graphic depiction of mass consumption. In "The Consumer's World: Place as Context," Sack (1988) explores the role of consumption in the geography of modernity. Central to the essay is the metaphor of a loom of mass consumption, upon which threads from the intellectual realms of nature, meaning, and social relations are woven into the fabric of modern places. Figure 6.7*a* is a representation of part of the metaphor used by the author during preparation of the essay. Figure 6.7*e* is an adaptation of the author's sketch, designed and produced by DiBiase for publication. The sequence, from *a* through *e*, illustrates the metamorphosis of the diagram as the concepts it summarizes became more concrete and its intended audience expanded. A measure of the importance of the figure to the essay is the fact that it is referred to nine times in eighteen pages of text.

Presentation

Once investigators are convinced that their particular understanding of a problem is correct, they have an obligation to present the ideas for peer scrutiny. At this stage, visualization tools provide a means of presentation. As in the case of the gentrification frontier map (fig. 6.6), the analytical visualization tool doubles as a presentation device. In other instances, however, the presentation is generated as a summary of a complex concept or pattern. Presentational visualization was used recently by Shannon and Pyle (1988), who examine theories concerning the potential African origin of AIDS and postulate various pathways for diffusion out of Africa. A flow map based on a global map projection, with Africa at the center, allows the reader to follow the authors' interpretation (fig. 6.8).

Most graphics published with geographic research are intended as presentation devices, but some also served as tools for visual thinking earlier in the research sequence. In an exploration of the role of visualization tools in visual thinking DiBiase (1989) examined illustrations appearing in three major geo-

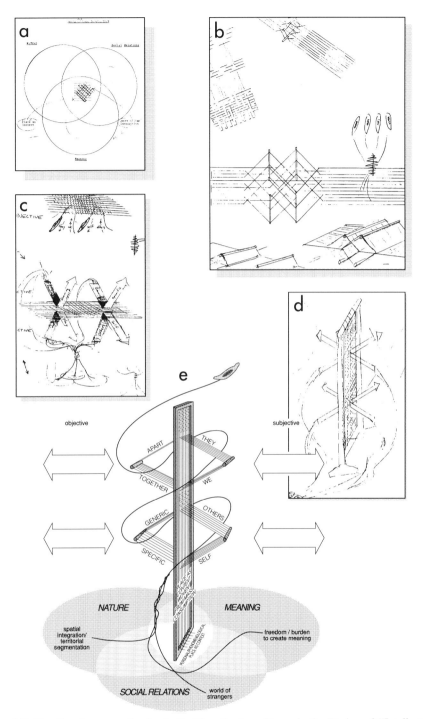

Fig. 6.7. The Development of a Synthetic Visualization: Stages in the Design of "Intellectual Surface and the Loom of Consumption." SOURCE for 6.7*e*: Sack 1988:645.

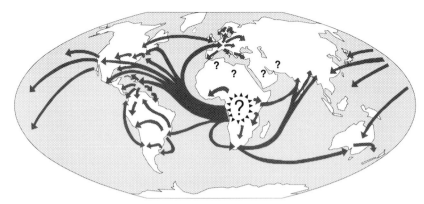

Fig. 6.8. Postulated Diffusion of AIDS during the 1970s and 1980s. SOURCE: Modified from Shannon and Pyle 1988.

graphical journals. He found that 78 percent of graphs, 58 percent of maps, and 53 percent of photoimagery was interpreted as serving an exploratory or confirmatory role in addition to its presentation function.

WHY VISUALIZATION TOOLS HELP

Evidence exists that mental visualization has contributed to major scientific discoveries. Shepard (1978), for example, cited Faraday's and Maxwell's use of imagery in their initial conceptions of electric and magnetic fields, Einstein's emphasis on visual over verbal thinking, and Kekule's visual images that led to his discovery of the structure of benzene. Not everyone finds mental visualization an automatic process; concrete representations are often required as a catalyst. Visualization tools seem to prompt the process by removing some of the information processing load from conscious thought and shifting it to preconscious visual processing (Friedhoff and Benzon 1989).

Larkin and Simon state that "diagrams and the human visual system provide, at essentially zero cost, all the inferences we have called 'perceptual' " (1987:92). This zero cost information seems to be responsible for what appeared to Faraday's contemporaries as intuition (Shepard 1978). Arnheim has referred to intuition as a component of visual thinking that operates like a "gift from nowhere" (1985:79). Geographers need this gift in order to grapple with the complex multidimensional nature of the worlds they study.

Change in Perspective

Most data collection techniques produce discrete pieces of information, often in tabular form. But geographers—with their interest in spatial distributions, regions, and interactions among places—are often best served by visualization tools that present information from an overhead perspective. Maps provide a means for combining such piecemeal observations into a single overhead perspective.

Remote sensing also provides an overhead perspective, a maplike view of the earth that permits patterns to be visualized in their correct positions and as geographic units in context with other patterns. In 1956 A. E. Gutkind wrote of the impact that conquest of the air had on our ability to observe the earth's geographic patterns:

> A new scale in time and space has been added to our mental and material equipment. Before the conquest we were winding our way like worms through narrow passages and seeing only more or less unrelated details. Today we can look at the world with a God-like view, take at a glance the infinite variety of environmental patterns spread over the earth and appreciate their dynamic relationships. (1956:1)

Although these words were published when aerial photography was the only kind of aerial imagery, they are even more true today.

Synoptic View

Maps and composite images both extend vision to areas much larger than are seen directly. The extended view expands visualization to patterns and relationships that might be overlooked entirely if one were not able to step back and see the big picture. Low-relief landforms of the midcontinent are revealed as never before, and previously unidentified lineaments become apparent in a shaded relief map derived from a detailed digital elevation matrix (fig. 6.9). The map is for the geographer what the telescope is for the astronomer.

Maps that provide such a synoptic view can mislead. They depict regions with apparent precision but they may be products of approximations based on incomplete or imprecise data. Prior to the advent of satellite remote sensing and other digital data collection aids, knowledge of the earth's major biomes as reflected in apparently precise boundaries around map regions was based largely on point observations of climate and vegetation extended over time and distance by rudimentary interpolation and extrapolation from known to unknown areas.

Viewing the Nonvisible

In addition to simple overhead photography, remote sensing imagery can depict spectral regions outside the range of human vision. Portions of the reflective

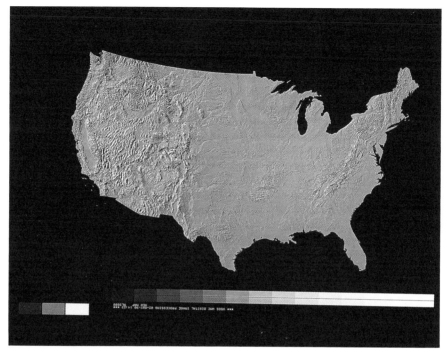

Fig. 6.9. Shaded Relief Map of U.S. Topography from Digital Elevation Models. SOURCE: Pike and Thelin 1991.

infrared spectrum, for example, convey information concerning the internal structure of plant leaves, revealing stress, disease, or the approach of maturity. Emissive infrared radiation conveys information concerning distributions of soil moisture and thermal properties of subsurface soil and rock. Because these phenomena cannot be seen directly and can be measured with conventional instruments only at points, remotely sensed images greatly enhance human abilities to visualize nature's complex geographic patterns.

Maps, graphs, and other forms of representation extend human vision even beyond the limits of reflected or emitted radiation. Any data that are spatially referenced can be mapped. Omernik and Powers (1983), for example, combined alkalinity measurements from approximately 2,500 streams and lakes with information on soils and geology to create a synoptic view of total alkalinity of surface water in the United States. Over the past century and a half, geographers and others have developed an impressive array of symbolization methods for depicting the nonvisible. Many of these are adaptations of symbol types such as contour lines or three-dimensional views that were originally devised for terrain representation (fig. 6.10). Others were developed exclusively for representation of abstract concepts such as migration space (fig. 6.11).

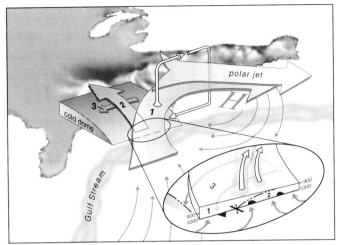

Fig. 6.10. Visual Synthesis of the Meteorological Process Contributing to an East Coast Cyclogenesis Event. (Created by David DiBiase for Nelson Seaman of the Department of Meteorology, Pennsylvania State University.)

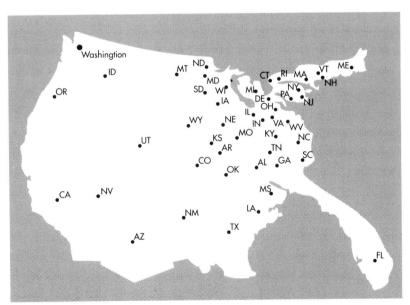

Fig. 6.11. Continuous Area Cartogram of In-Migration by State in the United States. The area of each state on this map is proportional to its number of in-migrants rather than to its geographic area. SOURCE: Modified from Plane 1984:253.

Resolution and Filtering

Geographic patterns and processes exist at different scales. Visualization tools help in the selection of appropriate scales for analysis. Because detail can be beneficial, remote sensing platforms continue to be developed with greater and greater resolution. For mapping with census data, patterns hidden in aggregate data (census tract totals), sometimes appear when we map things at a higher resolution (census blocks). Lewis (1976), for example, was able to demon-

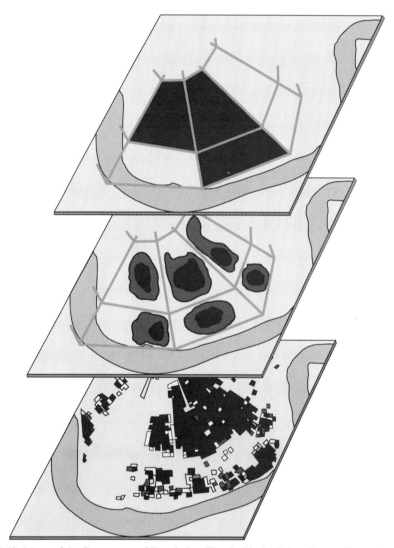

Fig. 6.12. Maps of the Percentage of Population That Is Black, Using Census Tracts (*top*) and Census Blocks (*bottom*) as Enumeration Units. The middle illustration depicts Lewis's interpretation of the census block data. SOURCE: Modified from Lewis 1976:46–47.

strate an unexpected multinucleated pattern of black population in New Orleans ringed by rich whites with homes on the great boulevards of the city. The pattern is not visible if the more highly aggregated census tract data are mapped (fig. 6.12).

High resolution is not always useful. It can actually conceal patterns and processes that operate at regional rather than local scales. In some instances coarser resolution permits analysts to detect patterns that are obscured by detail. The lineaments in the midcontinent United States, for example, were first noticed in the composite digital terrain model described earlier (fig. 6.9). Data-filtering procedures are routinely applied to remotely sensed images to enhance analysts' ability to detect patterns. Filtering is a means of selecting specified kinds of features from the more varied, complex patterns within an image or map. Some filters are designed simply to improve the visual qualities of an image by improving contrast, sharpening edges of image features, or removing extraneous features. Other filters operate as masks that are systematically moved over an image to select certain kinds of patterns or to screen out others. Digital filters, for example, can detect subtle changes caused by shadowing related to topographic differences or vegetation patterns and thereby identify lines that are not reliably identified by human interpreters.

Recognizing the value of separating global trends from local detail, geographers have applied an analytical visualization technique called trend surface analysis. The procedure allows identification of systematic patterns present at various scales of analysis. At a state level, for example, a quadratic polynomial trend surface can provide a clear depiction of the major thrusts of settlement in New York (fig. 6.13).

For statistical maps, arguments for removing detail have been put forward as a justification for grouping a large number of data values into a small set of

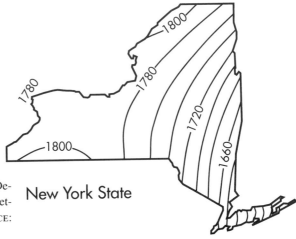

Fig. 6.13. Trend Surface Map Depicting the Major Trends of Settlement in New York. Source: Monmonier 1990:36.

New York State

data classes (fig. 6.14). Unlike spatial filtering of remotely sensed images or trend surface analysis, classification of quantitative data for mapping is usually a spatial process that may or may not enhance pattern strength. Some researchers have advocated data-grouping procedures that adapt standard statistical classification routines to include spatial proximity of data units (Monmonier 1972). Others (Tobler 1973, for example) have argued that human pattern analysis abilities are so powerful that analysts should provide data that are as complete as possible and leave classification to human vision (see fig. 6.15).

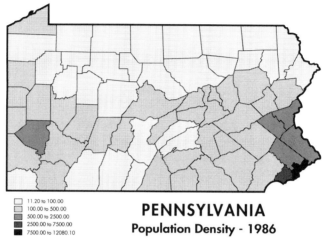

PENNSYLVANIA
Population Density - 1986

11.20 to 100.00
100.00 to 500.00
500.00 to 2500.00
2500.00 to 7500.00
7500.00 to 12080.10

Fig. 6.14. A Typical Population Density Map. Density is divided into five categories; with increased darkness corresponds to increasing density.

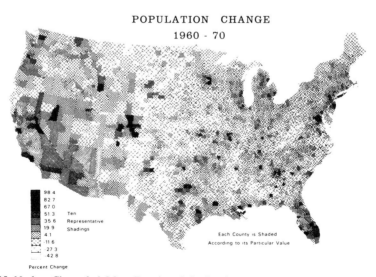

POPULATION CHANGE
1960 - 70

98 4
82 7
67 0
51 3   Ten
35 6   Representative
19 9   Shadings
4 1
-11 6
-27 3
-42 8

Percent Change

Each County is Shaded
According to its Particular Value

Fig. 6.15. N-class Choropleth Map. Density of shading is proportional to density of population.
SOURCE: Peterson 1979:23. Copyright © 1979 by the American Congress on Surveying and Mapping.

Transformations

Visualization tools generally involve some form of data transformation. Areal photographs are rectified, one-dimensional tables are transformed into two-dimensional graphs, and spherical geographic coordinates are transformed into flat maps. Whenever a view of any significant portion of the earth's surface is required, the transformation between spherical earth and planar representation will exert a profound influence on the viewer's understanding of geographic arrangement.

Although the earth is a continuous surface, maps have borders and sometimes breaks or interruptions. Where these borders and interruptions occur will influence judgments of spatial proximity (fig. 6.16). All flat maps distort area, angles, or both, and different projections favor different properties (fig. 6.17). Each projection leads to a different image of geographic relationships among phenomena depicted.

Tools developed to help geographers visualize relationships also allow them to transform complex thematic data into more easily interpreted depictions. For example, data on fiscal transfers between regions of the country can be transformed into a connectivity matrix. It is not until the matrix is further transformed into a map, however, that the pattern of interregional fiscal flow becomes apparent (fig. 6.18). In addition to transforming nonspatial information into visual representations, transformations between visual forms are used to examine information from multiple perspectives. A two-dimensional isoline representation, for example, can be transformed into a three-dimensional depiction in order to allow an analyst to visualize a surface more easily (fig. 6.19).

Multidimensional Representations

All maps are by definition multidimensional representations. Often two, three, or more data dimensions are added to the two spatial dimensions of the map base in order to examine geographic patterns of data. Remotely sensed images are similarly multidimensional; they present information for a variety of phenomena, including soils, vegetation, drainage, and transportation in a two-dimensional format.

Multiple phenomena can be examined with a single image because multispectral scanners collect data in several regions of the electromagnetic spectrum; that is, the same region of the earth is recorded on several spectral channels. Sometimes the width and position of these channels are defined arbitrarily, but often they are specifically tailored to record radiation reflected or emitted by specific features of interest. Because colors to which the human visual system is sensitive can be reproduced by only three primary colors, visual analysis of information derived by multispectral scanners uses only three channels of data at a time. Therefore, one problem is the task of selecting the set of three channels that best displays the features of interest.

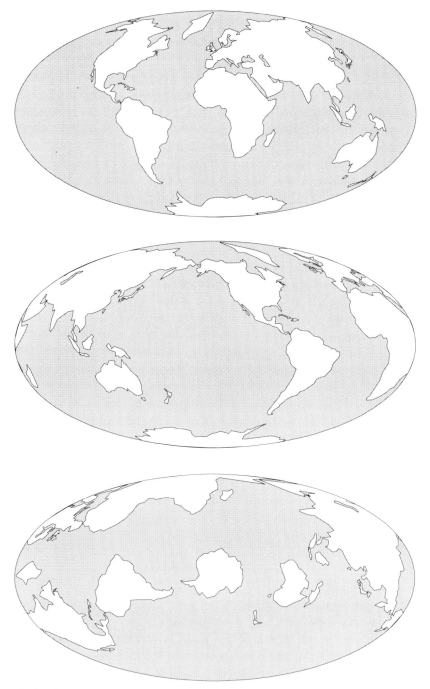

Fig. 6.16. Influence of Choice on Where to Interrupt the World. *Top*, a conventional Europe-centered depiction; *center*, an emphasis on the Pacific rim; and *bottom*, a view of the interconnected world oceans.

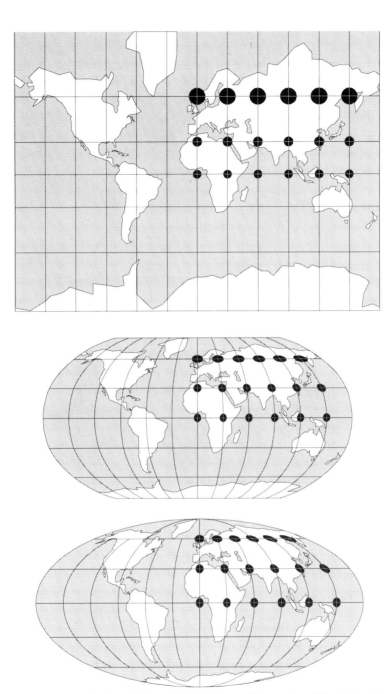

Fig. 6.17. Graphic Depictions of Size and Angle Errors Inherent in a Map Projection. *Top*, a conformal projection with area distorted as indicated by the changed size of the distortion ellipses; center, both size and angles distorted in the Robinson projection, a compromise projection, selected by the National Geographic Society for use in political reference maps; *bottom*, distortion diagrams indicating angular but not size distortion on the Mollweide projection, one of several popular equal-area projections depicting the entire world. SOURCE: Produced with the *WORLD* map projection package, Regents of the University of Minnesota.

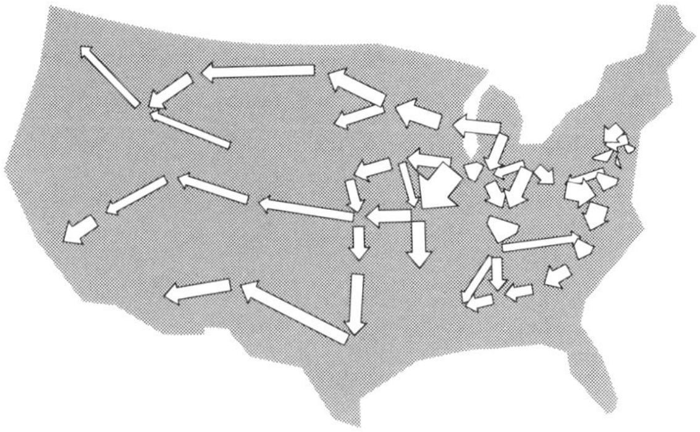

Fig. 6.18. Map of the Fiscal Flow Surface of the United States. Generated from data for net financial transactions by state. SOURCE: Modified from Tobler 1981.

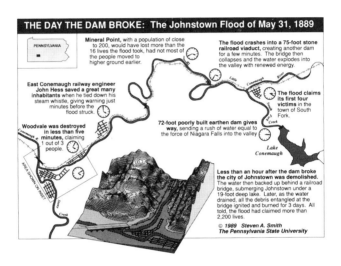

Fig. 6.19. Map of the 1889 Johnstown Flood. SOURCE: Smith 1989.

Removing Temporal Limits

Maps and graphs have long been used to depict change. With maps, there are two traditional strategies: single static maps that depict limited aspects of change in or across space and multiple static maps that can be visually compared. The latter have been called *chess maps* by Monmonier (1990) because of their similarity to illustrations found in daily newspapers that depict snapshots of chess boards over a sequence of moves. The chess map strategy is typical of many atlases. Recently, an entire atlas was designed around this concept to display voting patterns by congressional district in the United States from 1789 to 1989 (Martis 1989). This series of maps allows researchers to visualize the changing political landscape of the United States.

Single maps that depict change fit into two categories: those representing movement from place to place and those showing change over time in particular places. Maps of movement have been devised that use a series of point symbols to depict changing location, such as the migrating center of gravity of the U.S. population (see fig. 6.11), and line symbols to depict routes, such as those traveled by goods and people, diseases, money, and storms (See figs. 6.8, 6.18, 6.20). The advance of a frontier, as illustrated for gentrification, may also be viewed as an advancing wave and represented with isarithmic lines (see fig. 6.6). Change maps are used to depict variation of some attribute of a place over time. Temporal symbols, what Monmonier (1990) has termed *glyphs*, are used to indicate change at points in one or more variables (fig. 6.19). For areas, choropleth maps typically are used to depict change over a fixed time period (fig. 6.15).

An intriguing representation of temporal change leading to significant insight was developed by Stoddart, Reed, and French (1989). They investigated the accretion of salt marshes by examining tidal action across a marsh. By plotting tidal velocity against flood level, they were able to identify asymmetries between flood and ebb tides along with marked velocity pulses for tides that exceeded creek banks (fig. 6.21). The initial discoveries prompted further research that led to the conclusion that sediment for accretion on the salt marsh comes primarily from the eroding marsh edge, that sediment having been transported onto the marsh during over marsh tides.

Photography and other forms of remotely sensed images have added dramatically to scientists' ability to monitor change. Use has been made of photographic time sequences throughout geographic research. Veblen and Lorenz (1988), for example, used photographic comparisons in an analysis of vegetation changes in Patagonia. They matched photos taken near the turn of the

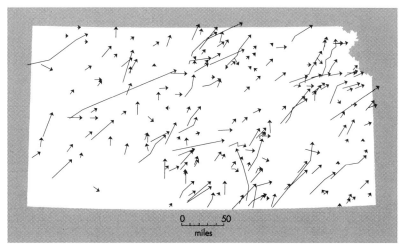

Figure 6.20. Map of Tornado Paths across Kansas, 1950–1970. SOURCE: Eagleman, Muirhead, and Willems 1971:42.

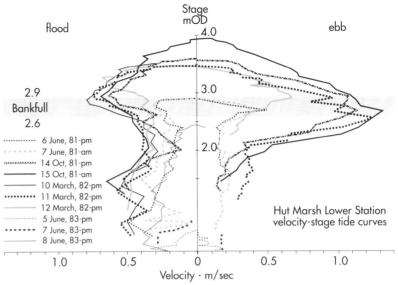

Fig. 6.21. Flood- and Ebb-Tide Velocity-Stage Curves for Tides of Varying Amplitudes, June 1981 to June 1983. Plots demonstrate the increase in velocity when tides overflow the banks and the velocity asymmetry for flood versus ebb tide. Source: Stoddart, Reed, and French 1989:234.

century with modern photos to support a contention that sharp fire boundaries result from differences in susceptibility of natural vegetation to fire. Another intriguing example of paired photographs, this time to record changes in the cultural landscape of the United States, was presented by Thomas and Geraldine Vale (1983). Their photos match those published by George Stewart in his 1953 book, *U.S. 40.* Overall, their comparison from coast to coast along the route of U.S. 40 led to the unexpected conclusion that the U.S. landscape is not changing as rapidly as generally thought. Their evidence suggests that, rather than becoming more standardized, the American landscape "remains a study in diversity."

In a similar manner, remote sensing imagery from aircraft or satellites extends the reach of human visualization over time. Phenologic changes in vegetation—seasonal changes in growth, flowering, maturity, and senescence of plants—observed over broad regions provide botanists and agronomists with the basis for delineating climatic regions, forecasting crop yields, and monitoring agricultural developments. Patterns such as tidal currents or snow cover change so rapidly that they often escape examination if not recorded as remotely sensed images. As remote sensing technology has advanced and greatly increased both the temporal and spatial scope of its representations, the geographer's ability to visualize the earth's broad-scale patterns of land use, vegetation, and climate has increased accordingly (fig. 6.22).

In figure 6.22, in the absence of remotely sensed imagery, humankind's

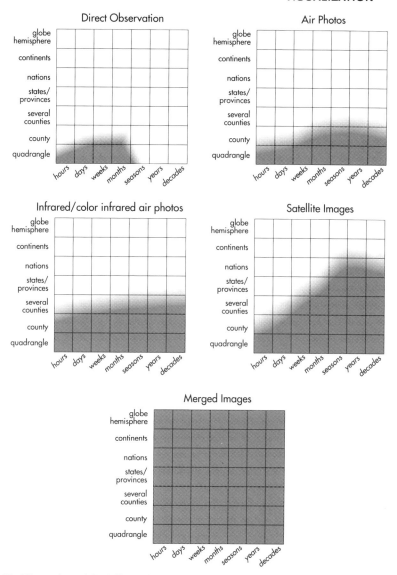

Fig. 6.22. Illustration of the Effect of Remotely Sensed Images on the Ability to Visualize the Earth. The horizontal axis in each graph represents time in an arbitrary scale of increasing intervals, from hours and days at one extreme, to years and decades at the other. The vertical axis represents space, also marked in an arbitrary interval from quadrangle-sized areas at one extreme to continents and hemispheres at the other.

direct vision is constrained (top left). The scope of the area one could examine would be limited to areas that could be viewed from a structure or a high hill. In this context, ability to examine changes over time and space is limited by the practical and conceptual problems of compiling and then interpreting graphs, diagrams, or maps. With aerial photography, vision expanded to permit examination of quadrangle- or county-sized areas at intervals of weeks to years, or even longer (top right). Mosaics of individual frames provided coverage of larger areas and were one means of expanding the area covered by an image, but practical difficulties of acquiring photographs with uniform illumination greatly restricted the availability, quality, and usefulness of such images. Later, routine availability of infrared films increased availability of high altitude photographs, extending the areal coverage of individual photographs and once more expanding human vision (center, left).

Range of vision again increased with the availability of satellite imagery, such as that of Landsat, Spot, and meteorological satellites (fig. 6.22, center right). Such satellites can show, as a single image, areas as large as one-hundred miles on a side, and their systematic coverage schedules have enabled compilation of image libraries that record images over periods of years and decades. The reach of human vision has been extended even further by the application of image-processing techniques to the analysis of data from meteorological satellites (bottom). Some of these techniques permit data from separate spectral channels to be processed to enhance their ability to reveal areas of vegetative cover. Other methods have permitted data from numerous satellite passes over periods of weeks to be combined into a single image, free of cloud cover, that can depict entire continents, or even the entire earth (except the polar regions), as a single image.

Composite images formed from data collected by meteorological satellites present entirely new images of vegetation and ecological patterns at continental scale. Such images have confirmed patterns defined by conventional climatic and ecological analysis and have also revealed spatial detail and year-to-year variation not observable by traditional methods. As a result, these images have modified the conceptual models by which we implicitly organize geographic information.

As a development complementary to the evolution of remote sensing technology, geographers and other earth scientists are developing comprehensive data bases with both space and time components. Data structures that put time and space on the same level are a prerequisite to visualization, analysis, and modeling of topics such as global climate change.

EMERGING TRENDS

The possibilities for development and use of visualization tools are increasing as rapidly as developments in computer graphics and geographic information systems. Increasing access to information and new visualization tools allow researchers to interact directly with the data. Profound changes in research methods are inevitable. For geography, several exciting trends have begun to emerge.

Exploration and Pattern Identification

The current resurgence of cartographic analysis for geographic exploration parallels a resurgence in the role of graphs in statistical analysis. Tufte's (1983) work on visual display of quantitative information, together with the efforts of Becker, Cleveland, and Wilks (1987) and Tukey (1977) to develop exploratory graphic techniques, has resulted in renewed appreciation for the power of graphic presentations to highlight both patterns and anomalies in data. Efforts in exploratory graphics have stimulated various geographic adaptations. Two innovative applications will serve as examples, one related to image analysis and the other to map analysis.

Analysts often use cospectral plots to identify land cover categories through their relative locations in spectral space. A common cospectral plot involves comparing brightness values in the red and the near infrared bands of a multispectral image. Regions with low reflectance on both of these axes are probably clear water while high reflectance on both indicates dry soil or concrete. The central diagonal is called the *soil brightness line* (fig. 6.23). Vegetation type and health can be identified by position in this cospectral space. Hodgson and Plews (1989) point out that a major limitation of cospectral plots is the restriction to comparison of only two bands at a time. They devise an N-dimensional display of cluster means in feature space for remote sensing analysis. This display makes use of several monocular depth cues—size, thickness, brightness and color—to depict more than two variables (fig. 6.24).

In a cartographic adaptation of exploratory data analysis, Monmonier (1989) suggests adding a map to a technique called *scatterplot brushing* (fig. 6.25). The result, *geographic brushing*, is an interactive graphic tool for exploring

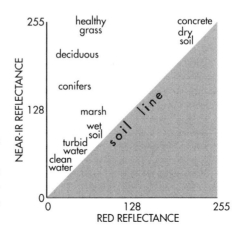

Fig. 6.23. A Standard Cospectral Plot with the Major Land-Cover Regions Labeled. Source: Modified from Hodgson and Plews 1989:613. Copyright © 1989 by the American Society for Photogrammetry and Remote Sensing. Used by permission.

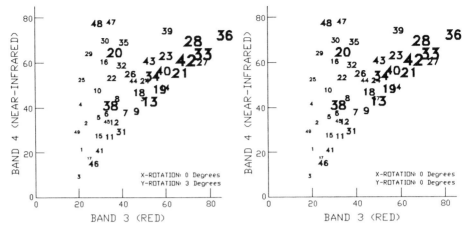

Fig. 6.24. A Monochromatic Three-Dimensional Stereo Cospectral Plot. Viewed under a pocket stereoscope, the images merge into one three-dimensional plot. Sᴏᴜʀᴄᴇ: Modified from Hodgson and Plews 1989. Copyright © 1989 by the American Society for Photogrammetry and Remote Sensing. Used by permission.

geographic correlation. With this technique, a moveable rectangular brush is used to select a group of points in one of an array of scatterplots in which each point represents a location. The graphics system highlights not only the selected points inside the brush but also the corresponding points in all the other scatterplots as well as their location on the map. The map permits the user to detect spatial patterns, and the geographic brush supplements the scatterplot brush by allowing the user to select places or regions on the map. This highly interactive graphics system, with a rapid response to movements of the brush, enables the analyst to explore statistical and geographical correlation among several variables simultaneously.

Animated Maps

Geographers are seldom interested in a static view of the world, but until recently, that was what visualization tools provided. If geographers wanted to examine change over time, static dance maps, change maps, and sets of chess maps (including sets of remotely sensed imagery) were the available display forms. Animation allows cartographers to incorporate time directly into visualization tools.

Map animations take several forms. The most common at present is computerized display of a rapid sequence of complete maps. The basic display strategy is derived from cinematographic methods that were first applied to map animation three decades ago (Thrower 1961). MacEachren and DiBiase (in press) used this technique to prepare a *map movie* that depicts AIDS distribution in Pennsylvania over time (fig. 6.26). Beyond sequenced maps, com-

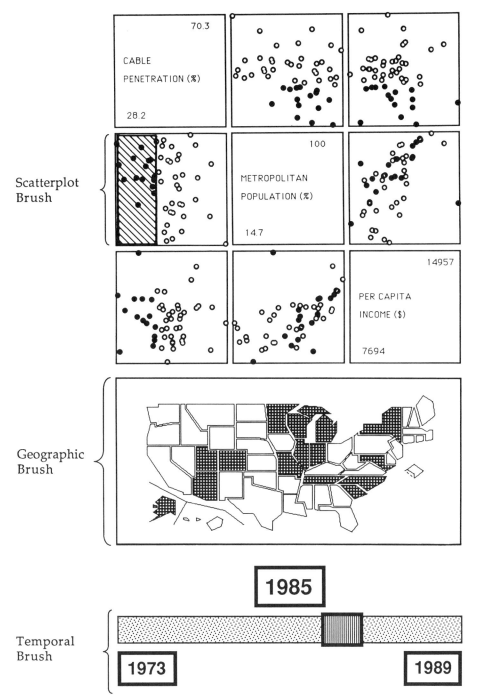

Fig. 6.25. A Scatterplot Matrix with a Scatterplot Brush, a Geographic Brush, and a Temporal Brush. SOURCE: Monmonier 1989:42.

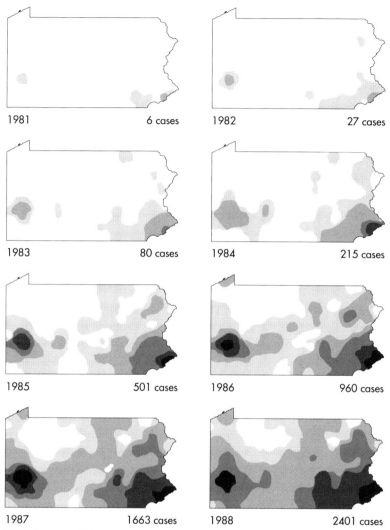

Fig. 6.26. A Sample of Frames from a Map Movie Depicting the Spread of AIDS in Pennsylvania between 1980 and 1988. SOURCE: MacEachren and DiBiase 1991:227.

puter animation techniques provide a variety of intriguing approaches to animation. These include what Gersmehl (1990) has called a *sprite*, a figure that moves across a background, such as depiction of a hurricane track, and *metamorphosis*, wherein change from one form to another is simulated, as in a depiction of continental drift.

Although an animated depiction avoids the static view of other representation modes, it still requires decisions on appropriate time intervals and spatial resolutions and whether these parameters should vary over the course of the temporal sequence. Monmonier (1990) has suggested that *graphic scripts* be developed for sequenced visualization of geographic data. They would be similar in concept to scripts used to organize traditional animations, and they could incorporate research results dealing with color choices, symbol sizes, and time sequences to enhance analysts' abilities to identify space-time relationships. Another important capability of computer-generated animations is the use of variable intensity. For presentational purposes, this capability allows dramatic fades and dissolves. Fading, fuzzy, or blinking symbols can play a role in visualization of data quality, a critical visualization issue that is just beginning to attract serious attention.

Visualization of Data Quality

Data quality is a critical issue because of the tendency of most people to treat both maps and computers as somehow less fallible than the humans who collect data. When a GIS is used to compile, analyze, and display information, the chance for unacceptable or variable data quality is high because the merged multiple layers of data vary in quality both for specified attributes and for locations. Cartographers have always worried about data quality. Thus far, however, only rudimentary steps have been taken to address the complex issues of visualizing data quality in multidimensional data displays that are used for image analysis and GIS applications. The importance of the topic is evidenced both by the decision of the National Center for Geographic Information and Analysis to make visualization of data quality the first visualization initiative undertaken by the new center and by the recent attention given to visualizing uncertainty by the U.S. Environmental Protection Agency.

Data quality involves more than locational verity. It is instructive to follow the framework of the proposed digital cartographic data standard (Moellering et al. 1988), which incorporates locational accuracy, attribute accuracy, logical consistency, completeness, and lineage. All data types, however, cannot be treated equally. For example, one can visualize the locational accuracy of discrete data (points and lines) by using error ellipses or epsilon bands (Perkal 1966). But error in continuous data, for example interpolation error on isarithmic maps, is not adequately portrayed by discrete ellipses. A significant research commitment is needed to develop and empirically evaluate other visual tools that assess data quality. The effort would contribute not only to geographic visualization, but to scientific visualization across fields. The cartographic traditions both of psychophysical testing and of attention to accuracy issues are strong foundations upon which to build.

User Interfaces and Pattern Recognition

How to visualize data quality is one of many issues encountered in the design of human-display interfaces for GIS. The success of the Macintosh graphical user interface and its dissemination first into SUN workstations and later into IBM software demonstrates that visualization is effective for cybernetic interfacing. The visual metaphor of the desktop breaks down human-machine barriers and makes increasingly complex operating systems transparent to nonsophisticated users. For GIS, graphical interfaces provide a visual command syntax for spatial tasks.

As an aid to visualizing multidimensional representations, Moellering (1989) has explored the applicability of stereoscopic technology that permits an image on a single monitor to appear three dimensional. A stereoscopic depiction of terrain allows other data distributions to be draped over the terrain representation, allowing the viewer to visualize the interrelationship of the topic mapped and terrain characteristics.

Although development of software and hardware continues to expand the capabilities of the visualization tool kit, these advances by themselves are not enough. Understanding how geographical patterns change as resolution or time changes is a key to designing a visualization interface that aids the researcher's identification of those patterns or anomalies. Both perceptual and cognitive factors influence whether or not a visual display, either diagrammatic or cartographic, will work. A concomitant need is to extend cartographic research dealing with legibility and effectiveness of printed maps to how humans perceive and interact with soft-copy maps on color monitors. There is, for example, little known concerning the use of color, contrast, or figure-ground for maps on computer screens, and virtually nothing is known about the variables of duration and frequency that influence interpretation of dynamic symbols. Other questions of interface design include: How many windows containing what kinds of information can a geographic researcher work with efficiently? What type of user manipulations are most appropriate? What numerical indicators of similarity, association, and correlation are useful as a complement to visual display? Research is also needed to determine whether the ideal design varies with the graphic savvy of the user, the phenomenon being observed, the user's interests and experience, or all those factors.

GIS Applications

Geographic information systems are becoming ubiquitous and powerful tools for geographic inquiry. All provide some method of graphic display. It would be very difficult to consider GIS without visual products, and to the extent that space and time are modeled with GIS tools, the real world is being explored by visual approximation. A GIS capable of combining images and maps, for example, allows visualization of topics such as vegetation changes through seasons or over years and in relation to drainage, slope, topography, population, fire, and other patterns. One of the most important functions for visualization in these systems will be the continuation of a role visualization played prior to computers. GIS can be thought of as a spatial form of data exploration and the models generated from GIS as a tangible expression of that data exploration. Visualization tools provide one means for exploring the reliability of those models; one can immediately *see* the implications of change in model parameters. Rather than simply calibrating a set of parameters and then using the model to derive a final answer, graphic displays of intermediate stages in the modeled scenario can be visually depicted. This depiction offers the potential to develop a better understanding of the processes at work. In addition, visual depiction of intermediate stages of modeled processes such as climatic change or disease diffusion allows an analyst to steer the model's course in direct response to the visual feedback. A scientist studying the potential for global warming, for example, should be able not only to view the time period at which the model starts and snapshots at various times in the future but also to notice spatiotemporal relationships while the model is progressing. Then "what if" questions can be asked, such as What if destruction of tropical rain forests was suddenly halted in 2003?

With the masses of data now available, it is more critical than ever to identify the appropriate questions to pose of the system and to determine the relevant scale at which to pose them. A typical past strategy was to attempt the development of graphically optimal representations, and much cartographic research over the past three decades has been devoted to the search for optimal symbolization. An obvious advantage of computer graphics linked to the GIS, however, is that the system is not limited to a single form of presentation. With GIS, alternate interpretations of information are available that may cause an investigator to consider other ways of addressing a problem.

Expert Visualization

The study of human strategies for visual search and pattern analysis may also suggest strategies for expert systems that address similar data base search and problem-solving tasks (see chapter 4). Data base search routines should be devised that can browse a data base with inferential queries and glean spatial patterns that are not explicitly encoded in the data base architecture yet are visually apparent in stored imagery. In a recent example, Penn State student Charles Roberts (1990) found that residential neighborhood types that were apparent to humans looking at high-resolution images were not detected by digital image analysis techniques. When these same tools were applied to lower resolution data, however, the neighborhood types were delineated correctly by the system.

Geographic phenomena look different at different scales (chapter 13). The complexities of geographic scale dependence continue to impede data encoding and pattern interpretation (Buttenfield 1990). Better understanding of the influence of how scale changes the appearance of geographic features and how that appearance is reflected in digital data bases must be achieved before effective geographic feature recognition algorithms can be developed. Using the power of vision, geographic knowledge, and practice in abstracting reality, cartographers are generally able to determine which geographic features are critical at a particular scale and the detail with which they should be depicted on a map. However the current inability to explicitly verbalize this cartographic process makes it difficult to determine the practical scale range of a particular cartographic data base or to automate the extraction and presentation of geographic patterns for analysis.

Electronic Atlases

As with other applications of the computer, initial attempts at electronic atlases have mimicked conventional atlases. Mainly, geographic data have been presented on conventional static maps. Electronic atlases have the potential to take the atlas from what has been primarily a spatial data storage device to an exploratory analytical tool. The Great American History Machine (GAHM) represents a first step toward this goal. The GAHM is an interactive historical atlas of the United States that allows students to access historical data, depict them on maps, manipulate symbolization, compare maps, and explore hypotheses (Miller and Modell 1988).

Monmonier (1990) has suggested *atlas touring* as a visualization support strategy for introducing a viewer to an electronic atlas. Atlas touring uses graphic scripts to generate a meaningful sequence of relevant views. Individual or simultaneous views might include maps, statistical diagrams, and text blocks with data definitions or verbalized interpretation. Tour variables and their treatment would be selected, in part, through a user profile that defines distributions, regions, trends, or relationships of particular interest to the user. Expert visualization can play a key role here in preliminary sifting of available data. A graphic script for access and display might be based on several standardized visual sequences with a common function. For example, a sequence addressing temporal trends might present a historical series of chess maps and change maps juxtaposed above a time-series graph that captures the developing trend for the entire region.

Although a series of views can be captured on video tape and shown to an audience, atlas touring is best used as a pump primer for interactive analysis. Programmed introductions seem most appropriate if the analyst is then encouraged to explore the data base and to return to particularly interesting views. Because of the limited capacity of short-term memory and the uncertain and slow process of capturing information for long-term memory, graphic scripts should be interruptible. The viewer should be able not only to pause and reflect or to fast-forward through boring sequences but to request more detailed sequences for selected distributions or relationships. Ideally, the user would toggle between atlas touring and high-interaction exploratory techniques such as geographic brushing.

Multimode Information Presentations

In the past, geographers have combined carefully crafted verbal descriptions with maps. Now both sound and tactile feedback can be added. Adding sound to maps might have an impact on cartography similar in scope to that of sound in the movie industry. One of the only multimedia mapping and GIS projects completed to date is the Domesday Project produced by a large team of British geographers (Openshaw and Mounsey 1987). This project involved creating a data bank of maps, images, photos, videos, and mappable data for Great Britain. Stored information is accessed from optical disks via a microcomputer. Users can call up locations randomly and zoom through a range of still representations—from a satellite mosaic to maps to photographs of particular streets—and access sequences of images with sound. Equally exciting multimedia possibilities involve use of sound overlays designed to accompany and explain map animations. Sound variables such as volume, duration, and pitch could be used to signal thresholds dealing with strength of relationships between variables, data quality, and speed of change or movement.

PROSPECT

Visualization is of growing importance to science in general. To these broader efforts, geographers offer an understanding of representation and a long history of use and development of representation, particularly of spatial representation. Cartographers, in particular, offer expertise in visual perception and its role in communication. In addition, cartographers and remote sensing specialists have worked with pattern recognition for years. Behavioral geographers and cartographers have developed models of how spatial information is acquired from maps and directly from the environment.

In working to advance the development of visualization tools both within the discipline and within science as a whole, geography draws from a wide cross section of fields. Both the history and the philosophy of science provide sources of information about how scientists have achieved insight, how psychologists have been learning about creativity and how to stimulate it, and how computer scientists are developing graphical interfaces, stereo displays, graphic coprocessors, and display algorithms that provide dramatic additions to our visualization tool kit. In addition, scientists in medicine, biochemistry, physics, chemistry, geosciences, meteorology, fluid dynamics, astronomy, and even political science (with its animated atlas of voting patterns) are developing computer visualization tools that can be applied to geographic problems. Statisticians, in complementary efforts, are developing exploratory graphic techniques that can be adapted to geographic applications.

Integration and analysis of data for such complex problems as geosphere-biosphere interaction, the impact of gentrification on homelessness, or geopolitical issues in a changing Europe will spur new visualization tools that provide both three-dimensional and dynamic depictions of available data and simulations of change. Emphasis will be on tools that help the analyst to identify patterns, anomalies, relationships, and hypotheses. To understand and react to global environmental crises or regional emergencies requires an efficiency of data processing that is only possible when human experts can interact with the computer to steer the system through the volumes of data available. What is needed to enable this steering is an intelligent map/graphic display that is an integral part of a GIS and that has the capability of responding to user queries (MacEachren 1987).

Geography has a history of bridging diverse worlds to synthesize information and to develop holistic approaches to problems. Development of scientific visualization tools and strategies will continue to offer opportunities to do just that.

NOTE

Most of the illustrations in this chapter were produced in the Deasy GeoGraphics Laboratory at Penn State. We are particularly grateful to Tom Davinroy of the lab for his work on a number of these illustrations and for managing the overall organization of illustrations for the chapter. We also thank Michael Hodgson and Reese Plews, Richard Pike and Gail Thelin, and Gary Shannon and Gerald Pyle for providing original prints, photographic negatives, or both, for illustrations from their research that are reproduced here.

REFERENCES

ABLER, RONALD F., JOHN S. ADAMS, and PETER R. GOULD. 1971. *Spatial organization: The geographer's view of the world.* Englewood Cliffs, N.J.: Prentice-Hall.

ARNHEIM, R. 1985. The double-edged mind: Intuition and the intellect. Chap. 5 in *Learning and teaching the ways of knowing, Eighty-fourth yearbook of the National Society for the Study of Education, Part 2*, ed. Elliot Eisner, 77–96. Chicago: University of Chicago Press.

BECKER, RICHARD, WILLIAM S. CLEVELAND, and ALLAN R. WILKS. 1987. Dynamic graphics for data analysis. *Statistical Science* 2:355–395.

BUTTENFIELD, BARBARA. 1990. Rules for automating feature distinctions. Paper presented at the symposium "Towards a Rule Base for Map Generalization," Syracuse, N.Y.

DIBIASE, DAVID W. 1989. Marketing cart lab services in a research-oriented university. Paper presented at the annual meeting of the North American Cartographic Information Society, Ann Arbor, Mich.

―――. 1990. Scientific visualization in the earth sciences, *Earth and Mineral Sciences* (Bulletin of the College of Earth and Mineral Sciences, Pennsylvania State University) 59, 2:13–18.

EAGLEMAN, J. R., V. U. MUIRHEAD, and N. WILLEMS. 1971. *Thunderstorms, tornadoes and damage to buildings*. Final research report to the University of Kansas. Environmental Control Administration, Department of Health, Education, and Welfare Grant #EC00303.

FRIEDHOFF, R. M., and W. BENZON. 1989. *Visualization: The second computer revolution*, New York: Harry N. Abrams.

GERSMEHL, PHILIP J. 1990. Choosing tools: Nine metaphors for map animation. *Cartographic Perspectives* 5:3–17.

GOODCHILD, MICHAEL F., and DONALD G. JANELLE. 1988. Specialization in the structure and organization of geography. *Annals of the Association of American Geographers* 78:1–28.

GUTKIND, A. E. 1956. Our world from the air; Conflict and adaptation. In *Man's role in changing the face of the earth*, ed. William L. Thomas, 1–44. Chicago: University of Chicago Press.

HODGSON, MICHAEL E. and REESE W. PLEWS. 1989. N-dimensional display of cluster means in feature space. *Photogrammetric Engineering and Remote Sensing* 55:613–619.

HOOVER, JOSEPH. 1989. *AIDS in Pennsylvania*. University Park: Deasy GeoGraphics Laboratory, Pennsylvania State University.

Larkin, J. H., and H. A. Simon. 1987. Why a diagram is (sometimes) worth ten thousand words. *Cognitive Science* 11:65–99.

LEWIS, PIERCE. 1976. The stages of metropolitan growth. In *New Orleans: The making of an urban landscape*, 31–66. Cambridge, Mass.: Ballinger.

MCCORMICK, B. H., T. A. DEFANTI, and M. D. BROWN. 1987. *Visualization in scientific computing*. Report to the National Science Foundation by the panel on Graphics, Image Processing, and Workstations. Baltimore: ACM SIGGRAPH.

MACEACHREN, ALAN M. 1987. The evolution of computer mapping and its implications for geography. *Journal of Geography* 86, 3:100–108.

MACEACHREN, ALAN M., and DAVID W. DIBIASE. In press. Animated maps of aggregated data: Conceptual and practical problems. *Cartography and geographic information systems* (18:221–229).

MACEACHREN, ALAN M. and JOHN H. GANTER. 1990. A pattern identification approach to cartographic visualization. *Cartographica* 27, 2:64–81.

MARTIS, KENNETH C. 1989. *The historical atlas of political parties in the United States Congress, 1789–1989*. Ruth Anderson Rowles and Gyula Pauer, cartographers. New York: McMillan.

MILLER, DAVID W., and JOHN MODELL. 1988. Teaching United States history with the Great American History Machine. *Historical Methods* 21, 3:121–134.

MOELLERING, H. 1989. A practical and efficient approach to the stereoscopic display and manipulation of cartographic objects. *Auto-Carto 9*, 1–4. Proceedings of the Ninth International Symposium on Computer-Assisted Cartography, Baltimore.

MOELLERING, H., L. FRITZ, T. NYERGES, B. LILES, N. CHRISMAN, C. POEPPELMEIER, W. SCHMIDT, and R. RUGG. 1988. The proposed national standard for digital cartographic data. *American Cartographer* 15:9–142.

MONMONIER, MARK 1972. Contiguity-biased class-interval selection: A method for simplifying patterns on statistical maps. *Geographical Review* 62 203–228.

———. 1989. Geographic brushing: Enhancing exploratory analysis of the scatterplot matrix. *Geographical Analysis* 21:81–84.

———. 1990. Strategies for the visualization of geographic time-series data. *Cartographica* 27, 1:30–45.

OMERNIK, JAMES M., and CHARLES F. POWERS. 1983. Total alkalinity of surface waters—A national map. *Annals of the Association of American Geographers* 73:133–136 plus map supplement.

OPENSHAW, S. and H. MOUNSEY. 1987. Geographic information systems and the BBC's Domesday Interactive Videodisc. *International Journal of Geographical Information Systems* 1:173–179.

PERKAL, J. 1966. An attempt at objective generalization. In *Michigan inter-university community of mathematical geographers*, ed. John Nystuen. Trans. W. Jackowski. Discussion paper 10. Ann Arbor: University of Michigan Department of Geography.

PETERSON, M. 1979. An evaluation of unclassed crossed-line choropleth mapping. *American Cartographer* 6:21–38.

PIKE, RICHARD J. and GAIL P. THELIN. 1991. Mapping the nation's physiography by computer. *Cartographic Perspectives* 8:15–24.

PLANE, DAVID A. 1984. Migration space: Doubly constrained gravity model mapping of relative interstate separation. *Annals of the Association of American Geographers* 74:244–256.

ROBERTS, CHARLES. 1990. Personal communication.

SACK, ROBERT D. 1988. The consumer's world: Place as context. *Annals of the Association of American Geographers* 74:642–664.

SHANNON, GARY W., and GERALD F. PYLE. 1988. The origin and diffusion of AIDS: A view from medical geography. *Annals of the Association of American Geographers* 79:1–24.

SHEPARD, R. N. 1978. The mental image. *American Psychologist* 33:125–137

SMITH, NEIL, BETSY DUNCAN, and LAURA REID. 1989. From disinvestment to reinvestment: Tax arrears and turning points in the East Village. *Housing Studies* 4:238–252.

SMITH, STEVE. 1989. The day the dam broke: The Johnstown flood of May 31, 1889. University Park: Deasy GeoGraphics Laboratory, Pennsylvania State University.

STEWART, G. S. 1953. *U.S. 40, cross section of the United States of America*. Boston: Houghton Mifflin.

STODDART, DAVID R., DENISE J. REED, and JONATHAN R. FRENCH. 1989. Understanding salt-marsh accretion, Scolt Head Island, Norfolk, England. *Estuaries* 12:228–236.

THROWER, NORMAN J. W. 1961. Animated cartography in the United States. *International yearbook of cartography* 1:20–30.

TOBLER, WALDO. 1973. Choropleth maps without class intervals, *Geographical Analysis* 5:262–265.

———. 1981. Depicting fiscal transfers. *Professional Geographer* 33:419–422.

TUFTE, EDWARD. 1983. *The visual display of quantitative information.* Cheshire, Conn.: Graphics.

TUKEY, JOHN W. 1977. *Exploratory data analysis.* Reading, Mass.: Addison-Wesley.

VALE, THOMAS R., and GERALDINE R. VALE. 1983. *U.S. 40 today: Thirty years of landscape change in America.* Madison: University of Wisconsin Press.

VEBLEN, THOMAS T. and DIANE C. LORENZ. 1988. Recent vegetation changes along the forest/steppe ecotone of northern Patagonia. *Annals of the Association of American Geographers* 78:93–111.

YARNAL, BRENT, and HENRY F. DIAZ. 1986. Relationships between extremes of the southern oscillation and the winter climate of the Anglo-American Pacific coast. *Journal of Climatology* 6:197–219.

Analysis

Michael F. Goodchild

────────────○────────────

THE SEARCH FOR PATTERN

Science seeks to make sense out of an apparently chaotic and unpredictable world by finding comparatively simple theories and laws that can explain natural and social behavior. Much scientific activity is now driven by economic motives; the control that comes from understanding natural and social processes ultimately leads to economic and social benefits. But simple curiosity also attracts many people to scientific careers and helps explain the altruistic nature of much basic science.

Geography's domain is the surface of the earth, the processes that shape it, and the activities that take place upon it. Geographers study the world of people and the spaces in which they live. People are naturally curious about the world, and in one sense geography is one of the most natural of the sciences in that the human and physical systems of the earth often excite human curiosity, if only in the fascination with maps and travel.

Sometimes the compulsion to explain arises from the unexpected simplicity of patterns. What determines the elegant form of river meanders, which take the same form in the wiggles of the smallest stream and in the massive, smooth curves of the lower Mississippi? What explains the symmetry of a distant view of Chicago's downtown skyline, with its smooth rise to a single peak? Why is the Atlantic coast from Charleston to Cape Hatteras composed of three huge but almost perfectly smooth scallops? Geometric form is a fertile source of possible explanations, because similarity of form between two widely different phenomena suggests some degree of similarity in their causative processes.

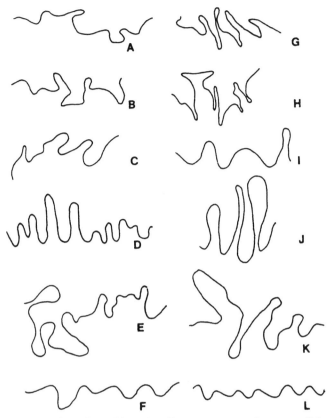

Fig. 7.1. Some Meandering Lines. Lines *A* to *E* represent natural streams; *F* is viscous fluid flow on an inclined plane; *G, H* and *K* are mountain roads; *I, J* and *L* are paths taken by a moth in a wind tunnel. SOURCE: Mark 1985:49.

Consider meander forms, for example (fig. 7.1). A road makes loops that resemble river meanders as it climbs a high pass to satisfy two constraints: an upper bound on gradient and a lower bound on turning radius. Rivers are also loosely constrained by turning radius, but what is the equivalent of a constraint on gradient?

The objective of science is to discover theories or laws that explain patterns and behavior in the natural and social worlds. The process of theory formation operates simultaneously in two directions—by induction, when observations suggest theories, and by deduction, when putative theories are tested by experiment. Induction works best when it deals with strikingly simple patterns, although the processes that produce such patterns may be extraordinarily complex. On the other hand simple processes often manifest themselves in complex patterns that are made confusing by the influence of extraneous factors. For example, the naked eye sees only confusion in the traffic patterns of major metropolitan areas, or in the branching of stream networks. In these cases we need a new perspective—a filter or a reworking of the data—before the patterns that are diagnostic of underlying laws become clear. The naked eye is simply too unselective and unfocused.

The term scientists give reworking or filtering scientific data is *analysis*. Analysis can range from simple display of information in the form of maps, tables, or graphs to complex arithmetic manipulation, but its objective is always the same—to present information about the world in a way that suggests explanation (induction), or to confirm some previously suspected explanation (deduction). In its simplest and most unstructured form, analysis consists of procedures learned as early as elementary school and by now largely intuitive. In more complicated and formal expressions, analysis may extend to exploratory data analysis (EDA). (Possible applications of EDA to geographical data will be explored later in this chapter.) Techniques of statistical analysis lie at the most structured end of the analytical spectrum. Many are highly complex and far from intuitive. An analyst's tool kit now contains many different and varied methods, and finding the right technique has become a major part of the analyst's task.

Although the purpose of analysis may be deductive—to test some previously established theory—its techniques are inherently general, and not specific to any particular theory. *Modeling*, on the other hand, begins with theory and uses arithmetic techniques to predict the consequences of the theory in a real setting. If the predictions are correct or fall within some accepted limits of accuracy, the theory is confirmed. The difference between *analysis* and *modeling* is not sharply defined, and the two terms are sometimes used almost interchangeably (see chapter 8).

The world is no longer as simple as it seemed in the early days of geographical analysis. Geographers no longer believe that the mere conjoining of objective analysis and raw geographical reality will yield powerful theory. Perhaps the simple problems have all been solved, or perhaps geographers were naïve to believe that the scientific ideal of neutral, dispassionate investigation of an objective reality was possible in the study of a phenomenon as intricate as the earth's surface.

In human geography the concept of dispassionate observation seems particularly unreasonable, and the positivist tradition of empirically verifiable theory has been increasingly rejected in the past two decades as a basis for theory building in human geography, mirroring trends in other social sciences, particularly sociology (Gregory 1978). At the extreme, those who reject positivism argue that observations tell us as much about the observer as about the observed, that science often has conservative political motives, and that critique is the major tool of analysis. In less extreme forms, positivist tools of analysis are retained, but interpretation is placed in a wider, less rigid context (Billinge, Gregory, and Martin 1984).

Fig. 7.2. The Snow Map of Cholera Incidence in the Area of Broad Street, London, in 1854. The contaminated water pump is located at the center of the map, just to the right of the D in BROAD STREET. SOURCE: Gilbert 1958.

Map Analysis

Despite its simplicity, the map is still a powerful tool in the geographical analyst's kit bag. The Snow map (fig. 7.2) remains the best known example of the power of maps to suggest simple yet powerful explanations. The map shows the residential locations of cholera victims in a section of London during an outbreak in September, 1854 that took more than five hundred lives. The strong symmetry of the pattern immediately suggests a cause located at the center of the cluster. To complete the explanation we need to know that drinking water was suspected as the carrier (deductive reasoning), that one of the local water pumps was located near the center of the cluster, and that the spatial extent of the cluster coincided with the area served by the well (its radius is similar to the distances traveled for water).

The informal, intuitive nature of much geographical analysis is worth em-

Automobile Sales in the coterminous United States

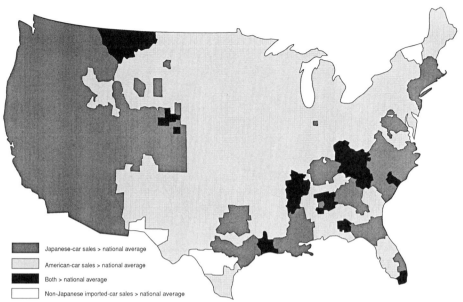

Fig. 7.3. Sales of Japanese and American Cars Relative to National Averages, by ADI, for 1989 and 1990 Models. SOURCE: Weiss et al. 1990:69.

phasizing because it is so simple and pervasive, and yet often overlooked. The August 1990 issue of the *Atlantic Monthly* included a map (fig. 7.3) showing areas of the United States where sales of Japanese and American cars exceeded their national averages of 23 percent and 64 percent respectively. Despite national advertising, sales of Japanese cars are higher around their ports of entry— Seattle, Washington; Long Beach, California; Houston, Texas; Jacksonville, Florida; Norfolk, Virginia; Newark, New Jersey; and Cranston, Rhode Island—than elsewhere. The map also shows the historic influence of Detroit and the broad economic disparities between the coasts and the interior— affluent, educated white-collar people are more likely to buy Japanese cars. The author of the accompanying discussion noted, "In an era of mass communications and glib evocations of a 'global village' it is sometimes easy to overlook the imperatives of geography," even though they become obvious once the data are displayed in geographical perspective (Weiss 1990). This simple, intuitive analysis nevertheless depends on an extensive knowledge of the geography of the United States. Note that the television market areas used as reporting zones are small enough to reveal patterns and effects that might have been missed had the data been presented for states.

Hochberg and Miller (1989) have drawn attention to a striking example of geographic perspective in the work of Mokyr (1983) on the Irish potato famine of the 1840s. Mokyr's table (7.1), lists upper bounds on computed "excess death rates" for Irish counties in alphabetical order. This lengthy and admi-

**Table 7.1. Upper Bounds on Average Annual "Excess Death Rates,"
by Irish County, 1846–1851**

County	Rate (*per 1,000 population*)
Antrim	20.3
Armagh	22.2
Carlow	8.8
Cavan	51.8
Clare	46.5
Cork	41.8
Donegal	18.7
Down	12.5
Dublin	0.7
Fermanagh	39.1
Galway	58.0
Kerry	36.1
Kildare	12.0
Kilkenny	18.1
King's	24.9
Langford	26.7
Limerick	20.9
Londonderry	10.1
Louth	14.6
Mayo	72.0
Meath	21.2
Monaghan	36.0
Queen's	29.1
Roscommon	57.4
Sligo	61.1
Tipperary	35.0
Tyrone	22.3
Waterford	30.8
Westmeath	26.3
Wexford	6.6
Wicklow	14.6

SOURCE: *Mokyr 1983:267.*

rable volume of economic historical analysis is remarkable for the absence of a single map—instead, the data are discussed in terms of the hierarchical grouping of counties into the four historic provinces of Ireland. Hochberg and Miller show that when the data are mapped (fig. 7.4), the eye is immediately struck by the simple core/periphery pattern centered on Dublin, and by its implications.

Today the straightforward causative mechanisms of diseases like cholera are well understood but the same techniques still help epidemiologists explain more difficult problems. Leukemia takes much longer to develop than cholera, so the location where the patient contracted the disease, because of migration, is not necessarily the current residential location. Long journeys to work may confuse patterns of incidence for diseases contracted in the work place. The incidence of Snow's cholera was sufficiently high that the uneven distribution of population within the study area could be ignored. But for a low-incidence disease like leukemia, uneven population distribution, uneven distribution of age and sex cohorts within the population, and the high mobility of late twentieth-century society must be considered.

Openshaw's recent work on leukemia clusters in the northeast of England illustrates the difficulties of using map analysis in this context (Openshaw et al. 1987, 1988; Openshaw 1988b). Statistical theory tells us the probability that a given number of cases will arise by chance in a given number of people, based on the average incidence in the population as a whole. When the probability is sufficiently low for an observed number of cases, that number is said to be statistically significant, suggesting the presence locally of a specific causal factor. But even though the likelihood is small, there is a chance of being wrong, a chance that is multiplied many times if large numbers of tests are performed—if an entire map is scanned looking for apparent clusters (fig. 7.5).

Successful detection of cancer clusters in the 1980s is far more difficult than was identification of the offending well in the 1850s. Openshaw has made good progress in harnessing the analytic power of computers to automate the search and testing process. He suggests this kind of automated geographical analysis is valuable for three reasons: because of the difficulty and complexity of the analysis involved, because of the social obligation to test for such things continuously as soon as data become available, and because of the very low cost of present-day computing.

But what would a cancer cluster mean if one were unambiguously identified? An above-average incidence of leukemia over an area of, say, a few city blocks would suggest some causative factor present in the same area. Exposure at the workplace would not show up as a local residential cluster unless all those exposed at work lived in the same neighborhood. Exposure through water or atmospheric contamination would not show up, either. It is difficult to imagine many causative factors that would appear as simple clusters, except for leakage

Average Annual Excess Death Rates in Ireland, 1846-51, by County (per 1000)

Fig. 7.4. Upper Bounds on Average Annual "Excess Death Rates" by Irish County, 1846–1851. SOURCE: Mokyr 1983:267; redrawn from Hochberg and Miller 1989.

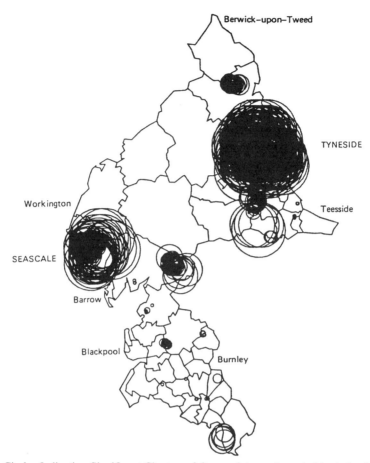

Fig. 7.5. Circles Indicating Significant Clusters of Cases of Acute Lymphoblastic Leukemia in the North of England, 1968–1985. SOURCE: Openshaw et al. 1987:348.

of a gas into basements from the surrounding subsoil. Because of the ease with which a map can suggest the presence of clusters, and because of the difficulties of confirming them through analysis and making sense of them etiologically, the drawbacks of map analysis may in some cases outweigh the benefits. Despite these difficulties, a map of individual cases of cancer is an enormously powerful argument in the hands of a local public interest group, which emphasizes the power of the map as a tool of analysis.

In looking for cancer clusters, the eye can estimate densities from patterns of points. This kind of analysis, in which a map of discrete events becomes a map of continuous variation in density, has broad applications (Silverman 1986). It has been used to identify trade areas from point maps of customer locations (Huff and Batsell 1977; O'Kelly and Miller 1989) and to create species range maps (Averack and Cawker 1982). Its validity lies in the ability of the human eye and mind to recognize areas of high density, and to find correlates—in the form of features or variables that coincide with areas of high density—that can explain them. When performing these often intuitive operations, analysts rely on knowledge of local conditions or on maps of other data for the same area (showing buried hazardous waste, for example). In Snow's case this was within the capabilities of the unaided eye and mind. But in Openshaw's analysis of low density leukemia cases, variations in population density and demographic mix, the sheer volume of data involved, and the need to carry out rigorous statistical testing combined to require high-speed computation. Ultimately, though, it is still the eye and mind that interpret the results, however they have been processed. As always, analysis massages data by a series of objective and logical steps into a form that is more suitable for interpretation and explanation.

Order in Chaos

Visitors approaching the Earth from space would see signs of order long before they touched down. They might overlook the complementary shapes of Africa and South America and their implications for plate tectonics, or the circular symmetry of settlement on the slopes of Mount Egmont in New Zealand, but they would certainly be struck by the circular irrigation patterns in North Africa and by the regularity imposed on western North America by the Public Land Survey System. Other features would seem hopelessly disordered at all scales. How, for example, would an extraterrestrial, or an earth scientist for that matter, explain the complexity of the Aegean Sea shoreline?

Faced with apparently complex patterns and behavior, geographers search for simplicity and the results have often been satisfying. There appears to be little general order in the sizes of each nation's cities, for example, but the German geographer Felix Auerbach observed that when cities are ranked by population and their populations are plotted against rank on double logarithmic paper, the points form a straight line (Haggett 1972). This rank-size relationship suggests no obvious explanation, although it has an impressive degree of generality. It has even been proposed as a planning tool (Berry 1961).

Another analysis that finds order in apparent chaos is the law of stream numbers (Horton 1945; Abrahams 1984). Stream networks are like trees—the outlet of a river is the base of the trunk and the stems branch at junctions and terminate at leaves. A numbering system devised by Strahler (1952) assigns an integer to each link in a stream basin. The source streams (which begin at leaves) are numbered one, two ones join to form a two, two twos form a three, and so forth. When the number of streams of a given order is plotted against stream order (using a logarithmic scale for the number of streams, but not stream order) the result is usually close to a straight line. Like the rank-size relationship, the law of stream numbers suggests no obvious explanation of the phenomenon it describes. Geographical analysis is replete with examples where straightforward data manipulation produces strikingly simple patterns with no obvious explanations.

The history of the spatial interaction model is somewhat more complex. The notion that masses of humanity might interact in a way analogous to the gravitational attraction between heavenly bodies has roots in the nineteenth Century. Calvert (1856) proposed a force of social attraction between major cities proportional to the product of their populations and inversely proportional to the square of the distance between them. Extensive research prior to 1969 on migration, social interaction, shopping trips, and so forth created a quandary. The *gravity model* seemed to fit interaction data surprisingly well, but it was difficult to find an explanation to replace Calvert's groundless speculation. However, since then research has produced numerous plausible explanations (Neidercorn and Bechdolt 1969; Wilston 1970; Fotheringham and O'Kelly 1989).

The forms and patterns observed in physical and human systems result from processes operating on complex initial conditions, subject to complex external factors. To understand the form of a hillside, one needs to know not only the processes that shape it but also the form of the hill before the action of those processes and the external factors that influenced their action. Of these, erosional processes are the analytical objective, yielding the laws or theories of geomorphology; the initial conditions and externalities are, by comparison, incidental. Ideally, researchers would prefer methods of analysis that remove incidentals, leaving only the processes themselves. In practice, analytic techniques remove some incidentals and leave others to be explained by uncertainties or imperfections in the operation of processes. Conventional analyses of spatial interaction data remove the effects of the sizes and spatial arrangement of places (Haynes and Fotheringham 1984).

Order in Complex Systems

The previous discussion was based on the belief that the processes that shape the physical and human worlds are simple, but that the patterns that result are complex because of the influence of initial conditions and other external factors. Yet despite its complexity, the world reveals surprising degrees of order and symmetry, as in the rank-size relationship. What mechanisms produce order in an apparently chaotic world?

Retailing has undergone enormous change in the past few decades as a result of innovations in transportation technology, consumer behavior, urban planning, and restructured retail industries. A *conservation principle* operates in retailing, however, that keeps some quantities constant when other aspects of the system change. Consumer spatial behavior in the Netherlands over the past few decades, for example, suggests that while consumers have changed the distances they travel to shop, their travel speed, their mode of transport, and the number of stores they visit, the quantities they have conserved despite those changes are the total time they devote to shopping and their frequency of trips, both of which have remained remarkably constant (Hupkes 1982).

The concept of *equilibrium* provides a powerful paradigm for understanding order in complex systems (chapter 11). If a process operates for sufficient time, the influence of initial conditions and externalities will disappear, leaving a pattern that is an outcome of the process alone. New external influences may disturb the equilibrium temporarily, but it will ultimately reestablish itself.

On a uniform landscape, the equilibrium pattern of retail centers is hexagonal—each place occupies the center of an identical, regular hexagonal trade area (Christaller 1933; Lösch 1954). Much effort was expended in the 1950s and 1960s searching for evidence of hexagonal patterns in suitable areas (Berry and Parr 1988). Unfortunately the economic advantages of a precise hexagonal network are few (Goodchild 1972), making it unlikely that equilibrium can evolve in the face of initial conditions and externalities, such as changes in the economics of retailing. Moreover, a real system can never escape the influence of certain initial conditions that are not met, particularly the assumption of a uniform landscape. If the underlying distribution of population is not uniform, the pattern will not be hexagonal. Analysis of the average number of edges of each center's trade area (Haggett and Chorley 1969) consistently produced a result very close to 6.0, which appeared to be evidence that real systems bore some resemblance to a purely hexagonal network, until it was noted that six edges is a necessary consequence of one of Euler's theorems that is applicable to *all* boundary networks irrespective of the processes that cause them (Getis and Boots 1978).

This example introduces an important point. Because analytic techniques are not tied to particular models, users may need high levels of expertise to interpret results correctly. An analysis that appears to a naïve user to confirm the existence of a nearly hexagonal network may do nothing of the kind. One way to safeguard against such misinterpretation is to repeat the analysis on a variety of patterns, including some which are known to have no hexagonal tendencies. If one takes the simple expedient of analyzing the average number of edges per polygon in the network visible on the side of a polystyrene coffee cup, one will find the answer similarly close to 6.0 and draw the obvious conclusion that the technique reveals nothing about hexagonality of the pattern. Analysis can reveal the appearance of order, structure, and simple process in complex systems but equally, as just noted, the results of analysis can be misinterpreted.

INTERPRETATION OF PATTERN

It seems unlikely that perfect order will ever be found in any geographic pattern. Although perfection may exist at atomic or molecular levels, there is simply too much complexity at the geographic range of scales. So geographers must deal with patterns that approximate the ideal, such as degrees of hexagonality. But this condition raises a constant specter of misinterpretation: How can geographers be sure that the degree of perfection observed is any greater than it would have been anyway? Although in Haggett and Charley's 1969 study the average number of edges was close to 6.0, the finding turned out to be no closer or more significant than it would have been in any pattern, however produced. A polystyrene cup, for example, displays the effects of a random process of boundary formation, one that differs from the retail center pattern only by the complete absence of any economic or behavioral pattern-forming processes. The statistician formalizes this kind of counterproposition as a null hypothesis (H_0).

In the absence of either a null hypothesis or of a clear idea of what a neutral or random pattern would produce, analysis will often mislead. What would someone expect to find purely by chance in the case of the size distribution of cities or the numbers of streams of different orders in a stream network? In a tree network with a given number of leaves created by a random process in which all alternative arrangements of junctions are equally likely to occur, the most likely distribution of numbers of streams by order is the Horton law of stream numbers (Shreve 1966, 1967; Abrahams 1984). In other words the law confirms nothing about the processes operating on the landscape except that they are sufficiently complex that any arrangement is equally likely. The action of geological and geomorphological processes is more clearly confirmed by deviations from the Horton law than by adherence to it. In effect, the law is its own null hypothesis. Similar interpretations of the rank-size relationship and the gravity law have been proposed (Curry 1964; Wilson 1970).

Geographers work in complex worlds where processes which may themselves be simple operate under complex conditions and produce patterns that are never perfect. They cope with that complexity by allowing a level of uncertainty or error to creep into tests of laws, models, and theories, and they find satisfaction whenever the world comes reasonably close to predictions. But there is always the possibility that their ideas of reasonably close are too wide, no better than what would have occurred if the law had been incorrect or unknown. Unfortunately, many apparent laws inferred from the appearance of order turn out on closer inspection to be no more than null hypotheses. This is not to say that they are of no value. The rank-size relationship has little value as a planning tool if it is no more than what should be expected from a random configuration of cities. But the gravity law, even though it is no more than the most likely aggregate pattern of randomly acting individuals, helps retailers predict consumer travel. They use it to estimate the effects of changes in population distributions or alterations in traffic patterns that result from changes in road networks.

Unfortunately the formulation of an appropriate null hypothesis is far from

easy in many cases. The rank-size rule is normally expressed as a double logarithmic plot, suggesting that the statistical technique of regression might be
used, as it contains its own null hypothesis and testing procedure (Clark and
Hosking 1986). However the null hypothesis of regression requires that the
axes of the graph represent independent measures of two different variables,
whereas in the case of the rank-size relationship, one variable (rank) is derived
from the other (size); independence does not exist. In fact it is impossible to
change the population of any city significantly without also changing its rank.

Geographers continue to wrestle with these and related problems. What, for
example, is the appropriate null hypothesis for topography? How would the
physical landscape look in the absence of some suspected geomorphological
process or effect, or in a hypothetical state before the action of such processes?
What is the appropriate null hypothesis for a boundary network? How would a
map look if it were divided into random counties or trade areas, without the
influence of spatial-political or spatial-economic processes (Pielou 1965)?

The days of ransacking reality in search of simple patterns are over; geographers no longer believe that analysis will demonstrate essential simplicity in
the arrangement of the earth's surface or of human activities upon it. They may
still believe in the essential simplicity of certain processes, but at the scale of
geographic observation, even simple propositions such as the Navier-Stokes
equations produce complex solutions (Scheidegger 1970), and it seems unlikely that human behavior will ever be reduced to simple principles unless
they are hedged with substantial uncertainty.

Given these reservations, what role does analysis play in modern geography? Current practice seems to fall into several distinct categories:

1. Identifying local departures from normality. The Openshaw
   work cited earlier (Openshaw et al. 1987; 1988; Openshaw
   1988b) falls into a class of applications in which analysis is used
   to isolate instances or places where behavior differs from expectations. The problems of evaluating normality may be sufficiently complex to require sophisticated computers.
2. Evaluating the strength of effects. Although the mechanisms that
   affect systems may be known in general terms, the specifics are
   often missing. We may know, for example, that distance affects
   shopping behavior, but not the specific magnitude of its influence or its relative influence compared to other factors in a given
   area. We may suspect that a certain factor is important, but not
   know the direction of its effect.
3. Predicting needs and outcomes. Studying correlations and relationships that have nothing to do with understanding or explaining the variables involved may nevertheless be useful for
   practical reasons. For example, it may be useful to collect and

analyze data on the relationship between city population and number of fire stations, not because of any natural curiosity about the processes involved but in order to compare a given city's level of fire service to a norm.

4. Exploring data. Despite the importance of deduction, the mechanisms that affect many types of phenomena are so complex as to defy attempts to describe them deductively. In such cases the inductive approach is clearly more appropriate. So little is known, for example, about the etiology of some forms of cancer that it is worthwhile analyzing relationships for which no obvious deductive basis exists, such as water hardness or altitude.

THE TOOLS OF ANALYSIS

Geographers employ standard analytical tools, but some unique issues arise in applying them to spatial data and problems.

Classification

The simplest form of analysis is classification, the act of placing events, objects, or observations into categories. The number of categories to be used, their representative archetypes, and possible hierarchical relationships among categories require decisions that will reflect the analyst's view of the world. In science generally, the Linnaean hierarchical system of biological classification is perhaps best known, but regions are geography's own distinctive contribution. Although simple in concept, geographers from Hartshorne (1939) to Hart (1982) have celebrated the power of regional classification to yield insights into complex geographic patterns and processes. While numerical taxonomy tries to reduce classification to an objective process by minimizing within-class variation, regional geographers have identified many types of regions (formal, functional, nodal, and equitable), and have devised many analytic techniques for objective delineation of each. Nevertheless, the sheer complexity of regional systems often defies formal analytical treatment.

Equiprobability

In the absence of other information analysts assume that all possible states of a system are equally likely. This elementary proposition can be dressed up as the principle of insufficient knowledge, but it merely states an obvious truth. The proposition is nevertheless sufficient, as we have seen, to provide a theoretical basis for the gravity model and the law of stream numbers. It is also the basis of the null hypothesis, and it closely approximates what we intuitively mean by random. Analysis that determines the extent to which a system is not in its most probable state, or the extent to which certain states are not equally likely, can be useful, as in the example of cancer clusters (Openshaw et al. 1987, 1988; Openshaw 1988b). Geographers make use both of statistics that measure the relative likelihood of an observed state of a system and compare it to the most likely state, and of the principle that a closed system will tend through time to attain its most likely state—maximum entropy.

Dimensional Analysis

A great deal can often be learned from a simple analysis of the structure of suspected relationships. Consider the problem of analyzing the number of trips made annually by consumers from a neighborhood to a shopping center. It might be supposed that the number of trips (T) depends on the population of the neighborhood (P), the floor area of the shopping center (A), and the distance between them (D). One possibility would be:

$$T = aP + bA - cD$$

where a, b, and c are constants to be determined. If two neighborhoods are merged, however, one would expect the number of trips for both to equal the sum of the trips from each individually, which would not result from applying this model. It was essentially this type of analysis that led Huff (1963) to use the gravity model to predict shopping behavior.

Dimensional analysis is often used to analyze the complex patterns produced by the behavior of viscous fluids, such as sand dunes, riverbed ripples, and offshore bars. Suppose, for example, that a pattern has a characteristic length, such as the distance between successive ripples (Goodchild and Ford 1971). One might expect the length to be influenced by the density, viscosity, and velocity of the fluid. Each of these parameters has its own characteristic dimensions of measurement, which are some combination of length, mass, and time units in each case. So a simple analysis of the dimensions can lead to intelligent guesswork about how the parameters are related, without any knowledge of the processes involved.

Linear Models

Statistical analysis offers a vast and rich tool kit of methods based on assumptions of linearity; that is, that the relationship between two variables, x and y, can be expressed as

$$y = a + bx$$

where a and b are constants. In some cases these methods are applied in the knowledge or expectation that the relationship is truly linear, but more often the analysis is carried out in the absence of any theory or specific expectations. When the variable y is thought to be affected by more than one x, the tool kit offers multiple regression, which assumes that all influences are linear, and that the effects of each x combine additively:

$$y = b_0 + b_1 x_1 + b_2 x_2 + \ldots$$

More complex situations are covered by a lengthy list of techniques, all based on the linear model: factor analysis, canonical correlation, discriminant analysis, logit regression, and so forth.

The linear model has been helpful over the last three decades as a means of analyzing data in the absence of well formulated theory. Its use embeds some degree of prior understanding of the processes operating, but in general it has been used in exploratory hypothesis generation rather than for hypothesis testing. The linear model succeeds because linearity is a reasonable first approximation of the behavior of y over the range of x values observed in most data sets. A more complex function may improve the degree of fit, but usually has no conceptual justification. For example one could develop a program to test large numbers of functions to find the one that best fits (Openshaw 1988a), but there would usually be no way to explain *why* one function provided a better fit than any other. The linear model conforms to the principle of Occam's razor and to the principle of insufficient knowledge. There are instances, nevertheless, where the linear model is clearly inappropriate. For example, the variable x may be circular, such as a compass bearing ($x = 360$ and $x = 0$ have the same effect). Age often has similar effects—research on the spatial distributions of people in urban areas has repeatedly shown that young and old have similar spatial distributions and are strongly differentiated from the middle-aged population.

Problems of Borrowed Tools

Few statistical techniques were developed explicitly for spatial analysis. In fact, rearranging the locations of the cases in most standard techniques would cause no change in results. A number of spatial pattern analysis techniques have enjoyed periods of extensive use in geographical research, particularly point pattern analysis (Boots and Getis 1988). Because these techniques assume that observed patterns are generated by stochastic processes, they suffer from two problems. First, since the hypothesized process is statistical, tests are generally weak, and vastly different processes can lead to the same statistical model. Second, the models describe only the simplest situations, and it is often difficult to modify them to allow for departures from their basic assumptions.

More general problems arise when standard statistical techniques are applied in spatial settings. Many techniques rely on the assumption of independence of observations—no observation should provide information about any other.

Spatial series are strongly autocorrelated; that is, dependence exists between observations that are proximal. "The first law of geography is that everything is related to everything else, but near things are more related than distant things" (Tobler 1970). Spatial analysts have developed methods for measuring the degree of spatial dependence present in data and the scale over which such dependence operates, and there is an extensive literature on spatially dependent processes (Griffith 1987; Goodchild 1988; Odland 1988). It has proven much more difficult to adapt statistical techniques such as regression for spatial analysis (Clifford and Richardson 1985) or to provide simple, easy-to-use implementations. As things stand, the geographic analyst must still rely on statistical packages that deliver the orthodox, nonspatial view.

Many data used to analyze the spatial patterns produced by human activities are available for aggregate units or reporting zones such as counties, municipalities, and census tracts. Definitions of those zones affect analysis in ways intuition suggests will be similar to the effects of sampling (Openshaw and Taylor 1979). But whereas it is usually possible to quantify sampling errors (for example by replication of an experiment), the ways that reporting-zone definitions affect the outcome of analyses are usually unquantified and unknown. In those cases where these definitions have been evaluated, the errors induced by them are often much larger than intuition would suggest.

NEW TECHNOLOGIES

Geographic Information Systems

The first burst of interest in geographical analysis came in the late 1950s and the 1960s when digital computers were beginning to affect academic research. In the 1980s and 1990s, geographical analysis was given renewed impetus by great interest in the technologies known as geographic information systems, or GIS (Burrough 1986; Aronoff 1989; Star and Estes 1990). GIS is often seen as the technology of delivery for spatial analysis, just as statistical packages deliver the technology of statistical analysis. Techniques that have long been buried in the literature of quantitative geography are now being rediscovered because of the practical motivation provided by GIS.

These systems are changing the role of geographical analysis and spatial data. Whether they represent a new paradigm remains to be seen (Hay 1989). Analysis will certainly become more data driven because of the overwhelming advantages offered by data that are already in digital form over data that must be digitized. Already, systems built around the TIGER (topologically integrated geographic encoding and referencing) system street files of the U.S. Bureau of the Census and the DEM (digital elevation model) topographic data of the U.S. Geological Survey have spawned computer systems designed to capitalize on the analytic capabilities these data sets support.

In the 1960s and 1970s it was comparatively easy to analyze the nonspatial aspects of data because packaged statistical software was readily available. Processing and analyzing spatial components were much more difficult, and a significant split developed between quantitative geography and the more spatially oriented fields of cartography and remote sensing. GIS has fostered a reemergence of the spatial aspects of quantitative data, and it has put pressure on theorists to deliver the missing components of spatial analysis, such as the spatial form of regression.

The use of high-precision processors for spatial data also raises serious questions about the quality of most spatial information. Small-scale maps of world soils, for example, are useful ways to deliver generalized information about the geography of the great soil groups, but they become hopelessly inaccurate when they are digitized and displayed at high precision. The prominence of GIS has led to renewed interest in the nature of spatial information, and to recognition of the unique problems of describing its errors and inaccuracies (Goodchild and Gopal 1989).

Exploratory Geographical Analysis

Proponents of exploratory data analysis (EDA) argue that the vitally important process of exploring data can be formalized and improved by a set of simple but rigorous methods (Tukey 1977). In essence, graphic displays allow users to perceive patterns and symmetries in data that are not otherwise apparent. Faced with a mass of data, EDA allows the user to extract information that is more meaningful or more indicative of underlying processes than the raw data themselves.

As shown in the discussion of cancer clusters earlier in this chapter, spatial data present special problems of perception and intuition, suggesting that the value of a spatial form of EDA may be greater than that of Tukey's nonspatial form. The eye is not particularly good at estimating densities from sparse point patterns, at removing the effects of varying population densities, or at integrating densities over extended areas. Recently several tools have become available that hint at the range of capabilities that might be built into a system for exploratory spatial analysis (ESA). The spatial integration function used to estimate potential markets for proposed retail facilities, for example, could be supported as a real-time function on the current generation of desktop computers. The Great American History Machine project (see chapter 6) of the Department of History at Carnegie Mellon University illustrates the power of computers to simultaneously explore both the spatial and temporal aspects of data. Its different screen windows for spatial and temporal series link them so that a selection in one (pointing to a location) produces an immediate update in the other (display of the time series for the selected place). Systems like this try to overcome the inherently high dimensionality of spatial data by taking advantage of the computer's ability to provide several views at once.

Although many analytic techniques present a simpler view of the world, real geographical distributions have the annoying habit of revealing more detail, apparently without limit, as they are examined more closely. Coastlines become more contorted, islands appear in lakes, and small vortices appear within the simple structures of atmospheric disturbances. Maps are of little help in coming to grips with the problems of scale and resolution, as they portray the world at fixed scale and suggest fixed scales of analysis. But given a suitably constructed data base, an ESA system could zoom and pan at will, with appropriate aggregation and disaggregation of data. Such an ESA system could bridge the conceptual gaps between micro- and macro-scale analysis and modeling.

One of the most suitable applications for ESA is global science. The need to project the earth's curved surface onto a flat sheet has created great problems in cartography and consequently in attempts to model and analyze processes on the globe. It is difficult, for example, to subdivide a projected globe into finite elements that can be used in modeling atmospheric processes without creating severe distortion. In principle, the capabilities of current hardware are sufficient to support a global science work station that would constantly display the earth in orthographic projection, with continually updated aspect. Data densities of perhaps 10^5 or even 10^6 finite elements can now be "draped" over the sphere and rotated or browsed in near-real time. Geographers stand on the verge of being able to work with the globe itself, thereby escaping the long-standing limitations imposed by fixed projections.

NEW DIRECTIONS

The geographer's view of the world has always been colored by the data available for analysis and by the ways those data have been presented. Presentation of data by reporting zones introduces its own biases, as noted above. Cartographic presentation, despite its power to suggest causes and correlates, imposes a view of the world that is constant in scale and organized in a Euclidean space. One of the major intellectual breakthroughs in GIS has been the development of hierarchical spatial data structures such as the quadtree that are based on entirely new ways of viewing spatial distributions and that have no roots in conventional display (Samet 1984). Interestingly, the quadtree's hierarchy is in many ways analogous to other breakthroughs where scale is similarly seen as a variable rather than as a constant attribute of data. The fractal literature introduced the notion that there might be system to the behavior of spatially distributed phenomena across different scales, and that measures of geographic features at one scale might be predictable from measures at other scales (Mandelbrot 1977; 1982). That perspective provides a conceptual and mathematical framework that may permit GIS to escape the limitations of fixed scales. Fractals and scaling phenomena represent a radical departure for a discipline that has traditionally begun with assumptions of spatial uniformity. How, for example, might central place theory have developed if the underlying assumption was that the distribution of human population on the face of the earth is self-similar (appears the same at all scales) rather than uniform?

The most important contribution of GIS to spatial analysis has been to draw attention to aspects of spatial analysis that are incidental and that confuse interpretation, such as data accuracy, the influence of reporting zones, and scale. All are potentially more avoidable in a GIS environment than in the traditional, manual environment of spatial analysis. Frame-independent spatial analysis, a body of methods whose results are independent of the spatial frame—would yield additional progress (Tobler 1989). Spatial analysis is at somewhat of a disadvantage because it lacks sound underpinnings. Unlike disciplines such as psychology, geography lacks a comprehensive theory of spatial information or spatial relationships. GIS provides a candidate theory in its relational data model, but implementation is still incomplete and unsatisfactory. The natural sciences and remote sensing see spatial variation as continuous and amenable to associated techniques such as spectral analysis. Architecture and urban planning see the world as populated by objects. Geographical analysis lies in the awkward area where those two views of the world—continuous variation versus objects—compete, each being more suitable for certain purposes. Thus GIS remains, to some extent, split between the corresponding raster and vector views of the world. No system yet provides complete interchangeability between the two data models.

ANALYSIS AND UNDERSTANDING

The objectives of spatial analysis seem clear: to filter or process raw data so that they are more suggestive, easier to interpret, and more helpful to understanding the processes that operate on the surface of the earth. Spatial analysis has a well-defined tool kit that was developed in the early days of the quantitative revolution, although gaps exist in certain areas, especially in the very simplest forms of ESA. The results of any analysis will always admit a variety of interpretations, correct and false, and analysis, with its roots in induction, must compete with more deductive methods in the development of true understanding.

Despite the simplicity and elegance of many spatial patterns, analysis of static forms leads inevitably to ambiguities when inferences are made about processes. Yet the study of form in and of itself has legitimate purpose, and the form versus process debate will continue in geography. Much can be learned about the world by observing its forms and patterns, even when such observation does not contribute directly to their explanation; analysis plays a useful role as formalized description.

New technology and the interest it generates in the world of applications provide renewed impetus to spatial analysis and at the same time a significant change of perspective that is related to new viewpoints in science generally—a computationally intensive form of investigation developed because most of the simple problems have been solved. Chaos, for example, is a radical departure from previous ways of looking at complex systems (Gleick 1987). The roles of simulation and visualization provide simple examples of these changes. In his 1977 and 1982 works Benoit Mandelbrot presented striking illustrations of simulated landscapes that were generated by statistical models that made no attempt to incorporate real geological or geomorphological effects. The same technology now provides realistic simulations of trees, clouds, and numerous other complex natural phenomena. The test of realism is purely visual but the technology and the plausibility of its results raise awkward questions. Although there is no attempt to explain or understand, these techniques are capable of generating convincing simulations of real forms that conventional analyses of process cannot yet produce.

Tools of analysis will continue to improve, providing new power to visualize, simulate, and display information in its true spatial context. These changes offer exciting possibilities for analysis, suggesting new roles for simple methods of exploring and presenting data that may reverse the trend toward more and more mathematical sophistication. The broader implications of such new tools are profound. Geographers live in worlds of limited data where methods of socioeconomic data gathering are little changed from those used a hundred years ago and where labor-intensive methods of data collection continue to lose ground. Geography's worlds change with increasing speed. New analytical tools cannot reverse those trends, but they will provide a means of maintaining balance between data and geographers' abilities to figure out what they mean.

NOTE

The National Center for Geographic Information and Analysis is supported by the National Science Foundation, Grant SES 88–10917.

REFERENCES

ABRAHAMS, ATHOL D. 1984. Channel networks: A geomorphological perspective. *Water Resources Research* 20:161–188.

ARONOFF, S. 1989. *Geographic information systems: A management perspective*. Ottawa: WDL Publications.

AVERACK, R. and K. CAWKER. 1982. Climatic controls on the range limits of *Celtis occidentalis* in southern Ontario: A discriminant analysis. *Ontario Geography* 20:21–31.

BERRY, BRIAN J. L. 1961. City size distribution and economic development. *Economic Development and Cultural Change* 9:583.

BERRY, BRIAN J. L., and J. B. PARR with B. J. EPSTEIN, A. GHOSH and R.H.T. SMITH. 1988. *Market centers and retail location: theory and applications*. Englewood Cliffs, New Jersey: Prentice-Hall.

BILLINGE, MARK, DEREK GREGORY and R. L. MARTIN 1984. *Recollections of a revolution: Geography as spatial science*. New York: St. Martin's.

BOOTS, BARRY N., and ARTHUR GETIS 1988. *Point pattern analysis*. Scientific Geography Series no. 8. Newbury Park, Calif.: Sage.

BURROUGH, PETER A. 1986. *Principles of geographical information systems for land resources assessment*. Oxford: Clarendon.

CALVERT, G. H. 1856. *Introduction to social science: A discourse in three parts*. New York: Redfield.

CHRISTALLER, WALTER. 1933. *Die zentralen orte in Süd-deutschland*. Jena, Ger.: Fischer.

CLARK, WILLIAM A. V., and P. L. HOSKING. 1986. *Statistical methods for geographers*. New York: Wiley.

CLIFFORD, P., and S. RICHARDSON 1985. Testing the association between two spatial processes. *Statistics and Decisions* (supplement) 2:155–160.

CRAIG, R. L. 1980. A computer program for the simulation of landform erosion. *Computers and Geosciences* 6:111–142.

CURRY, L. 1964. The Random spatial economy: An exploration in settlement theory. *Annals of the Association of American Geographers* 54:138–146.

FOTHERINGHAM, A. STUART, and M. E. O'KELLY. 1989. *Spatial interaction models: Formulations and applications*. Amsterdam: Kluwer Academic.

GETIS, ARTHUR, and BARRY N. BOOTS. 1978. *Models of spatial processes*. New York: Cambridge University Press.

GILBERT, E. W. 1958. Pioneer maps of health and disease in England. *Geographical Journal* 124:172–183.

GLEICK, J. 1987. *Chaos: Making a new science*. New York: Viking.

GOODCHILD, MICHAEL F. 1972. The trade area of a displaced hexagonal lattice point. *Geographical Analysis* 4:105–107.

———. 1988. *Spatial autocorrelation; Concepts and techniques in modern geography*. CATMOG No. 47. Norwich: Geo Books.

GOODCHILD, MICHAEL F. and D. C. FORD. 1971. Analysis of scallop patterns by simulation under controlled conditions. *Journal of Geology* 79:52–62.

GOODCHILD, MICHAEL F., and S. GOPAL, eds. 1989. *The accuracy of spatial databases*. Basingstoke, Eng.: Taylor and Francis.

GREGORY, DEREK. 1978. *Ideology, science and human geography*. London: Hutchinson.

GRIFFITH, D. A. 1987. *Spatial autocorrelation: A primer*. Resource Publications in Geography 1985–4.Washington, D.C.: Association of American Geographers.

HAGGETT, PETER. 1972. *Geography: A modern synthesis*. New York: Harper & Row.

HAGGETT, PETER, and RICHARD J. CHORLEY. 1969. *Network analysis in geography*. London: Edward Arnold.

HART, JOHN FRASER. 1982. Presidential address: The highest form of the geographer's art. *Annals of the Association of American Geographers* 72:1–29.

HARTSHORNE, RICHARD. 1939. The nature of geography: A critical survey of current thought in the light of the past. *Annals of the Association of American Geographers* 29, 2–4:171–645.

HAY, A. M. 1989. Commentary. *Environment and Planning A* 21:709–710.

HAYNES, KINGSLEY E., and A. STUART FOTHERINGHAM. 1984. *Gravity and spatial interaction models*. Scientific Geography Series 2. Beverly Hills, Calif.: Sage.

HOCHBERG, L., and D. W. MILLER. 1989. Regional boundaries and urban hierarchy in pre-famine Ireland: A preliminary assessment. Paper presented at the annual meeting of the Social Science History Association, Washington, D.C.

HORTON, R. E. 1945. Erosional development of streams and their drainage basins: Hydrophysical approach to quantitative morphology. *Geological Society of America Bulletin* 56:275–370.

HUFF, DAVID L. 1963. A probabilistic analysis of shopping center trade areas. *Land Economics* 39:81–90.

HUFF, DAVID L., and R. R. BATSELL. 1977. Delimiting the areal extent of a market area. *Journal of Marketing Research* 14:581–585.

HUPKES, G. 1982. The law of constant travel-time and trip rates. *Futures* 14:38–46.

LÖSCH, AUGUST. 1954. *The economics of location*. New Haven: Yale University Press.

MANDELBROT, BENOIT B. 1977. How long is the coast of Britain? Statistical self-similarity and fractional dimension. *Science* 156:636–638.

———. 1982. *The fractal geometry of nature*. San Francisco: Freeman.

MARK, DAVID M. 1985. Fundamental spatial patterns: The meander. *Ontario Geography* 25:41–53.

MOKYR, J. 1983. *Why Ireland starved: A quantitative and analytical history of the Irish economy, 1800–1850*. London: Allen & Unwin.

NEIDERCORN, J. H., and B. V. BECHDOLDT, JR. 1969. An economic derivation of the "gravity law" of spatial interaction. *Journal of Regional Science* 9:273–282.

ODLAND, JOHN. 1988. *Spatial autocorrelation*. Scientific Geography Series no. 9. Newbury Park, Calif.: Sage.

O'KELLY, MORTON E., and H. MILLER. 1989. A synthesis of some market area delimitation models. *Growth and Change* 20, 3:14–33.

OPENSHAW, STANLEY. 1988a. Building an automated modelling system to explore a universe of spatial interaction models. *Geographical Analysis* 20:31–46.

————. 1988b. Leukaemia patterns in northern England: A new method of finding cancer clusters. *Northern Economic Review* 16:52–69.

OPENSHAW, STANLEY, M. CHARLTON, A. W. CRAFT, and J. M. BIRCH. 1988. An investigation of leukaemia clusters by use of a geographical analysis machine. *Lancet,* 6 Feb., 272–273.

OPENSHAW, STANLEY, M. CHARLTON, C. WYMER, and C. CRAFT. 1987. A Mark I geographical analysis machine for the automated analysis of past data sets. *International Journal of Geographical Information Systems* 1:335–358.

OPENSHAW, STANLEY, and PETER J. TAYLOR. 1979. A million or so correlation coefficients: Three experiments on the modifiable areal unit problem. In *Statistical applications in the spatial sciences,* ed. N. Wrigley, 127–144. London: Pion.

PIELOU, E. C. 1965. The concept of randomness in the patterns of mosaics. *Biometrics* 21:908–920.

SAMET, H. 1984. The quadtree and related hierarchical data structures. *Computing Surveys* 16:187–260.

SCHEIDEGGER, A. E. 1970. *Theoretical geomorphology.* New York: Springer Verlag.

SHREVE, R. L. 1966. Statistical law of stream numbers. *Journal of Geology* 74:17–37.

————. 1967. Infinite topologically random channel networks. *Journal of Geology* 75:178–186.

SILVERMAN, B. W. 1986. *Density estimation for statistics and data analysis.* London: Chapman and Hall.

STAR, J., and J. E. ESTES. 1990. *Geographic information systems: an introduction.* Englewood Cliffs, N.J.: Prentice-Hall.

STRAHLER, ARTHUR N. 1952. Hypsometric analysis of erosional topography. *Geological Society of America Bulletin,* 63:1117–1142.

TOBLER, WALDO R. 1970. A computer movie simulating urban growth in the Detroit region. *Economic Geography* (supplement) 46:234–240.

————. 1989. Frame independent spatial analysis. In *Accuracy of spatial databases,* ed. Michael F. Goodchild and S. Gopal, 115–122. Basingstoke, Eng.: Taylor and Francis.

TUKEY, J. W. 1977. *Exploratory data analysis.* Reading Mass.: Addison-Wesley.

WEISS, M. J., CLARITAS CORP., and R. L. POLK & CO. 1990. At last count: Japanese and American cars. *Atlantic Monthly,* Aug., 69.

WILSON, ALLAN G. 1970. *Entropy in urban and regional modelling.* London: Pion.

Modeling

Cort J. Willmott and Gary L. Gaile

· · · ———————————————— ◯ ———————————————— · · ·

REALITY, MODELS, AND KNOWLEDGE

Knowledge of the world depends on models, whether they are idiosyncratic interpretations, culturally based stereotypes, or sets of differential equations that describe environmental systems. Perceptions are influenced by biology, intellectual potential, previous learning, and the stimuli of the moment. Knowledge, perceptions, and understanding arise largely from the ways people filter information: from the ways individuals separate the important from the unimportant, the rule from the exception, the stereotype from the atypical, the signal from the noise. All such evaluations are based on the existence of one or more mental models. On the personal level, modeling processes may be complex, hidden, or not well understood; they also may be illogical and unamenable to articulation or precise specification. On the scientific level, models should be as explicit as possible, and they may be precisely stated and highly quantitative. Scientific modeling often begins with theory and uses mathematical techniques to explore the implications of the theory in real settings.

Abstractions of Reality

Models never contain all the detail of the real systems they represent. All geographic systems and phenomena consist of basic patterns or trends (signals), and of deviations (noise) that may mask those signals. By highlighting the regular behavior of systems—the rules rather than the exceptions—models reduce complexity and reveal system structures and functions. Viewing systems in terms of signal and noise also implies that noise—whether it takes the form of unsystematic error, important anomalies, or trivial bias—can only be identified and evaluated in the context of a model.

Abstract model building requires symbolic representation using building blocks that include the alphabets, words, and phrases of language, as well as mathematics and pictures. *Linguistic* models rely on words and pictures; metaphor is an essential component of such models (Buttimer 1990). Linguistic models may be eloquent, entertaining, or evocative depictions of concepts, places, or processes. Such models do not cleanly separate signal from noise, and the relative importance of model components is often vague. Linguistic models, as a result, are difficult to verify, replicate, and extend. *Mathematical* models permit verification, replication, and extension, and they are internally consistent. Relationships among model components are explicitly specified according to mathematical rules, and therefore the path from premises to conclusions is well marked. This chapter focuses primarily on mathematical models and their uses in geography.

History and Background

Richard Chorley and Peter Haggett (1967) extolled the many virtues of models in geography and the model-based paradigm. They also presented an array of modeling applications to topics ranging from geomorphology to the locations of settlements. Chorley and Haggett foresaw the central role that modeling would come to play in geography, and the number and variety of models geographers would develop. But factors about which Chorley and Haggett could only speculate in 1967 also have profoundly affected modeling in geography and in science generally. Advances in computing, evidence that chaos may exist within deterministic systems (Lorenz 1964), and the epistemological restructuring of geography (MacMillan 1989) have coevolved to produce a diversity of models that would have surprised even the most expansive thinkers of the 1960s. Perhaps the 1960s were halcyon years for statistically based modeling. But as statistical methods became generally accepted, debate turned to applied, technical, and epistemological issues. Geographers also sought a wider array of relevant and challenging problems, and concern for structure grew into concern for process. With greater understanding of the scientific method, its inadequacies became more apparent, and a critical literature began to develop within human geography.

Dissimilar theoretical bases also spawned divergence in the types of models developed by physical and human geographers. Within the physical sciences, well-accepted laws and theories about small-scale aggregate behavior have existed for centuries: contributions in mechanics by Galileo and Newton in the sixteenth and seventeenth centuries immediately come to mind, as do later extensions by Navier-Stokes to hydrodynamics. Such laws provide insights about the order, causes, and effects of events, as well as knowledge of what later came to be called feedback. Within human affairs, invariant laws of behavior exist at neither the individual nor the aggregate level. Changes of state within models that account for individual behavior must therefore be handled by assigning probabilities to possible outcomes from each individual's actions. Each action by each individual at each time and place then can be simulated. Aggregate models of human actions are frequently modeled as one or more statistical functions among independent and dependent variables that may have little or no theoretical link but that exhibit some empirical correlation.

Prelude to Modeling

One of two cornerstones of the scientific method is *deduction*. A set of hypotheses is constructed on the basis of another set of simpler assumptions taken as premises. Deductive reasoning develops preliminary hypotheses before data are collected and analyzed to empirically verify them. Brian Hanson (1990), for example, formulated a useful theory of icecap thermal changes—a logical sequence of hypotheses about glacial movements that took place over thousand-year time scales—well in advance of the availability of the data to verify the theory. *Induction,* the other cornerstone of science, explores data in the search for hypotheses (see chapter 7). Inasmuch as more than one process can produce the same data structure, inductive reasoning may yield false hypotheses or conclusions. Yet it is often used when meaningful hypotheses cannot be deduced. Geographers can empirically identify higher probabilities of residential relocations by collecting and analyzing mobility and related data (White et al. 1989), for example, even though they may not be able to deduce the equations that govern residential mobility. Exploratory data analysis (EDA) is a more generic form of inductive inquiry (chapter 7).

Deduction and induction often involve the assignment of names or numbers to phenomena, depending on the nature of the phenomena, how well analysts understand them, and the context in which they occur. Such assignments are defined on the basis of their information content. When type or kind is all that can be discerned, the assignment is *nominal.* If the rank, order, or relative position that the event takes in a set of events can be ascertained, an *ordinal* assignment can be made. Land-use categories, for example, are nominal, and individual preferences for different landscapes can be ranked and are therefore ordinal. *Interval* assignment is possible when events can be placed on a number line and the distances among them expressed as real numbers. *Ratio* assignment requires that the phenomenon being represented have a meaningful zero value. Fahrenheit temperature is an interval-scale assignment whereas Kelvin-scale temperature is a ratio-scale designation. When an event has magnitude specifiable on an interval or ratio scale and direction (such as wind velocity), a *vector* assignment may be appropriate. An even higher-level designation, a *tensor* assignment, is needed to describe events manifest as stresses and strains.

The way that events and processes are quantified significantly affects the degree to which analysts can model them. It is desirable to make the highest-level assignments possible, because they contain all the information captured by any lower-level assignment and more. Level of quantification governs model sophistication or realism because low-level assignments preclude certain mathematical operations. Ordinal observations, for example, cannot be multiplied or divided meaningfully, and ratios cannot be constructed usefully from interval scale numbers. Scalar arithmetic on vector observations is also inappropriate (Klink and Willmott 1989) even though commonly employed.

Classifications of Models

Models may be palpable or abstract. Palpable models are scaled-down representations of real systems whereas abstract models are constructed with symbols. With the rapid expansion of digital computation, palpable models have declined in importance, although they remain invaluable for certain purposes (Oke 1981). Abstract (especially mathematical) representations of geographic systems have found increasing favor among geographers.

A mathematical model can be classified according to (1) the way basic changes within the system modeled are thought to occur; (2) the overall structure of the model; or (3) the model's intended use. When changes from one state to another within a geographic system are thought to be predictable without error and they are specified that way within a model, the model can be said to be *deterministic*. Deterministic models arise from the belief that salient changes of state are governed by known, specifiable, and invariant rules. When individual changes of state cannot be predicted without error, probabilities are used to predict each change of state (Kirkby et al. 1987). Such models are said to be *stochastic* or probabilistic; they have been used to simulate the spread of an innovation or of a disease, for example. They treat individual exchanges as random processes that occur in accord with probabilities.

Purely deterministic or stochastic models, however, have been outnumbered overwhelmingly in geography by *parameterized* models. Parameterized models usually specify changes of state in the aggregate; that is, they are usually statistical summaries whose parameters have been estimated from empirical observations. They also may be integrals of governing differential equations. Regardless of their genesis, they are deterministic in function. Once the parameters have been obtained, dependent variables are calculated solely from independent variables. Any uncertainty that may exist is not incorporated into the model itself.

Geographic models also can be classified according to their overall structure. Models that embody correlative relationships can be considered *morphological* (or structural). Intrinsic relationships among cause and effect or pertinent sequences of events typically are not apparent in morphological constructions. When the order of events or theorized cause-and-effect chains are explicitly represented within a model, the model can be thought of as a *cascade*. Cascading models trace flows such as energy, mass, ideas, or money through the modeled system; they mimic transfer processes within, and to and from, the system of interest. An important refinement of cascading models is the specification of *feedback*. Feedback mechanisms allow models to regulate flow rates in response to changes in pertinent internal or external conditions. Negative feedback dampens flow rates while positive feedback enhances them. Cascading models that contain salient feedback mechanisms are called *process-response* models; generally they have the most realistic overall structures (Terjung 1982). For decades, geographers have quantitatively described usually static morphological relationships among geographic variables; few geographers have tried to model flows through geographic systems, and even fewer have formulated process-response models.

A third way that models can be categorized and evaluated is on the basis of their purposes. Distinguishing among models on this basis can be problematic because a model may have more than one purpose. Nevertheless, most geographic models can be classed as *exploratory, descriptive, predictive,* or *explanatory*. Exploratory and descriptive models usually have statistical bases. They are formulated to identify and describe the morphologies common to geographic processes and phenomena. Predictive models also are commonly based in statistics, but their sole purpose is to estimate one or more dependent variables as accurately as possible from a set of independent variables. Predictive models seek to minimize a cost-accuracy ratio, where cost is defined to include the overall effort necessary to use the model. Explanatory models, by contrast, minimize a cost-realism ratio, where realism is the fidelity with which the model mimics a system of interest. Explanatory models realistically describe and explain system structure and function. They are based more on theory than on data, and they are often the only means of answering questions such as "What will happen in response to human-induced or natural event *x*?" when insufficient observations of *x* have been made and its effects have not been (sufficiently) observed. While explanatory models can be used to predict, their main value lies in their contributions to explaining complex geographic systems.

MODELS FIT TO DATA

Models whose functional forms are established independently of the problem of interest (but whose parameters have been estimated from data thought to represent the problem of interest) comprise the largest group of geographic models. Their functional forms are usually derived from statistics as are the criteria and algorithms for estimating their parameters. In a few instances, functional forms have been borrowed from allied disciplines and used as a kind of quantitative metaphor. Consider, for example, gravity models (Sheppard 1984). Simple fit models include the linear regression model, and attendant parameter-estimate techniques are commonplace in the literature (Gaile and Grant 1989; Gaile 1990; Slocum 1990). Such models describe variance and covariance patterns within and among data or, in other words, structural or morphologic relationships.

Models fit to data serve four purposes. First, they may be used to explore data in order to discover intrinsic or important structure. Identification of structure then may lead to testable hypotheses. Second, when hypotheses already exist and can be given a functional form, their efficacy can be tested by examining the degree to which the hypotheses statistically explain variance/covariance patterns in the data. Third, models may be fit to data in order to estimate a rule for predicting dependent variables from independent variables. Fourth, fit models can reduce the complexity of more detailed process-based models. As process-based geographic models (discussed below) have been simple relative to process-based models in several other disciplines, simplification of more complex process-based models has been rare in geography.

Functional Forms and Parameter Estimation

The functional form of a model should follow from the model's purpose and from an understanding of the variability within and correlation among the data used. Model form also must allow its parameters to be estimated from the data. While the model's purpose and an understanding of the data should be the primary considerations, the existence of parameter-estimation procedures for linear or intrinsically linear functions has strongly influenced model development in geography. Intrinsically linear models can be made linear (usually through a simple power transformation), and therefore even untransformed linear models are intrinsically linear. A discussion of models fit to data, then, must focus on intrinsically linear models.

Many examples of the development and use of intrinsically linear models exist in geography. Economic geographers, for example, have made extensive use of the gravity model to characterize flow between nodes (Wheeler and Mitchelson 1989). Geomorphologists similarly have used power transformations to estimate parameters for stream-channel equations (Magilligan 1985), and climatologists have estimated the exponent in Beer's law to obtain solar irradiation at the bottom of a turbid atmosphere (Cerveny 1989).

The gravity model is widely used in geography. Inspired by Newton's law of universal gravitation, the gravity model may be written

$$\hat{I}_{ij} = GM_i^{\alpha}M_j^{\gamma}d_{ij}^{-\beta} \tag{1}$$

where $\hat{I}_{ij}$ is the estimated number of exchanges from location i to location j, M_i is a measure of the propensity of an exchange from i to j to occur, M_j is a measure of the propensity of j to attract an exchange from i, and d_{ij} is some measure of the separation between i and j. The model also contains four parameters (G, α, γ and β) that are estimated from data. When a logarithmic transformation is applied, the intrinsically linear form, equation (1), becomes linear; that is,

$$ln(\hat{I}_{ij}) = ln(G) + \alpha\ ln(M_i) + \gamma\ ln(M_j) - \beta\ ln(d_{ij}). \tag{2}$$

In this form, the gravity model is, in effect, a linear regression model of the generic form

$$\hat{Y} = \beta_0 + \beta_1 X_1 + \beta_2 X_2 + \beta_3 X_3 \tag{3}$$

and its parameters can be readily estimated using a solver such as the two-stage least-squares (2SLS) estimator or the ordinary least-squares (OLS) estimator (Sheppard 1984). A solver is a procedure for obtaining parameters that minimize some undesirable error, usually in estimating Y.

There is considerable controversy about which estimator is best (Sheppard 1984) and whether least-squares estimates in Y are appropriate at all (Mark 1984). Although the problem is often presented as a statistical issue, it arises primarily because geographic theory is insufficiently developed to select the best solver. Consider a simple bivariate case where $\hat{Y} = \beta_0 + \beta_1 X_1$ is estimated by three plausible solvers and graphed among the data used to estimate the parameters (fig. 8.1). Without geographic understanding of the processes the data represent, no adequate basis exists for selecting among these three or other linear solutions. The problem is amplified if the intrinsically linear model that has been selected is inappropriate. Parameter estimates also could be adaptive, that is, variable over the domain of the independent variable(s) or modifiable by new data. Once again, without geographic theory, there is no clear-cut way to estimate even a simple linear bivariate model from data.

Geographers, nevertheless, have estimated the parameters for a plethora of intrinsically linear models. Parameters, by and large, have been static rather than adaptive and the transformation presumed is usually logarithmic, even though optimizing transformations such as the Box-Cox transformation are available (Legates 1991). Spatial autocorrelation biases are usually ignored, although literature on the topic is rapidly developing (Odland, Golledge, and Rogerson 1989). Multiple-linear or polynomial regression has been preferred in estimating the best predictor of Y from data. Principal components or factor analysis has been used to identify the important variance-covariance (linear) structure when independent and dependent variables cannot or should not be identified (Legates and Willmott 1984). Sine-cosine series have been employed to examine those observations that are thought to oscillate (Rayner 1971; Willmott, Rowe, and Mintz 1985), while auto- or cross-covariance functions in-

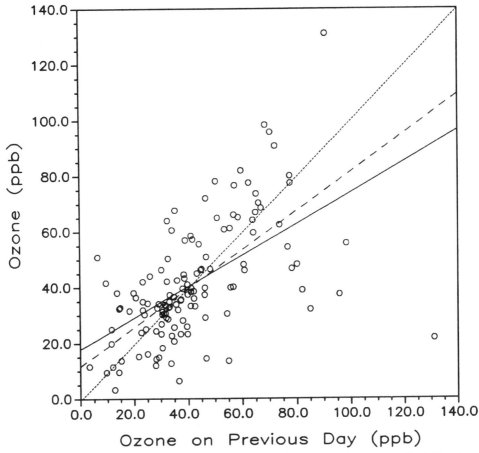

Fig. 8.1. Three Best-Fit Linear Models of the Relationship between the Previous Day's Hourly Maximum 0zone Concentration (*X*) and the Current Day's Concentration (*Y*) at a Station in Port Moody, British Columbia. SOURCE: Robeson and Steyn 1990. Intercept (β_0) and slope (β_1) parameters are alternatively estimated by ordinary least squares (solid line), least absolute deviations (dashed line), and orthogonal (to the line) least squares (dotted line).

creasingly have been applied to time or space series (Aguado 1982; Burt 1986). It is evident that geographers have identified a wide variety of model forms that can be made linear and for which parameters can be estimated from data.

While most geographic models are intrinsically linear, there are serious shortcomings. The system of interest may be intrinsically nonlinear, or the wrong intrinsically linear model may be selected. An inappropriate solver may produce inappropriate estimates of the parameters. More importantly, even when the model and solver are appropriate, models fit to data cannot explain system structure and function. Geographic theory, as opposed to statistical theory, holds greatest promise for improving the models geographers fit to data.

Data Collection

Data must be obtained to estimate model parameters, and special care assures that data adequately represent what the model purports to characterize. An unbiased sample design, one that has a good chance of yielding most of the information contained in all the data, should be developed prior to data collection. It should specify the limits of the sample in space and time, the number of observations to be collected, and the location in both space and time of the observation set. Over- or underspecification of a sampling domain may induce serious biases in parameter estimates.

Size (N) is an important element of sample design that is too often determined by time and resources rather than by sound science (Dixon and Leach 1977). With a good sample design, parameter reliability asymptotically increases with the number of sample observations. A problem faced by modelers is estimating in advance the minimum acceptable sample size. A small but reliable sample is normally desirable in light of data collection costs. Modern computers have so drastically reduced computational costs, however, that the expense of analyzing large samples seldom constrains modeling.

In geography especially, the locations of sample points are as important as sample size. When sample points do not represent variability adequately, parameter estimates will be biased correspondingly. For instance, a terrestrial rain gauge network may misrepresent terrestrial precipitation because of biases in weather-station siting (Willmott, Robeson, and Feddema 1991). Biased parameters also can arise if the sampling domain is not commensurate with the population domain. A truly representative sample is absolutely prerequisite to reliable estimation of model parameters.

We believe that in geography purely statistical considerations such as randomness should be subordinated to considerations specific to the areally distributed and referenced data that geographers study. Such considerations provide better problem-specific guidance on what is truly representative of geographic phenomena. When statistical issues are subordinated, representative samples can sometimes be culled from data that were collected for other purposes, such as the census. But when statistical considerations remain paramount (significance-testing assumptions, for example), such samples of convenience preclude meaningful estimates of probabilities from existing data (Morrison and Henkel 1970; Freedman, Pisani, and Purves 1978). When geographers use existing data or collect geographically meaningful samples, they should question uncritical adherence to classical rules of inferential statistics and let their sampling be guided by geographic understanding.

Evaluation of Model Fit

Models fit to data may be evaluated by the degree of correspondence between the models and the data to which they are fit using two specific criteria:

(1) graphic comparison of the model's function(s) and data; and (2) calculation and interpretation of diagnostic statistics.

Although geographers have infrequently attempted to graph the correspondence between models and data, such displays are especially revealing of model fit or performance (Willmott 1984). Scatter plots between X_i and Y_i (fig. 8.1) can be especially useful for interval or higher-level observations. Two-dimensional residual plots or maps of residuals also can reveal autocorrelation patterns and other anomalies that reduce model effectiveness. A variety of other graphic forms can be developed for special purposes. Given their expertise at cartographics (chapter 6), geographers should be at the forefront of using graphics to illuminate error patterns that would not be apparent in a table of diagnostic statistics. Graphics should accompany most presentations and interpretations of model fit or performance.

Diagnostic statistics should accompany such graphics and, at a minimum, should include dimensional and dimensionless indices of model fit. Fit statistics should describe the extent to which the solver minimized its undesirable error. The fit standard error (SE) and the coefficient of determination (r^2) usually are appropriate, although other statistics such as the mean absolute error (MAE) also may be helpful (Willmott 1984). Because fit statistics (r^2s and SEs) normally exaggerate performance in the model's favor, resampled estimates of r^2 and SE are preferable (Efron and Tibshirani 1986; Michaelsen 1987). It follows that geographers should replace traditional fit diagnostics with resampled, computer-intensive estimates. Several computer-intensive procedures permit straightforward evaluations of parameter or statistic confidence bounds (Willmott et al. 1985; Efron and Tibshirani 1986). Classical hypothesis-testing methods have become largely irrelevant as well as fraught with problems (Morrison and Henkel 1970). In spite of their current popularity in geography, they need not be used.

Geographical interpretation of model fit often depends on the degree to which geographic factors have been included in the model. Gravity and spatial interaction models (Haynes and Fotheringham 1984) and Casetti's (1972) expansion method, for example, explicitly incorporate distance as a variable, and therefore their geographic interpretation is relatively straightforward. Other models that use areal units as cases may not incorporate spatial factors; it may then be necessary to use spatial autocorrelation measures to elicit a geographical understanding of model fit (Odland 1988).

MODELS BASED IN PROCESS

Models based in process take their mathematical forms primarily from geographic rather than statistical understanding. Particularly in physical geography, differential equations represent fluxes that are clearly distinguished from state variables, initial conditions, and boundary conditions (Terjung 1982). Process-based models take cascading or process-response forms, and they are explanatory in that they are meant to represent and explain both system structure and behavior. Among their most important advantages relative to models fit to data is that they can incorporate feedback. Process-based models have two important uses in geography. They are employed to interpolate or extrapolate events when observations are unavailable, and they are used to examine the sensitivity of a modeled system to changes in input variables, boundary conditions, or initial conditions. In both instances, absence of data usually precludes using models fit to data.

Conceptualization

Conceptualization of a system of interest is the first and most important step in developing a valid process-based model (fig. 8.2). Few geographers have paid adequate attention to conceptualization because linguistic models and models fit to data, unlike process models, can be developed without careful attention to such basics as units, domain, boundary conditions, minimum resolvable scales, and the links among state and flux variables (Chorley and Haggett 1967; Terjung 1976, 1982; Strahler 1980). Successful process-based models require clear specification of these and other salient components of the system of interest. A modeler must, for example, specify whether relationships among variables are correlative, ordered, or cause and effect, and how feedbacks are coupled to the variables (see chapters 10–13 and 15 for aspects of model conceptualization such as scale and resolution).

Operationalization

Once a system of interest has been conceptualized, the model often must be reduced to a more tractable form (see fig. 8.2). This reduction requires parameterization of small-scale or ill-defined processes and specification of differential (flux) equations with appropriate initial and boundary conditions. While differential equations typically represent fluxes, in some instances probabilities may determine flow rates and directions (Kirkby et al. 1987), for example, in stream network development models that use random-walk formats and in disease diffusion models. A set of equations from which calculations can be made typically emerges from parameterization and specification of a conceptual model.

Geographers have long used one such set of differential equations (the climatic water balance) to solve water resource and climatological problems (Willmott, Rowe, and Mintz 1985). The climatic water balance estimates soil

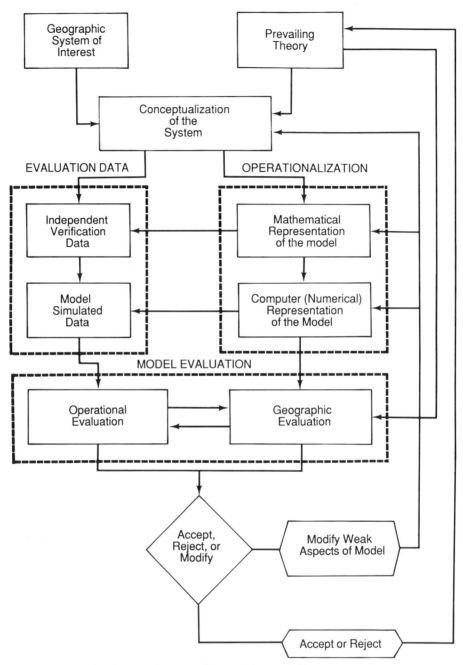

Fig. 8.2. Schematic Diagram of the Path from Problem Definition and Prevailing Theory to the Development of a Process-Based Geographic Model.

moisture—the water available to plants—assuming mass continuity and negligible horizontal moisture transfer within the soil. It can be written as

$$\partial w^s/\partial t = P^s - M \tag{4a}$$

and

$$\partial w/\partial t = P^r + M - E - S \tag{4b}$$

where w^s is the water available in the snow pack (if one exists), t is time, P^s is the snowfall rate, M is the snow melt rate, w is the soil moisture, P^r is the rainfall rate, E is the evapotranspiration rate, and S is the rate at which surplus water runs off or percolates to a depth below the root zone.

Changes in soil moisture (the important state variable) occur in direct response to net imbalances between incoming $(P^r + M)$ and outgoing water $(E + S)$. Each flux (independent variable) is estimated from separate expressions with additional independent variables. Feedbacks, such as the partial dependence of E and S on w, and mass continuity, require repeated estimation until all terms in equation (4) balance. Equations (4a) and (4b) also should be solved simultaneously to obtain physically realistic estimates of soil moisture (w). Unless climate is changing, any long-term integration of equations (4a) and (4b) must approach zero. In other words, physical understanding of system structure and function guides both the specification and solution of equation (4).

Process-based models are often too complex to be solved without digital computers, especially if they include feedbacks. Therefore the mathematical form of the model must be translated into a numerical form (a computer program) to perform the calculations. Differential equations usually become finite-difference (Rayner 1984) or finite-element (Hanson 1990) equations that can be integrated approximately by computers. When the fluxes are guided by probabilistic expressions, the integration may be a simple summing or statistical operation. Numerical models have found wide acceptance in physical geography, especially in climatology (Oliver et al. 1989). Geographers have used them to simulate urban climates (Terjung and Louie 1974), agricultural productivity (Hayes et al. 1982), available solar energy (Cerveny 1989), and glacier dynamics (Hanson 1990), in addition to soil moisture budgets. Within human geography, an increasing number of model forms use computational formats (Odland, Golledge, and Rogerson 1989). Ratick and Osleeb (1983), for instance, developed a computer-based coal supply and transportation model that minimizes total system costs. Computational process models also are being applied to growing numbers of problems (Odland, Golledge, and Rogerson 1989). These models rely heavily on computerized solvers.

Efficacy and Tuning Problems

Models containing feedback typically must adjust solution sets until they are plausible according to predetermined criteria. A number of coupled, related-rate problems are solved until the modeled system achieves an equilibrium. Within Ratick and Osleeb's (1983) system, for example, overall minimum cost is the desired equilibrium, and other aspects of the system are adjusted accordingly. Within Hanson's (1990) glacier system, all estimated fluxes together must conserve heat, mass, and momentum. For many simple systems, a single equilibrium or solution set is easily found; the system may also be robust inasmuch as models of such systems repeatedly and consistently converge on the same solution set. Some ostensibly deterministic systems, however, may not converge on a unique solution. Modeled climates, for example, may exhibit irregular behavior because of numerical instabilities or the existence of more than one solution set that satisfies all theoretical and numerical convergence criteria (Lorenz 1964). Problems associated with indeterminate, nonexistent, or plural equilibrium solutions have far-reaching implications for the future of process-based modeling in geography.

Another problem associated with numerical modeling is tuning; that is, parameter adjustment until model estimates fit observations well. When parameters are iteratively adjusted as opposed to being fixed analytical solutions or summaries of observations, their values depend on the collective values of all other parameters. There is no clear-cut geographic interpretation of tuned parameters or of the relative contributions of the various processes a model embodies. Once tuned, a numerical model's predictions may compare well with observations, but its explanatory power is reduced to little more than a multi-parameter model fit to data. Geographers should resist the temptation to tune numerical models to improve prediction.

EVALUATION OF MODEL PERFORMANCE

An important issue associated with model building is determining how well models work. Evaluations of model performance should consider the model's consistency with prevailing theory as well as its ability to accurately and precisely estimate real-world events. Theoretical consistency is a scientific (for our purposes, geographic) criterion whereas the empirical validity is an operational consideration (fig. 8.2). Within any particular model evaluation effort, the mix of geographic and operational evaluations should reflect the model's purpose. If a model is intended to minimize the cost/realism ratio, geographic criteria should dominate. If it is meant to minimize the cost/accuracy ratio, operational concerns should be paramount. We focus here on operational evaluation of geographic models because it is possible to outline widely applicable procedures for examining model precision and accuracy. Emphasis on the operational should not be interpreted as suggesting that geographic evaluation is unimportant, but geographic evaluation is model-specific and therefore not amenable to general discussion.

Many geographers fail to adequately estimate and report how well their models perform, whether they use models fit to data or process-based models. Their oversight arises from the misconception that the fit diagnostics discussed earlier are satisfactory indices of precision and accuracy and from infrequent demands by reviewers and editors for detailed explication of model performance. As a result, less competitive or redundant formulations have become commonplace in the literature. Such models do not serve geography well. From the literature alone, these models are difficult or impossible to verify or to compare with other models for purposes of selecting a good formulation.

Fit diagnostics (for example, r^2 and SE) are usually reported for models fit to data, but they are inappropriate performance measures. These indices are biased in the model's favor because the real-world observations to which the model-predicted values are compared influence estimation of the model's parameters. Such comparisons are therefore based on circular reasoning. Model predictions should be compared to independent observations *not* used to estimate model parameters, whether a fit or process-based model is evaluated. Several resampling procedures provide a useful compromise for models fit to data (Michaelsen 1987). Most fit diagnostics also assume that the solver removes any systematic error, especially in the mean. Since this removal ordinarily does not occur in comparisons with independent observations, statistics that are insensitive to differences in observed and model-predicted means and variances (such as r^2) are inappropriate. Furthermore, it is risky to report statistical significance of fit diagnostics, especially when the statistics are inappropriate or biased (Morrison and Henkel 1970).

Comparisons of model predictions (P_i) with independent real-world observations (O_i) in order to evaluate model performance or to compare alternate formulations should be based on graphical and statistical summaries of the difference variable, $D = P - O$, where P and O are arrays whose elements are pair-wise model-predicted and empirical observations, respectively (Willmott 1984; Willmott et al. 1985). The elements of D, P, and O may be either scalars or vectors. Necessary assumptions include: pairwise comparability of P_i and O_i; independence between P and O; and the condition that O is observed without error. Any bias known to be contained in O should be estimated and removed before P and O are compared.

Whether evaluating model fit or performance, an important first step is the graphic comparison of P and O. Scatter plots, such as those used to compare four potential evapotranspiration models (fig. 8.3) are quite useful, although time- or space-series plots and residual maps may help illuminate error patterns. Graphic comparisons among scalar observations are illustrated by Robeson and Steyn's (1990) comparison of three statistical air pollution models. When the elements of P and O are vectors, performance evaluation graphics

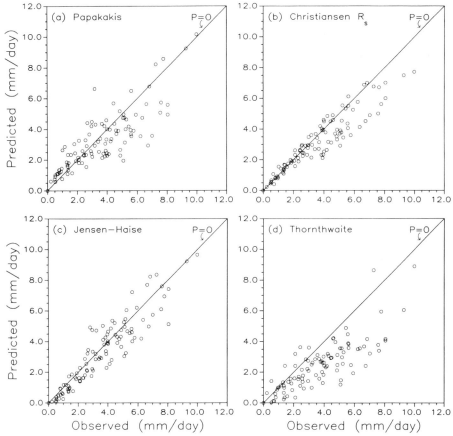

Fig. 8.3. Scatter Plots of Monthly Lysimeter-Derived (Observed) Potential Evapotranspiration at Ten Stations versus Four Model-Predicted Variables for Those Same Months and Station Locations. The model-predicted variables were generated by the models of Papadakis, Christiansen, Jensen-Haise, and Thornthwaite. SOURCE: Willmott 1984.

are illustrated by a comparison of two models of off-shore wind velocity (Willmott et al. 1985) and an examination of the simulated wind field within the Fraser Valley in British Columbia (Steyn and McKendry 1988). In-depth examinations of model-performance graphics exist in several of these articles.

An array of diagnostic statistics that describe aspects of P, O, and D should accompany graphic comparisons. It is especially important to quantify meaningfully the precision and accuracy with which P approaches O. Univariate statistics of P and O, for example, their means ($\bar{P}$ and $\bar{O}$) *and standard deviations (s_p and s_o)* as well as N, should also be reported so the characteristic levels of P and O can be appreciated. More importantly, decompositions of D are needed to assess precision and accuracy. Willmott (1984), for instance, recommends that overall accuracy be described by the root mean square error

(RMSE) or the MAE as well as by the dimensionless index of agreement (d). The dimensioned statistics (RMSE and MAE) arise from the solution to

$$E^{1/\gamma} = [N^{-1} \sum^{n}_{i=1} | P_i - O_i |^{\gamma}]^{1/\gamma}, \qquad \gamma > 0 \qquad (5)$$

with $\gamma = 2$ and $\gamma = 1$, respectively. A corresponding d statistic is then

$$d_{\gamma} = 1 - (E/E_p), \qquad 0 \le d \le 1 \qquad (6)$$

where $E = (E^{1/\gamma})^{\gamma}$ and E_p is the maximum possible value that E can attain (Willmott 1984). The index of agreement, d, may be interpreted as the degree to which the potential variability around $\bar{O}$ is explained by P, and it varies between zero and one, one indicating perfect prediction.

Average error also may be partitioned to obtain estimates of the systematic (linear) and unsystematic components of error. For the case $\gamma = 2$, the partition is

$$RMSE = [RMSE^2_s + RMSE^2_u]^{0.5} \qquad (7)$$

where $RMSE_s$ is the systematic component of $RMSE$ and $RMSE_u$ is the unsystematic component. Unsystematic error ($RMSE_u$) can be interpreted as a measure of model precision whereas $RMSE_s$ suggests the level of error reduction that may be possible through respecification of model parameters.

When univariate and difference statistics are reported together, they provide a reasonably complete and easily interpretable summary of operational performance. Willmott (1984), for example, was able to demonstrate the relative accuracy of the Christiansen and Jensen-Haise potential evapotranspiration models as well as the relatively high level of precision associated with Thornthwaite's low-cost model, which requires only one independent variable, air temperature. Potential evapotranspiration is the evapotranspiration that would occur at a place if its vegetation experienced no water stress. The statistics indicate that all four models tend to underpredict (table 8.1). Robeson and Steyn (1990) found it difficult to distinguish among three models that forecast ozone concentrations, although a time-dependent regression model was slightly more accurate than the others. Legates and Willmott (1986) were able to show that bidirectional hermitian-spline interpolation was demonstrably more accurate than cubic-spline or linear-based methods. If reliability of performance statistics is of interest, it may be evaluated through computer-intensive resampling procedures such as the bootstrap (Efron and Gong 1983; Willmott et al. 1985; Efron and Tibshirani 1986). A comprehensive evaluation of model performance should, at a minimum, include the statistics discussed here, graphics, and an informed geographic evaluation of the model's worth.

Overview

Within the two large and overlapping classes of models fit to data and models based in process, there exists a large and growing number of model forms. De-

Table 8.1. Quantitative Measures of Evapotranspiration Model Performance

	Summary univariate measures					Difference measures					
	$\bar{O}$	$\bar{P}$	s_o	s_p	N	MAE	$RMSE$	$RMSE_s$	$RMSE_u$	d
Papadakis	3.65	3.30	2.17	1.84	104	0.96	1.28	0.74	1.05	0.89
Christiansen R_s	3.65	3.18	2.17	1.77	104	0.65	0.94	0.69	0.64	0.94
Jensen-Haise	3.65	3.38	2.17	2.20	104	0.69	0.89	0.30	0.84	0.96
Thornthwaite	3.74	2.44	2.13	1.56	101	1.44	1.81	1.56	0.92	0.79

SOURCE: *Willmott 1984.*

Note: The terms N and d are dimensionless while the remaining measures have the units mm day^{-1}.

terministic models have arisen primarily in physical geography from certainty about cause-and-effect relationships among basic within-system exchanges. Physical systems often are continuous and physical geographic models are therefore commonly based on differential equations. Physical geographic models are frequently cast as cascading or process-response models. When relationships among exchange processes are uncertain, probabilistic links frequently have been specified, giving rise to stochastic models. Human geographic models, because they are usually based on individual or aggregate probabilities, are overwhelmingly statistical in nature. They also tend to be discrete because the basic operational unit is often the individual. Aggregate representations of within-system exchanges, for deterministic or stochastic exchanges, have resulted in parameterized models, the most common form in geography. Morphologic constructions have been developed to describe system structure or covariance, and cascading and process-response models depict linked sequences of events and feedback. Most geographic models have been developed to describe and predict empirical regularities; a much smaller number of models have attempted to explain system structure, function, and behavior.

Explanatory (process-response) models remain poorly developed in geography relative to models fit to data, a lacuna that retards development of explicit, verified theory in the discipline. Geographers should subordinate their overriding sensitivity to statistical issues (Odland et al. 1989) to concerns about modeling intrinsically geographic systems and processes. Model efficacy and operational performance also have received too little attention in geography, which makes it difficult to judge the success of individual models or the state of the modeling literature. To understand geographic dynamics, the task ahead is clear. Geographic theory will be greatly advanced by the construction, evaluation, and application of realistic, process-based geographic models.

MODELING AND THE 'ISMS'

Scientific method underlies most modeling, and it is the established operating procedure in the physical and natural sciences. It permits verification, replication, and extension of research results previously achieved by others. Scientific method also provides a set of agreed-upon criteria for judging whether an explanation is satisfactory or plausible.

While scientific method remains accepted in physical geography, critical social theorists have raised challenges to the use of formal modeling in human geography, challenges presaged by the rejection of positivist research by some of its foremost advocates, notably David Harvey, Gunnar Olsson, Allen Scott, Edward Soja, John Hudson, and William Bunge. Radical geographers challenged positivism along with the existing economic order. Scientific method has always been less applicable in human than in physical geography. It is particularly difficult to simplify, classify, and measure complex social relationships, and identifying functional and membership laws among social groups remains elusive (Johnston 1989). Human geographers continue to be engaged in a lively epistemological debate in which there is little consensus except for considerable negativism toward logical positivism.

Advocates of critical social theory have extended the early debates and diverging points of view to produce today's panoply of paradigms. Many contemporary human geographers embrace critical social theory, including humanism and structuralism, and its challenges to positivist orthodoxy. Giddens's (1984) theory of structuration has won many adherents in geography by emphasizing agency-structure, time-space, and context-concept relationships (Palm 1986). Theoretical realism, an epistemology with close parallels with positivism, has found its share of advocates in geography (Bhaskar 1975; Johnston 1989; Lawson and Staeheli 1990). Major works on postmodernism by Soja (1989) and Harvey (1989a) have highlighted an outlook that calls for deconstructing existing knowledge. Soja goes a step further by characterizing geography as the Phoenix that will rise from the ashes of philosophical anarchy (cf. Buttimer 1990).

Contemporary social theories disdain strict empiricism or positivism and formal modeling (Peet and Thrift 1988). Critics focus on what models do not do; they "cannot replace evocative description" or "answer the important questions" (Flowerdew 1989, 252). These critiques of positivism cannot be dismissed, but neither should they be accepted in total. Critical social theorists often fail to come to grips with physical geography, wherein positivism has been especially fruitful. They also find positivistic methods useful while rejecting the philosophy that underlies them.

Modeling is an inescapable facet of human thought and discourse, whether it is positivistic or critical, and models should be appreciated in their full variety, from the positivistic formulations of physical geography to Marxist dialectics of historical materialism (Harvey 1989b). Rejection of modeling risks a retreat into neoexceptionalism while other sciences continue to advance by constructively criticizing and improving their models.

NOTE

We received thoughtful comments on an earlier draft from Brian Hanson, Kathy Klink, and Scott Robeson, and we are grateful for their assistance. We also thank Peter Rogerson for his useful suggestions and Jeff Osleeb for supplying pertinent papers. We are most appreciative to the editors and other chapter authors for enhancing this presentation.

REFERENCES

AGUADO, EDWARD. 1982. A time series analysis of the Nile River low flows. *Annals of the Association of American Geographers* 72:109–119.

BHASKAR, R. 1975. *A realist theory of science*. Brighton, England: Harvester.

BURT, JAMES E. 1986. Time averages, climatic change, and predictability. *Geographical Analysis* 18:279–294.

BUTTIMER, ANN. 1990. Geography, humanism, and global concern. *Annals of the Association of American Geographers* 80:1–33.

CASETTI, EMILIO. 1972. Generating models by the expansion method: Applications to geographic research. *Geographical Analysis* 4:81–91.

CERVENY, RANDALL S. 1989. Shadowing of nonpolluted locations by urban pollution. *Annals of the Association of American Geographers* 79:242–256.

CHORLEY, RICHARD J., and PETER HAGGETT, EDS. 1967. *Models in geography*. London: Methuen.

DIXON, C., and B. LEACH. 1977. *Sampling methods for geographical research*. CAT-MOG no. 17. Norwich, England: GeoAbstracts.

EFRON, BRADLEY, and GAIL GONG. 1983. A leisurely look at the bootstrap, the jackknife, and cross-validation. *American Scientist* 37:36–48.

EFRON, BRADLEY, and R. TIBSHIRANI. 1986. Bootstrap methods for standard errors, confidence intervals, and other measures of statistical accuracy. *Statistical Science* 1:54–77.

FLOWERDEW, ROBIN. 1989. Some critical views of modelling in geography. In *Remodelling geography*, ed. B. MacMillan, 245–252. Cambridge: Basil Blackwell.

FREEDMAN, DAVID, ROBERT PISANI, and ROGER PURVES. 1978. *Statistics*. New York: W. W. Norton.

GAILE, GARY L. 1990. Whither spatial statistics? *Professional Geographer* 42:95–100.

GAILE, GARY L., and R. GRANT. 1989. Trade, power, and location: The spatial dynamics of the relationship between exchange and political-economic strength. *Economic Geography* 65:329–337.

GIDDENS, A. 1984. *The constitution of society: Outline of the theory of structuration*. Cambridge: Polity.

HANSON, BRIAN. 1990. Thermal response of a small ice cap to climatic forcing. *Journal of Glaciology* 36:49–56.

HARVEY, DAVID. 1989a. *The condition of postmodernity*. Oxford: Basil Blackwell.

———. 1989b. From models to Marx: Notes on the project to "remodel" contemporary geography. In *Remodelling geography*, ed. B. MacMillan, 211–216. Cambridge: Basil Blackwell.

HAYES, JOHN T., PATRICIA A. O'ROURKE, WERNER H. TERJUNG, and PAUL E. TODHUNTER. 1982. A feasible crop yield model for worldwide international food production. *International Journal of Biometeorology* 26:239–257.

HAYNES, KINGSLEY E., and A. STUART FOTHERINGHAM. 1984. *Gravity and spatial interaction models*. Scientific Geography Series 2. Beverly Hills, Calif.: Sage.

JOHNSTON, RONALD J. 1989. Philosophy, ideology, and geography. In *Horizons in human geography*, ed. Derek Gregory and R. Walford, 48–66. Totowa, N.J.: Barnes and Noble.

KIRKBY, MIKE J., PAM S. NADEN, TIM P. BURT, and DAVE P. BUTCHER. 1987. *Computer simulation in physical geography*. New York: John Wiley and Sons.

KLINK, KATHERINE, and CORT J. WILLMOTT. 1989. Principal components of the surface wind field in the United States: A comparison of analyses based upon wind velocity, direction, and speed. *International Journal of Climatology* 9:293–308.

LAWSON, VICTORIA A., and LYNN A. STAEHELI. 1990. Realism and the practice of geography. *Professional Geographer* 42:13–20.

LEGATES, DAVID R. 1991. An evaluation of procedures to estimate monthly precipitation probabilities. *Journal of Hydrology* 122:129–140.

LEGATES, DAVID R., and CORT J. WILLMOTT. 1984. On the use of factor analytic techniques with geophysical data. *Modeling and Simulation* 15:417–426.

———. 1986. Interpolation of point values from isoline maps. *American Cartographer* 13:308–323.

LORENZ, E. N. 1964. The problem of deducing the climate from the governing equations. *Tellus* 16:1–11.

MACMILLAN, BILL, ED. 1989. *Remodelling geography*. Cambridge: Basil Blackwell.

MAGILLIGAN, FRANCIS J. 1985. Historical floodplain sedimentation in the Galena River Basin, Wisconsin and Illinois. *Annals of the Association of American Geographers* 75:583–594.

MARK, DAVID M. 1984. Some problems with the use of regression analysis in geography. In *Spatial statistics and models*, EDS. Gary L. Gaile and Cort J. Willmott, 191–199. Dordrecht: D. Reidel.

MICHAELSEN, JOEL. 1987. Cross-validation in statistical climate forecast models. *Journal of Climate and Applied Meteorology* 26:1589–1600.

MORRISON, DENTON E., and RAMON E. HENKEL. 1970. *The significance test controversy—a reader*. Chicago: Aldine.

ODLAND, JOHN. 1988. *Spatial autocorrelation*. Newbury Park, Calif.: Sage.

ODLAND, JOHN, REGINALD G. GOLLEDGE, and PETER A. ROGERSON. 1989. Mathematical and statistical analysis in human geography. In *Geography in America*, eds. Gary L. Gaile and Cort J. Willmott, 719–745. Columbus, Ohio: Merrill.

OKE, TIMOTHY R. 1981. Canyon geometry and the nocturnal urban heat island: Comparison of scale model and field observations. *Journal of Climatology* 1:237–254.

OLIVER, JOHN E., ROGER G. BARRY, WALTRAUD A. R. BRINKMANN, and JOHN N. RAYNER. 1989. Climatology. In *Geography in America*, eds. Gary L. Gaile and Cort J. Willmott, 47–69. Columbus, Ohio: Merrill.

PALM, RISA. 1986. Coming home. *Annals of the Association of American Geographers* 76:469–479.

PEET, RICHARD, and NIGEL THRIFT, EDS. 1988. *The new models in geography*. Hemel Hemstead, England: Allen & Unwin.

RATICK, SAMUEL J., and JEFFREY P. OSLEEB. 1983. Optimizing freight transshipments: An evaluation of East Coast coal export options. *Transportation Research A* 17A:493–504.

RAYNER, JOHN N. 1971. *An introduction to spectral analysis*. London: Pion.

———. 1984. Simulation models in climatology. In *Spatial statistics and models*, eds. Gary L. Gaile and Cort J. Willmott, 417–442. Dordrecht: D. Reidel.

ROBESON, SCOTT M., and DOUW G. STEYN. 1990. Evaluation and comparison of statistical forecast models for daily maximum ozone concentrations. *Atmospheric Environment* 24B:303–312.

SHEPPARD, ERIC. 1984. The distance-decay gravity model debate. In *Spatial statistics and models*, eds. Gary L. Gaile and Cort J. Willmott, 367–388. Dordrecht: D. Reidel.

SLOCUM, TERRY A. 1990. The use of quantitative methods in major geographical journals, 1956–1986. *Professional Geographer* 42:84–94.

SOJA, EDWARD W. 1989. *Postmodern geographies: The reassertion of space in critical social theory*. New York: Verso.

STEYN, DOUW G., and IAN G. McKENDRY. 1988. Quantitative and qualitative evaluation of a three-dimensional mesoscale numerical model simulation of a sea breeze in complex terrain. *Monthly Weather Review* 116:1914–1926.

STRAHLER, ARTHUR N. 1980. Systems theory in physical geography. *Physical Geography* 1:1–27.

TERJUNG, WERNER H. 1976. Climatology for geographers. *Annals of the Association of American Geographers* 66:199–222.

———. 1982. *Process-response systems in physical geography*. Bonn: Ferd. Dümmlers Verlag.

TERJUNG, WERNER H., and STELLA S-F. LOUIE. 1974. A climatic model of urban energy budgets. *Geographical Analysis* 6:341–367.

WHEELER, JAMES O., and R. L. MITCHELSON. 1989. Information flows among major metropolitan areas in the United States. *Annals of the Association of American Geographers* 79:523–543.

WHITE, STEPHEN E., LAWRENCE A. BROWN, WILLIAM A. V. CLARK, PATRICIA GOBER, RICHARD JONES, KEVIN E. McHUGH, and RICHARD L. MORRILL. 1989. Population geography. In *Geography in America*, eds. Gary L. Gaile and Cort J. Willmott, 258–289. Columbus, Ohio: Merrill.

WILLMOTT, CORT J. 1984. On the evaluation of model performance in physical geography. In *Spatial statistics and models*, eds. Gary L. Gaile and Cort J. Willmott, 443–460. Dordrecht: D. Reidel.

WILLMOTT, CORT J., STEVEN G. ACKLESON, ROBERT E. DAVIS, JOHANNES J. FEDDEMA, KATHERINE M. KLINK, DAVID R. LEGATES, JAMES O. O'DONNELL, and CLINTON M. ROWE. 1985. Statistics for the evaluation and comparison of models. *Journal of Geophysical Research* 90(c5):8995–9005.

WILLMOTT, CORT J., SCOTT M. ROBESON, and JOHANNES J. FEDDEMA. 1991. Influence of spatially variable instrument networks on climatic averages. *Geophysical Research Letters* 18:2249–2251.

WILLMOTT, CORT J., CLINTON M. ROWE, and YALE MINTZ. 1985. Climatology of the terrestrial seasonal water cycle. *Journal of Climatology* 5:589–606.

Communication

David A. Lanegran

An army marches on its stomach. Academic disciplines such as geography move forward by talking and writing. These information exchanges occur in the meetings, seminars, letters, electronic messages, journal articles, books, reports, and atlases that scholars and practitioners use to engage each other's excitement and interest. Oral and written discourse among established and apprentice scholars, dissemination of the discipline's insights to college and university students, and exchanges of problems and solutions between professional geographers and their clients link geography's inner worlds and convey its insights to society. Efficient and open communication is necessary for a discipline to survive and thrive. Disorderly as they often are, the talking and writing in which geographers engage constitute a social process of achieving consensus regarding the questions members of the community will ask and the methods to be used in answering them (Zelinsky 1975).

TALKING TO EACH OTHER

Meetings run the gamut from informal gatherings of a clutch of geographers interested in a narrow problem to national and international gatherings with casts of thousands. Geographers attend meetings more often than do their counterparts in other disciplines, perhaps because their meetings offer opportunities to explore new and distant places.

Annual Meetings

The annual meetings of the Association of American Geographers (AAG) and the National Council for Geographic Education (NCGE) afford scholars, advanced students, and practitioners formal opportunities to present the results of their research and thinking and to respond to the work of others. AAG meetings provide a hiring fair; the AAG's Convention Placement Service links employers and those seeking jobs. Corridor conversations and casual meetings in lobbies, lounges, and over meals also advance the discipline, for such sessions encourage frank discussion of questions, methods, and solutions that is often inhibited by formal presentation. They also serve as the arteries for the gossip that forms much of the lifeblood of any scholarly group. Some wags have proposed that formal sessions be dispensed with in favor of more time for informal interaction, but such a change is unlikely—many employers require presentation of a paper as a condition of financial support for attendance.

Two major changes in the structure of the annual meetings of the AAG were implemented during the last twenty years. An open program was adopted in 1973, and specialty groups were given program responsibilities in 1978. Annual meeting program time has been rationed in a variety of ways during the association's history. Before the minuscule AAG was democratized by its amalgamation with the much larger American Society for Professional Geography (ASPG) in 1948, only AAG members could present papers at the Association's meetings. After 1948, the new AAG's relaxed membership criteria enabled any interested scholar to participate. Formal screening of abstracts was begun in 1969 (Nelson and Tomica 1970), and meeting proceedings were published from that date. Growing numbers of papers made the process increasingly cumbersome, and allegations that papers were being rejected on grounds other than scholarship made the process increasingly contentious. Screening and the proceedings were abandoned after the 1976 meeting. Since then, any member or nonmember who submits an abstract in acceptable form has been placed on the meeting program.

Debate about the wisdom of an open program continues. Some geographers argue that open meetings enliven and improve communication among geographers. Critics contend that communication is inhibited by a glut of shoddily prepared presentations and claim that the absence of peer review suggests that geographers have no standards. In the early 1990s, 57 percent of those replying to a survey favored screening in principle, but many qualified their approval with concerns that political considerations would influence paper selection (AAG 1991).

For many years individuals or groups have organized sessions containing invited papers. Such informal screening was partially formalized after 1978, when the AAG Council recognized AAG specialty groups. The 1970s had seen the formation of interest or specialty groups within many scholarly societies.

Designed primarily to foster better communication among members with similar methodological, regional, or topical interests, AAG specialty groups were encouraged to sponsor sessions at annual meetings as one way of achieving that goal. The AAG Task Force on Communications that recommended specialty groups foresaw a limited number of specialty groups, more than ten but fewer than thirty, that would eventually plan up to 80 percent of the annual meeting paper sessions (Conzen 1978). That intention had not been realized by the early 1990s, when specialty groups typically sponsored about a third of annual meeting sessions.

Coincident with these changes, annual meeting attendance grew rapidly (table 9.1). Almost 3,500 geographers registered at the 1990 Toronto meeting, more than half the association's membership. No other scholarly society in the United States attracts as high a proportion of its membership to its annual meeting. AAG meetings are huge affairs, typically requiring the efforts of 1.5 person-years of AAG office staff time and hundreds of hours of volunteer time.

Table 9.1. Registration at AAG Annual Meetings, 1971–1991

Year	City	Paid registration
1971	Boston	1,100
1972	Kansas City	1,475
1973	Atlanta	1,650
1974	Seattle	1,330
1975	Milwaukee	1,880
1976	New York City	1,705
1977	Salt Lake City	2,055
1978	New Orleans	2,915
1979	Philadelphia	2,720
1980	Louisville	2,315
1981	Los Angeles	2,045
1982	San Antonio	2,085
1983	Denver	2,280
1984	Washington, D.C.	2,705
1985	Detroit	2,375
1986	Minneapolis	2,695
1987	Portland	2,500
1988	Phoenix	2,750
1989	Baltimore	3,115
1990	Toronto	3,471
1991	Miami	2,675

SOURCE: *AAG office records.*

In the early 1990s, a typical meeting consisted of 1,300 presentations organized into more than 200 sessions, not counting plenary papers and business
meetings.

Aside from their roles in the AAG's annual meetings, specialty groups have
gone far toward improving communication among like-minded geographers.
They were a response to perceived needs to create new modes of communication and to restructure the profession by emphasizing the importance of
specialization (Goodchild and Janelle 1988; Marcus 1988). Many of the 41
specialty groups (table 9.2) that existed in the early 1990s published their own
newsletters. When they were formed, specialty groups were deemed by the
AAG president to be "the most significant activity undertaken by the Association in the last decade," an innovation whose "results will influence our
goals, operations, and professional philosophies for many succeeding decades" (Marcus 1977).

Regional Meetings

When the AAG amalgamated with the ASPG in 1948 it acquired the ASPG's
regional divisions. Some regional groups had existed before the founding of
the ASPG in 1943. The Association of Pacific Coast Geographers (APCG)
dates from 1935. What is now the Southeastern Division of the AAG was
formed in 1947, and the once-independent regional divisions still retain distinctive personalities. The APCG maintains the independent ethos it developed
when the region was distant in travel time from the midwestern focus of the
AAG. The Southeastern Division runs a lively annual meeting. Its members
take great pride in rigorously screening all papers presented at the division's
meeting and assigning a discussant for every presentation.

The AAG's ten regional divisions differ in size, but all hold annual meetings
and several publish journals. Attendance at regional division meetings varies
from about a hundred for dispersed units such as the Great Plains/Rocky Mountain Division up to several hundred in populous regions such as the Southeastern Division. Although attendance at national AAG meetings has increased in
recent years, regional meeting attendance has stagnated. At one time graduate
students often presented their first papers at regional meetings, and senior
scholars maintained strong regional commitments. The ease of entry at national meetings may have diminished the attraction of regional meetings for
students and young practitioners, and scholars seem to focus more on national
and international contacts than on regional concerns.

Specialized Meetings

Several independent groups hold regular meetings. The Conference of Latin
Americanist Geographers, for example, holds annual meetings in the United

Table 9.2. AAG Specialty Group Membership, 1991

Group	Members
Africa	170
Aging and the Aged	45
American Ethnic Geography	new
American Indians	71
Applied	340
Asian	184
Bible	43
Biogeography	241
Canadian Studies	113
Cartography	467
China	121
Climate	289
Coastal and Marine Geography	132
Contemporary Agriculture and Rural Land Use	145
Cultural Ecology	211
Cultural Geography	216
Energy and Environment	228
Environmental Perception and Behavior	265
Geographic Information Systems	706
Geographic Perspectives on Women	153
Geography in Higher Education	174
Geography of Religion and Belief Systems	new
Geomorphology	339
Hazards	123
Historical Geography	377
Industrial Geography	192
Latin America	247
Mathematical Models and Quantitative Methods	252
Medical	137
Microcomputer	302
Political Geography	306
Population Geography	248
Recreation, Tourism, and Sport	214
Regional Development and Planning	319
Remote Sensing	344
Rural Development	180
Socialist Geography	133
Soviet, Central, and East European	151
Transportation	186
Urban	569
Water Resources	290

SOURCE: *AAG office records.*

States and Latin America. Founded in 1970, it serves as a medium of exchange for geographers interested in Latin America. An applied geography conference is held every year under the auspices of geographers at Kent State University and the State University of New York at Binghamton. Many participants are members of the AAG Applied Geography Specialty Group, but the conference is an independent enterprise that serves the needs of geographers practicing in government and the private sector. Although organized by geographers, these meetings attract participants from other disciplines. Conversely, geographers participate in the meetings of organizations such as the American Association for the Advancement of Slavic Studies, the Asian Studies Association, and so forth.

Since 1987, the AAG has cosponsored an annual GIS/LIS conference in collaboration with the American Congress on Surveying and Mapping, AM/FM (Automated Mapping/Facilities Management) International, the American Society for Photogrammetry and Remote Sensing, and the Urban and Regional Information Systems Association. The GIS/LIS conference caters to an interdisciplinary group of practitioners and scholars interested in geographic information and land information systems. Many of the 3,000 professionals and students who attend its meetings find them especially stimulating because of the variety of disciplinary and vocational perspectives that focus on GIS/LIS technology and questions.

Specialist meetings serve as media for professional and scholarly communication that the major geographical societies cannot easily provide. The tensions between specialization and community within geography will continue to generate a shifting array of specialty groups and independent meetings that create fission within the discipline at the same time that they serve the needs of specialists. Cosponsored meetings such as GIS/LIS will continue to arise as new specialties develop at the margins of existing disciplines and as new structures are needed to accommodate evolving needs to meet and talk.

International Meetings

The International Geographical Union (IGU) holds international congresses at four-year intervals. The IGU also sponsors regional conferences at four-year intervals between international congresses (table 9.3). The IGU executive establishes study groups and commissions that are close analogues of the AAG's specialty groups, although they have finite life spans. Commissions and study groups are reviewed at each congress. Commissions are extended or terminated on the basis of their activities in the previous four years. A study group may exist for only four years, at which point it will either become a commission or be dissolved. Members are appointed to study groups and commissions by the IGU executive upon the recommendation of national representatives. Commissions and study groups may have only ten full members, no two of whom may

Table 9.3. IGU Congresses and Regional Conferences, 1972–1998

Year	IGU Congress	IGU Regional Conference
1972	Montreal	
1974		Palmerston North
1976	Moscow	
1978		Lagos
1980	Tokyo	
1982		Sao Paulo
1984	Paris	
1986		Barcelona
1988	Sydney	
1990		Beijing
1992	Washington, DC	
1994		Prague
1996	Amsterdam	
1998		Lisbon

SOURCE: *IGU secretary general, personal communication, Sept. 1991.*

be from the same country; both may have unlimited numbers of corresponding members.

Geographers from throughout the world attend congresses and regional conferences. U.S. participation in congresses normally numbers several hundred and at regional conferences ranges from tens to more than a hundred, depending on location. Because IGU study groups and commissions hold symposia or seminars before each congress or regional conference, many geographers who attend IGU meetings participate in a premeeting symposium. In addition, IGU meetings are customarily followed by lengthy field excursions, so that full participation in a congress or regional conference often requires two to three weeks' time and several thousands of dollars. The logistics to be overcome in participating in IGU activities are often formidable. Geographers who participate regularly, however, argue that the rewards are commensurate with the difficulties. The great differences among national approaches to geography can be exceptionally stimulating, and the postmeeting field excursions offer unparalleled opportunities to learn about exotic places under expert guidance.

Organization Committees

The AAG and sister societies such as the NCGE boast numerous standing and ad hoc committees. Their members are charged to monitor specific programs, to formulate policy recommendations, or to solve problems. In the early 1990s, the AAG had seventeen standing committees, three ad hoc study groups, two elected committees, and two project steering committees, in addition to the annual meeting local arrangements and program committees. These committees engaged over 160 members in their deliberations and operations. Committees are a powerful communications medium. Bodies that appoint committees normally strive for balance with respect to gender, specialty, age, and regional representation, among other considerations. Whether they fulfill their nominal purposes or not, such groups bring geographers of diverse dispositions and interests into contact with each other, thereby fostering a greater sense of community and common purpose.

Informal Networks

The communications mechanisms outlined above are, for the most part, formal structures established specifically with communication in mind. They stand in contrast to the quiet conversations with colleagues and the exchanges of letters that have long enabled professionals and scholars to exchange ideas and gossip. Because such exchanges are either unrecorded or scattered among personal papers, it is difficult to estimate their effects beyond noting that they are undoubtedly of great importance. Such informal means of exchanging information have taken new and exciting turns in recent years. The consequences are difficult to foresee in detail but they seem portentous in outline.

Chief among the devices that have altered communication among geographers since 1960 are the telephone, facsimile transmission, and electronic mail. At first used only sparingly when a letter would be too slow, telephonic communication is now used for routine business and is probably less expensive than communication by letter. Facsimile (FAX) transmission use exploded in the United States in the late 1980s. Because it is almost instantaneous and does not demand that communicating parties be available simultaneously, it has become essential for scholarly communication. Electronic mail (E-mail) offers the speed of electronic communication and the advantages of a written text at costs much lower than telephone or FAX. A number of interconnected E-mail networks exist.

Many geographers in the United States use BITNET (Because It's Time Network). Begun in 1981 to connect the City University of New York and Yale University, by the early 1990s BITNET linked hundreds of academic institutions throughout the world (Lewis 1989). A growing share of the discipline's administrative and intellectual business is now conducted via BITNET and similar networks. E-mail also provides access to electronic bulletin boards (often referred to as *lists*) on which any subscriber can post and read messages. A mid-1991 tabulation of the lists accessible on BITNET alone numbered over 2,600 electronic bulletin boards. Bulletin boards of interest to geographers are devoted to cartography, climatology, geographic information systems, and graphics. Some lists are open; anyone can post a message to the bulletin board, which can then be read and responded to by any other subscriber. Some are screened, with an individual acting as a gatekeeper to screen messages that she or he deems to be irrelevant or inadequate. Some are refereed; messages posted to the list are reviewed by a panel before being placed on the bulletin board. Refereed lists serve as electronic journals, providing subscribers with research findings that have been approved by peers in their respective specialties.

WRITING FOR EACH OTHER

In contrast to the letters and digital messages exchanged among geographers stands the information shared more widely through the discipline's print media.

Geographical Journals

Most specialized geographic journals were started after 1960. According to Harris and Fellmann (1980), the number of geographic serial publications issued annually grew from 45 in the 1950s to more than 100 in the late 1970s. Not all this growth occurred in the United States, nor were all the new journals available to American scholars. Harris and Fellmann estimated that of the 3,335 serial publications they tabulated, only 1,089 were active in 1980, and of these, only 443 were readily available to the international research community. The accessibility of geographic journals to nongeographers is further limited because the most widely used abstract and citation services analyze fewer than 30 geography journals. Thus, scholars outside the discipline may have a restricted view of the research published by geographers (Harris and Fellmann 1980).

American geographers have created a division of labor among their journals. The *Annals of the Association of American Geographers*, published quarterly by the AAG, is the flagship geography journal in the United States. The *Professional Geographer* (also published quarterly) became an AAG journal when the AAG and the ASPG amalgamated. Originally focused on matters of interest to practitioners, it acquired a more scholarly character in the 1970s and 1980s. Its articles are shorter than those appearing in the *Annals*. The American Geographical Society (AGS) publishes the *Geographical Review*, whose nineteenth-century emphasis on exploration and commercial geography has become more scholarly since 1920. The journal is more likely to publish articles devoted to artistic and humanistic aspects of geography than the *Annals* or the *Professional Geographer*. The NCGE publishes the *Journal of Geography* containing articles devoted to the methods, substance, and theory of geographic education. Whereas the circulations of the AAG, AGS, and NCGE journals number in the hundreds or the thousands, the monthly *National Geographic Magazine*, published by the National Geographic Society (NGS), reaches eleven million subscribers. *National Geographic* provides the lay public with superb photographs of distant and familiar places in articles written to capture popular attention. NGS also publishes a scientific journal entitled *Research and Exploration* (formerly *National Geographic Research*), which contains articles by geographers but has been dominated by archeological and life science content.

Growth in the number and the circulation of geographic journals has led to speculation about their prestige and impacts. A 1983 survey of five hundred geographers in graduate programs produced a ranking of 34 journals (Lee and Evans 1984, 1985). The *Annals* ranked first among the top five and the *Geographical Review* second, followed by the *Professional Geographer*, *Economic Geography* (published quarterly by the Graduate School of Geography at Clark University), and the *Journal of Geography*. Specialized journals such as *Physical Geography, Urban Geography*, and *Historical Geography* ranked lower. State geography journals (for example, the *Pennsylvania Geographer*) were least popular. An age preference was evident in the responses. Older geographers preferred well-established journals that publish articles on a variety of topics, whereas younger geographers favored specialized journals. Physical geographers were devoted to their specialized publications and were less interested in human geography journals (Lee and Evans 1984:299).

The discipline's journals can inhibit rather than foster communication if they lead specialists to ignore research and teaching outside their primary interests. Specialization in geography increased in the 1960s and 1970s to the point where new journals were started because specialist groups were dissatisfied with traditional outlets. The first in the United States was *Geographical Analysis*, a rallying point for geographers of the theoretical and quantitative bent. *Antipode*, the journal of radical geography, was begun because its founders believed that established journals were not hospitable to their views (Peet 1985). Questions of where geography journals stand relative to those in related disciplines have also been raised. The small number of geographers means that geographers in some specialties can reach larger audiences by publishing outside of geography. An added incentive for doing so is evidence that articles in geographical journals are not cited as often as those appearing in journals in other disciplines (Turner 1988). As specialties continue to evolve and geographers continue to work with colleagues in other disciplines, publishing patterns will undoubtedly become more complex. A growing interest in citation analysis (Whitehand 1984, 1985; Turner and Meyer 1985; Wrigley and Matthews

1986; Bodman 1991; Curry 1991) will permit better measurement of the flow of ideas among geographers through the medium of journal articles.

Books and Monographs

Summarizing the hundreds of geographic books and monographs published in the United States over the last thirty years would be a daunting task. Geographers have written on local neighborhoods, cities, the United States, and systematic specialties, among other topics. Donald Meinig's *The Shaping of America: A Geographical Perspective on 500 Years of History* (1986) presented a broad view of the historical geography of the United States. It was widely and favorably received by historians as well as by geographers. Reviewers applauded Meinig's demonstration of the value of the geographical perspective and felt that the work would become a standard in the field. However, the book has been categorized as a geographic perspective on history, and not as an example of a genre of literature particularly associated with geographers (Cronan 1986). *The Shaping of America* is the first of a projected trilogy summarizing Meinig's lifelong research on American historical geography. (Chapter 3 contains a more detailed discussion of Meinig's volume.) *America's Northern Heartland*, by John R. Borchert (1987) of the University of Minnesota, won the AAG's Jackson Prize in 1989 for a serious geography of the American landscape that was designed to reach the general public. The book is an outstanding regional economic geography. Borchert applies the concepts of physical, historical, urban, and economic geography in an avowed spatial analysis of the Upper Midwest region of the United States. In physical geography, William L. Graf's book on Western rivers, *Fluvial Processes in Dryland Rivers* (1988), demonstrates the ways a geographic perspective enriches the understanding of regional differences among river systems.

Edward W. Soja's view of Los Angeles offers an American example of a geographic monograph rooted in a postmodern perspective. He uses his "free-wheeling essay on Los Angeles" to "both integrate and disintegrate" his introduction to postmodern geographies (1989:2), presenting the region as "a *prototopos*, a paradigmatic place . . . a *mesocosm*, an ordered world in which the micro and the macro, the idiographic and the nomothetic, the concrete and the abstract, can be seen simultaneously in an articulated and interactive combination" (1989:191). "It All Comes Together in Los Angeles" (1989:190) is the substantive paradigm of postmodernism, "a multi-layered geography of socially created and differentiated nodal regions nesting at many different scales around the mobile personal spaces of the human body and the more fixed communal locales of human settlements" (1989:8). (Chapters 12 and 14 include additional comments on Soja's monograph.)

Few attempts to collate information about books and monographs in geography have been produced, with the exception of *A Geographical Bibliography for American Libraries* (Harris and others 1985). Although I can adduce no numerical data to support the impression, more geographers may be turning to books as outlets for their analytical and synthetic energies than was common in the 1970s and 1980s. In the late 1980s, the discipline's leaders regularly bemoaned a perceived dearth of scholarly books in geography (Demko 1987; Jordan 1988). Although hundreds, if not thousands of books have been published in the last three decades, many are highly specialized and hold limited appeal for a broad audience of geographers. Because of their many divisions and specialties, not to mention some deterministic skeletons in their closets, geographers have often been hostile to grand works of synthesis such as those noted above. The majority of the books and monographs geographers write are addressed to specialist audiences rather than to the discipline as a whole, and it seems that few books are addressed primarily to the public.

Although publication of a doctoral candidate's dissertation is a nominal requirement for the doctorate in North American universities, most programs deem that test to have been met if a dissertation is available on interlibrary loan or on microfilm. Since 1948, the Department of Geography (now Program in Geography) at the University of Chicago has published the *Department of Geography Research Papers*. The series includes, but is not restricted to, all doctoral dissertations written by geographers who complete the doctorate in the Chicago program. The Chicago series is unique in formally publishing all department dissertations. A number of other monograph series published by various organizations have focused on geography, including several issued for different periods by the AAG, but none has published as consistently as the Chicago series.

Reference Works

A number of geographical dictionaries have been compiled in the last several decades (Harris 1986), the most recent of which are *Modern Geography: An Encyclopedic Survey* (Dunbar 1991), *The Encyclopaedic Dictionary of Physical Geography* (Goudie et al. 1985), and *The Dictionary of Human Geography* (Johnston, Gregory, and Smith 1986). Although the latter two are British works, they are widely used in the United States along with the earlier *Dictionary of Concepts in Human Geography* (Larkin and Peters 1983), which covers eighty-five basic concepts in detail and is largely based on American literature. Geography has evolved to the point where a specialized terminology exists, but consensus on the meaning of terms and concepts does not exist. Therefore, each entry in the dictionaries is signed to enable readers to evaluate the sources of definitions and explanatory essays. Gazetteers were once produced regularly by major publishing houses and government agencies, often at untold expense in human labor. Traditional printed compilations are now giving way to digital data bases of named features and their locations that can be corrected and updated much more readily than printed volumes.

Atlases

Many geographers believe that maps are the geographer's most distinctive communication medium (chapters 4 and 6). Although maps are used in many disciplines, they are employed most intensively in geography; maps and geography have always been inextricably linked. Collections of maps bound into atlases are an especially effective medium. They make it possible to supplement maps with text that offers exposition, interpretation, and synthesis. Throughout most of the history of geography, maps and atlases have stood alone. Only experts used them, and little explanation or interpretation was required. Since 1960, atlases have incorporated increased amounts of prose to embellish the maps they contain. This new genre of atlases constitutes a major development in geographic communication, and examples now appear regularly.

An early example was Norton Ginsburg's *Atlas of Economic Development* (1961), which presented a series of world maps of variables indicative of levels of economic development. The ratio of text to maps was unusually high for the time, owing both to the innovative variables used and to the techniques of analysis and display developed by collaborator Brian J. L. Berry. Other atlases devoted to regions and topics have appeared in subsequent years, including *A Comparative Atlas of America's Great Cities: Twenty Metropolitan Regions* (Abler 1976), *An Historical Atlas of South Asia* (Schwartzberg 1978), *This Remarkable Continent: An Atlas of United States and Canadian Society and Culture* (Rooney, Zelinsky, and Louder 1982), and *We The People: An Atlas of America's Ethnic Diversity* (Allen and Turner 1988).

Abler's *Comparative Atlas* was based largely on the 1970 Censuses of Population and Housing, and all maps are accompanied by essays. Cartographic symbols were designed to facilitate comparisons among the many variables and cities. The project that produced the atlas also generated monographs on each of the book's twenty metropolitan areas and thirteen monographs on urban policy issues (Adams 1976a, 1976b). Financial support for the AAG-sponsored project was provided by the National Science Foundation. The *Historical Atlas of South Asia* represented fourteen years of intense labor by a large staff. Because it contains 650 original maps compiled from sources drawn from throughout the world, it is far more than a compilation. It is a product of research as well as being a tool for research and teaching. David Sopher (1980) called the work a "towering scholarly achievement." *This Remarkable Continent* focuses on folk and popular culture. It was published by the North American Cultural Survey, an informal group of cultural and historical geographers, and was based on a wide variety of innovative and unusual data. The mixed nature and appeal of this new style of atlas is evident in a review of *We the People*, which judged it to be a "marvelous book [that] is at the same time a thematic atlas, a cultural geography, a social history and an encyclopedic reference work" (Bergen 1989:235). *We the People* suggests that geographers in the late twentieth century have adopted a pluralistic view of geography, a suggestion borne out by earlier works such as *The Women's Atlas of the United States* (Gibson and Fast 1986) and *Atlas of American Women* (Shortridge 1987). Both depict conditions faced by women. The former aims to provoke readers to include considerations of space and location in debates over women's issues. The latter focuses on spatial variations in the advances toward equality made by women in recent years.

The 1980s and 1990s saw a surge of state atlas making, usually centered in geography departments at state universities. Examples include California (Donley et al. 1979), Florida (Fernald 1981), Kentucky (Karan and Mather 1977), Massachusetts (Wilkie and Tiger 1991), North Carolina (Clay, Orr, and Stuart 1975), Oregon (Loy and Allan 1976), and Pennsylvania (Cuff et al. 1991). State or topical atlases are often huge undertakings. Gestation periods of eight or more years are not uncommon. They are also expensive, requiring their editors to spend more time raising money than they devote to atlas production. Despite such drawbacks atlases are widely acclaimed by scholarly colleagues and the public, and geographers increasingly find atlases attractive ways to present their insights.

Newsletters

Many AAG specialty groups issue newsletters. The AAG publishes a newsletter monthly except in July and September that is sent to all members and to institutional subscribers. *AAG Newsletter* content consists of exhortations and observations from the association's leadership, notices of forthcoming meetings, news of recent events, and the minutes of the AAG's council and executive committee meetings. Perhaps the most important items communicated via the *AAG Newsletter* are the employment listings contained in its "Jobs in Geography" section, which serves as the journal of record for advertising domestic and overseas positions available and positions sought.

The impact of electronic networks on media that have formerly been restricted to print is difficult to forecast. It seems unlikely that printed journals, books, and atlases will become obsolete in the near future. At the same time, journals are now being made available in some libraries on compact laser disks (CDROM) rather than in hard copy, and innovations such as E-Mail and its bulletin boards portend competition with traditional print forms. Anyone now planning a major atlas would have to consider dissemination on CDROM as well as in printed form. Further advances in computerized display technologies and in computer networking will doubtless offer geographers a wider array of communications media in the future, including digital access to journals, books, atlases, and newsletters.

COMMUNICATING WITH GEOGRAPHY'S CLIENTS

Students constitute the largest market for geography's insights. Accordingly, most American geographers spend the largest share of their time serving student needs in the nation's colleges and universities, and many members of the discipline have a keen interest in improving the geography taught in the nation's elementary and secondary schools. A third audience geographers address is the public and its leaders.

Colleges and Universities

In 1990, an estimated 586,000 college and university students took geography courses at more than 600 academic institutions in the United States (Walker 1991). What they learned ran the gamut from basic introductory material in survey courses to highly specialized topics for advanced undergraduate students. Most students take one or a few geography courses, but some elect to major in geography and complete baccalaureate degrees. Introductory college geography courses are a critical communication link for geographers, because very few students enter college intending to major in geography. When introductory courses are effective, students take more classes and some decide to major in geography. In recent years, the number of students completing bachelor's degrees in geography has hovered around 3,000 each year, down from the early 1970s when over 4,000 students per year majored in geography, but up from 1961 when the number of geography majors completing degrees exceeded 1,000 for the first time. Graduate training is available in approximately 135 institutions, of which 51 offer the Ph.D. In recent years the nation's production of Master's degrees in geography has averaged about 550 per year. Completed doctorates have numbered 120 to 130 per year recently, down significantly from the 200 per year produced between 1972 and 1975.

American colleges and universities are typically organized into divisions focusing respectively on the arts, engineering, the humanities and social sciences, and the natural sciences, or different combinations thereof. Geography fits neatly into none of these categories. Undergraduate students can often use cultural and human geography courses to satisfy institutional social science distribution requirements, and physical geography often meets laboratory science or natural science requirements. In a few places, courses on historical geography or landscape interpretation qualify as humanities offerings. Although geography and geographers are enriched by embracing the humanities, natural science, and social science ways of knowing the world, the fact that the discipline spans that breadth sometimes communicates more confusion than enlightenment to colleagues in other disciplines and to the public.

Attention to undergraduate instruction by the discipline's leadership has varied from time to time. The AAG established a Commission on College Geography (CCG) in 1965 (Commission on College Geography 1967). Supported by a series of grants from the U.S. Department of Education, the CCG was charged with strengthening geography at a time when the baby boom generation was crowding onto campuses across the country and geography enrollments were also booming. The CCG focused primarily on curriculum development and on providing supplementary materials devoted to substantive and methodological topics of current interest. The CCG was phased out in the mid-1970s when funds to support such efforts dried up as the post-Sputnik concern with education waned. In 1991, at the instigation of Susan Hanson, then AAG president, a second Commission on College Geography (CCG II) was appointed with the charge to "make geography programs at the college level more compatible with the broader educational needs of colleges and universities throughout the nation" (AAG Council minutes, 12 April 1991, Miami, Florida).

The messages geographers communicate via undergraduate courses and graduate programs defy summary in any detail. No compilation of course syllabi is available, and the use of even the most popular textbooks varies greatly. A thorough examination of both would most likely demonstrate diminished use of regional organization after 1960 and a corresponding increase in presentations that stressed systematic approaches to human and physical geography. A survey of the dozen college textbooks in use in the early 1990s leads to the conclusion that after flirting with quantitative and topical approaches to the study of society pioneered by economists and political scientists, geographers now prefer the narrative style common in anthropology, a mode exemplified by some of the selections highlighted in chapter 3 of this volume. Current textbooks are wonderfully illustrated with color photos, charts, and maps, but students are seldom challenged to make a map, graph a relationship, or perform a computation. Little evidence of geography's quantitative revolution appears in the textbooks used in the introductory geography courses taught in the early 1990s.

Elementary and Secondary Schools

The discipline's early leaders devoted a great deal of attention to the geography taught in elementary and secondary schools. Compared to other countries, assuring the presence and quality of geography was more difficult in the United States, where national curricula have been anathema and where almost all aspects of elementary and secondary education are controlled by the nation's thousands of independent school districts. Nevertheless, many academic geographers worked closely with state school officials and other bodies to ensure that sound geography was an integral part of the standard curriculum. American academic geography began to close in on itself in the 1920s, and one of the first casualties of that narrowing perspective was attention to geographic education. At the same time, a movement began to abandon traditional instruction in disciplines in favor of social studies instruction that presented a melange of anthropology, economics, geography, history, political science, and sociology—none of which were ever mentioned as components of the mixture. The consequences were devastating. By the 1950s, geography as a distinct subject had essentially disappeared from the country's schools. Although most teachers took a geography course during their college years, the courses they took after 1960 often stressed quantification and theory more than regional content. The messages transmitted by academic geographers since the 1920s were that geographers were not interested in collaboration with teachers, regional geography was not the core of the discipline, and the best geography was theoretical.

The intense interest in revitalizing education that followed Sputnik aroused some geographers to see how the discipline could benefit from the widespread research and experimentation with school curricula. With financial support from the National Science Foundation, the AAG launched the High School Geography Project (HSGP) in 1963. The project's leaders attempted to merge the latest thinking in education and geography by producing materials that explored major issues in students' social and physical environments. HSGP sought to create instructional units that would break tradition by covering topics rather than regions, and it produced a series of modules rather than a textbook. Guiding principles were: (1) to encapsulate selected ways the discipline looked at the world, (2) to emphasize questions geographers ask, and (3) to use methods geographers employ when seeking answers to questions. Units were field-tested by master teachers, and summer institutes were established to train teachers to use the materials. Although they were enthusiastically promoted and received, the HSGP instructional modules were not widely adopted. The absence of a true textbook placed HSGP products at a disadvantage in a marketplace dominated by texts, a drawback that was compounded by pitching the units to grades ten through twelve whereas most geography is taught in grades seven through nine. Teachers who mastered the materials in the HSGP summer institutes quickly became strong advocates for the materials and HSGP (Stoltman 1980). By the mid-1970s, however, funding for the summer institutes had dried up, and by 1985 the HSGP units had become collectors' items.

A new effort to transmit sound geography to the nation's students began in 1985 with the publication of *Guidelines for Geographic Education* by a joint committee of the AAG and the National Council for Geographic Education

(Joint Committee 1984). The *Guidelines* define five fundamental geographic themes that are attractive to social studies teachers, school administrators, and parents. Through 1991, over 70,000 copies of the *Guidelines* had been distributed to teachers and school administrators. The joint AAG-NCGE committee encouraged its parent bodies to establish an organization that would develop a plan to promote geographic education. The Geographic Education National Implementation Project (GENIP) was subsequently inaugurated with a steering committee consisting of representatives from the AAG, AGS, NCGE, and NGS. GENIP has concentrated on developing and disseminating information on ways the five themes from the *Guidelines for Geographic Education* can be incorporated into social studies classes or developed into independent courses. GENIP has supplemented the original *Guidelines* publication with *Guidelines for Geographic Education, 7–12* and an assessment of secondary school textbooks (St. Peter 1989).

AAG-NCGE efforts to upgrade geography education were greatly enhanced by the programs instituted by the National Geographic Society to combat geographic ignorance. During the early 1980s, NGS president Gilbert Grosvenor grew increasingly alarmed about the deplorable levels of geographic knowledge among the American public. In 1984 he launched a major effort to combat what came to be called *geographic illiteracy* (Grosvenor 1984). NGS established a geographic education program to: (1) increase public awareness of the need for improved geographic education, (2) train inservice teachers in geography, and (3) establish a system of state-based Alliances for Geographic Education. To further these ends, NGS has commissioned Gallup polls that document how little Americans know about world place location, lobbied Congress to create National Geography Awareness Week as a means of drawing media attention to geographic ignorance, and hosted congressional hearings on the need for geographic education. State alliances have used similar strategies to promote geographic education within states. At the national level, NGS's World Magazine Division initiated an annual geography bee that has proved immensely popular. NGS and its Geographic Education Foundation have eschewed direct involvement in academic geography and the publication of geography texts. Their programs make extensive use of the *Guidelines for Geographic Education* and its five fundamental themes. Nonetheless, the NGS's Education Media Division has experimented extensively with a variety of new methods for teaching geography, and it appears that the nation's most innovative technological research on geographic instruction is being conducted at NGS.

In the late 1980s and early 1990s, large corporations began to promote improved geographic competence on the part of the American public. NGS's education foundation received major contributions from several corporations and foundations, and CitiCorp began to sponsor the materials produced and distributed by NGS during National Geography Awareness Week. In 1989,

AAG and American Express Travel Related Services created a geography contest for secondary students modeled on the Westinghouse science competition. The American Express geography competition stresses geographic inquiry by awarding prizes for the best research papers produced by individual high school students or teams on geographic topics.

The programs undertaken by the geographical societies and their congressional and corporate supporters have begun to produce results. The National Council for Social Studies, which in 1970 had deemed the basic ideas taught in school geography out-of-date and flawed, concluded in 1989 that "because they offer the perspectives of time and place, history and geography should provide the matrix or framework for social studies" (National Commission on Social Studies 1989). More importantly, geography was designated one of the five core subjects proposed for inclusion in the curricula of all the nation's schools in the *America 2000* plan put forth by the National Governors Association and the U.S. Department of Education in 1991. Although it will take years of effort fully to implement the programs proposed in the *America 2000* plan, sound geography should once again be imparted to most of the nation's school students early in the twenty-first century.

The Public

Communicating geographical insights to the public has rarely been a priority in American geography. While many individual practitioners and scholars give occasional public lectures, few if any geographers see the public as the primary audience for their speaking and writing. *National Geographic Magazine* enjoys huge public success, to the tune of a circulation of eleven million and a readership of at least twice that number. Relations between academic geographers and the NGS have improved greatly in recent years, but many academic geographers remain uncomfortable with *National Geographic*'s emphasis on the unusual and the photogenic. While grateful for the publicity NGS has given to the discipline and for the largesse it has earmarked for improving geographic education, academics do not view *National Geographic Magazine* as a medium to convey their messages to the public.

Many geographers throughout the country are interviewed often for newspaper, radio, and television pieces on local and regional questions. Geographers concerned to reach a national audience continue to search for a channel to a society that exhibits great interest in many aspects of geography. Harm J. de Blij, professor of geography at Georgetown University and the University of Miami, has found one. He was appointed geography editor for ABC Television's *Good Morning America* program in August 1990. He appears regularly on the popular morning program to provide a geographic perspective on current affairs and events. His appointment to the position followed a chance appearance in mid-1989 that led to a five-segment series on geography broadcast during the week of 25–29 September. The five programs were devoted to geography as a discipline, physical geography, historical and environmental geography, population and urban geography, and a general overview. Strong popular response to the series led to further appearances and then to de Blij's appointment to a regular position with ABC (de Blij 1990). De Blij's appearances have given geography an enviable platform from which to demonstrate the value of geographic insights.

Other programs to educate the public that have been started in recent years hold the promise of enlarging geography's constituency beyond its traditional audiences of students and professionals in related disciplines. The American Geographical Society has enjoyed considerable success with its program that engages academic geographers to lecture on areas and regions traversed by cruise ships. The wealthy and influential individuals who patronize cruises often become enthusiasts for the discipline once they are exposed to first-rate lectures delivered by accomplished and dedicated teacher-geographers. AGS has also begun to hold seminars for business audiences in New York City at which geographers provide their perspectives on events and issues with commercial and industrial implications. These and other recent developments portend greater attention by academic geographers to sharing their analyses and thoughts with adult learners.

MISSING LINKS

Despite the gusto with which geographers communicate, some elaboration and restructuring of their networks would enhance the discipline's effectiveness and its position in American society. Organized geography has comparatively weak contacts with the students it has trained, with policy makers, and with the general public.

One link that could be stronger is that between academic geographers and practitioners. The year 1990 witnessed the 100,000th bachelor's degree in geography conferred in the United States since World War II. Few of the 100,000 have maintained their ties to the community, despite the fact that almost all of them remain in the labor force. Because of the penchant among academic geographers to write largely for each other, geography lacks a corps of self-identifying practitioners who maintain strong ties to the core discipline. Such groups do exist in economics and geology, for example.

Geographers have too often failed to share their understanding and recommendations with public administrators and political leaders. The intensive involvement of much of the discipline's leadership in public policy formulation and in major projects such as the Tennessee Valley Authority before World War II stands in sharp contrast to the limited impact geographers have had on public policy since 1945. Government officials, leaders in business, and the titans of industry need and want the advice and insights geographers can provide, and geography would benefit from the appreciation of the discipline that nonacademic decision makers would acquire were geographers to communicate more effectively and more often with such leaders.

The American public has a great thirst for geography. It is evident in the membership in the National Geographic Society, the numbers of viewers who watch geographic specials on public television, the sales of books written by James Michener (the greatest exponent of the geographic method of sequent occupance the world has yet known), and in the popularity of writers such as Barry Lopez, who can evoke haunting and vivid images of unfamiliar landscapes in his readers' minds with precisely selected words. Since Ellen Churchill Semple and Ellsworth Huntington, the discipline has had no figures whose writings have reached nongeographers and the public the way Margaret Mead (1936) popularized anthropology or Barbara Tuchman (1962) catered to the popular interest in vividly written history.

Since World War II, two distinct views of geography in the United States have resulted in a great deal of mis- and noncommunication among geographers and between geographers and American society. Most academic geographers have viewed geography as an abstract science to be communicated to other university geographers through scholarly journals and specialized books and monographs. Although based in places and spatial relationships, it has often been far removed from real places and people, and it has been more analytic than synthetic. The needs and wants of its own practitioners and of the public have not ranked high on the discipline's agenda. The public, geographers teaching in the nation's schools, and regional and national policymakers, on the other hand, need and want a geography that focuses on understanding and explaining the world, one that deals with real places, real people, real problems, and how they interact and change. Redirecting some of the discipline's effort toward meeting the needs of policymakers and the public would benefit both geography and American society.

NOTE

I am grateful to Ronald F. Abler and the AAG office for their assistance in preparing this chapter.

REFERENCES

ABLER, RONALD F., ed. 1976. *A comparative atlas of America's great cities: Twenty metropolitan regions.* Ki-Suk Lee, Chief Cartographer. Minneapolis: University of Minnesota Press.

ADAMS, JOHN S. 1976a. *Contemporary metropolitan America*, Vols. 1–4. Cambridge, Mass.: Ballinger.

———. 1976b. *Urban policymaking and metropolitan dynamics: A comparative geographical analysis*. Cambridge, Mass.: Ballinger.

ALLEN, JAMES P., and EUGENE JAMES TURNER. 1988. *We the people; An atlas of America's ethnic diversity*. New York: Macmillan.

America 2000: An educational strategy Sourcebook. 1991. Washington, D.C.: GPO.

Association of American Geographers. [AAG]. 1991. Survey of registrants, Eighty-seventh annual meeting of the Association.

BERGEN, JOHN V. 1989. *We the people: An atlas of America's ethnic diversity*. *Professional Geographer* 41:235–236.

BODMAN, ANDREW R. 1991. Weavers of influence: The structure of contemporary geographic research. *Transactions of the Institute of British Geographers*, n.s. 16:21–37.

BORCHERT, JOHN R. 1987. *America's northern heartland*. Minneapolis: University of Minnesota Press.

CLAY, JAMES W., DOUGLAS M. ORR, JR., and ALFRED W. STUART. 1975. *North Carolina atlas: Portrait of a changing southern state*. Chapel Hill: University of North Carolina Press.

Commission on College Geography. 1967. *New approaches in introductory college geography courses*. Washington, D.C.: Association of American Geographers.

CONZEN, MICHAEL P. 1978. The formation of specialty groups within the AAG. *Professional Geographer* 20:309–314.

CRONAN, WILLIAM. 1986. The shaping of America. *New York Times*, 17 Aug., 12.

CUFF, DAVID J., WILLIAM J. YOUNG, EDWARD K. MULLER, WILBUR ZELINSKY, and RONALD F. ABLER, eds. 1989. *The atlas of Pennsylvania*. Philadelphia: Temple University Press.

CURRY, MICHAEL R. 1991. On the possibility of ethics in geography: Writing, citing, and the construction of intellectual property. *Progress in Human Geography*, 15:125–147.

DE BLIJ, HARM J. 1990. Geography on "Good Morning America." *Focus* 40 (Winter): 32–37.

DEMKO, GEORGE. 1987. President's column. *AAG Newsletter* 21, 7:1.

DONLEY, MICHAEL W., ALLAN STUART, PATRICIA CARO, and CLYDE P. PATTON. 1979. *Atlas of California*. Portland, Ore.: Professional Book Center.

DUNBAR, GARY S., ed. 1991. *Modern geography: An encyclopedic survey*. New York: Garland.

FERNALD, EDWARD, ed. 1981. *Atlas of Florida*. Tallahassee: Florida State University Press.

GIBSON, ANNE, and TIMOTHY FAST. 1986. *The womens' atlas of the United States*. New York: Facts on File.

GINSBURG, NORTON S. 1961. *Atlas of economic development*. Chicago: University of Chicago Press.

GOODCHILD, MICHAEL F., and DONALD G. JANELLE. 1988. Specialization in the structure and organization of geography. *Annals of the Association of American Geographers* 78:1–28.

GOUDIE, ANDREW, eds. 1985. *The encyclopaedic dictionary of physical geography.* Oxford: Basil Blackwell.

GRAF, WILLIAM L., B. W. ATKINSON, K. J. GREGORY, I. G. SIMMONS, D. R. STODDART, and DAVID SUGDEN. 1988. *Fluvial processes in dryland rivers.* Berlin-Heidelberg: Springer-Verlag.

GROSVENOR, GILBERT M. 1984. The society and the discipline. *Professional Geographer* 36:413–418.

HARRIS, CHAUNCY D. 1986. Book reviews: Geographical dictionaries. *Annals of the Association of American Geographers* 76:284–311.

HARRIS, CHAUNCY D,. and JEROME FELLMANN. 1980. *International list of geographical serials.* Chicago: University of Chicago Press.

HARRIS, CHAUNCY D., SALVATORE J. NATOLI, RICHARD W. STEPHENSON, HAROLD A. WINTERS, and WILBUR ZELINSKY, eds. 1985. *A geographical bibliography for American libraries.* Washington, D.C.: Association of American Geographers and the National Geographic Society.

HILL, A. DAVID. 1981. A survey of the global understanding of American college students: A report to geographers. *Professional Geographer* 33:237–245.

JOHNSTON, RONALD J., DEREK GREGORY, and DAVID M. SMITH, eds. 1986. *The dictionary of human geography.* 2d ed. Oxford: Basil Blackwell.

Joint Committee on Geographic Education of the NCGE and AAG. 1984. *Guidelines for geographic education: Elementary and secondary schools.* Washington, D.C.: Association of American Geographers.

JORDAN, TERRY G. 1988. Critical mass. *AAG Newsletter* 23, 3:1.

KARAN, P. P., and COTTON MATHER, eds. 1977. *Atlas of Kentucky.* Lexington: University Press of Kentucky.

LARKIN, ROBERT P., and GARY PETERS, eds. 1983. *Dictionary of concepts in human geography.* Westport, Conn.: Greenwood.

LEE, DAVID, and ARTHUR EVANS. 1984. American geographers' rankings of American geography journals. *Professional Geographer* 36:292–300.

———. 1985. Geographers' rankings of foreign geography and non-geography journals. *Professional Geographer,* 37:396–402.

LEWIS, LAWRENCE. 1989. BITNET: A tool for communication among geographers. *Professional Geographer* 41, 4:470–479.

LOPEZ, BARRY H. 1986. *Arctic dreams: Imagination and desire in a northern landscape.* New York: Charles Scribner's Sons.

LOY, WILLIAM G., and STUART ALLAN. 1976. *Atlas of Oregon.* Eugene: University of Oregon Press.

MARCUS, MELVIN G. 1977. A letter from the president. *AAG Newsletter* 12, 7:1–2.

———. 1988. New twists on the horns of an old dilemma. *Annals of the Association of American Geographers* 78:540–542.

MEAD, MARGARET. 1936. *Coming of age in Samoa: A psychological study of primitive youth for Western civilization.* New York: Blue Ribbon Books.

MEINIG, DONALD. 1986. *The shaping of America: A geographical perspective on 500 years of history.* New Haven: Yale University Press.

MICHENER, JAMES A. 1974. *Centennial.* New York: Random House.

National Commission on Social Studies in Schools. 1989. *Charting a course: Social studies for the 21st century: A report of the curriculum task force.* Washington, D.C.: The Commission.

NELSON, CARNOT E., and KAZUA TOMICA. 1970. Impact of the proceedings on the annual meeting of the association of American geographers: A comparison of scientific information exchange at the 1967 and 1969 annual meetings. *Professional Geographer* 22:221–226.

PEET, RICHARD. 1985. Evaluating the discipline's journals: A critique of Lee and Evans. *Professional Geographer* 37:59–62.

ROONEY, JOHN F., WILBUR ZELINSKY, and DEAN LOUDER, eds. 1982. *This remarkable continent: An atlas of United States and Canadian society and culture.* College Station: Texas A&M University Press.

ST. PETER, PATRICE H. 1989. *Text assessments in geography.* Chicago: Geographic Education National Implementation Project.

SCHWARTZBERG, JOSEPH. 1978. *An historical atlas of South Asia.* Chicago: University of Chicago Press.

SHORTRIDGE, BARBARA GIMLA. 1987. *Atlas of American women.* New York: Macmillan.

SOJA, EDWARD W. 1989. *Postmodern geographies: The reassertion of space in critical social theory.* New York: Verso.

SOPHER, DAVID. 1980. *Historical atlas of South Asia:* A review. *Annals of the Association of American Geographers.* 70:28–291.

STOLTMAN, JOSEPH P. 1980. Round one for HSGP: A report on acceptance and diffusion. *Professional Geographer* 23:209–215.

TUCHMAN, BARBARA. 1962. *The guns of August.* New York: Macmillan.

TURNER, B. L. II. 1988. Whether to publish in geography journals. *Professional Geographer* 40:15–18.

TURNER, B. L. II, and W. B. MEYER. 1985. The use of citation indices in comparing geography programs: An exploratory study. *Professional Geographer* 37:271–278.

WALKER, WILMA J., ed. 1991. *Schwendeman's directory of college geography of the United States.* Richmond: Geographical Studies and Research Center, Eastern Kentucky University.

WHITEHAND, J.W.R. 1984. The impact of geographical journals: A look at the ISI data. *Area* 16:185–187.

———. 1985. Contributors to the recent development and influence of human geography: What citation analysis suggests. *Transactions, Institute of British Geographers* n.s. 10:222–234.

WILKIE, RICHARD W., and JACK TIGER, eds. 1991. *Historical atlas of Massachusetts*. Amherst: University of Massachusetts Press.

WRIGLEY, N., and S. MATTHEWS. 1986. Citation classics and citation levels in geography. *Area* 18:185–194.

ZELINSKY, WILBUR. 1975. The demigod's dilemma. *Annals of the Association of American Geographers* 65:123–143.

PART III

HOW GEOGRAPHERS THINK

Location, Place, Region, and Space

Helen Couclelis

The most fundamental concepts in the sciences are often expressed in ordinary language. In physics, it is *mass* and *energy*; in chemistry, the *elements*; *demand* and *supply* in economics; the *organism* in biology—and the list goes on. It is largely through such familiar-sounding concepts that the continuity between science and the everyday world is maintained, that the formal and formidable blends with the informal and familiar.

Geography too has its basic concepts, and they also are expressed in common English words: *location*, *place*, *region*, *space*. Of these, *space* is probably the most fundamental. But unlike $E = mc^2$, space resists definition in either formulas or words. A look at the dictionary suggests why:

> *space*: An interval between things, this regarded as empty of matter, sum of these as opposed to matter, this together with the room taken up by matter regarded as containing all things, any part of such s., regions beyond ken, a distance, an area, room available or required, a period or interval of time, ... (*The Pocket Oxford Dictionary*, 4th edition)

So space is both expanse and confine, both what is between things and what contains them, both empty of matter and defined by the presence of matter; space is even a period or interval of time! The broadness of the dictionary definition is characteristic of the difficulty of pinning down space. It is also indicative of the fact that, unlike most concepts developed to refer to some specific thing or property of the real world, space is part of the definition of that world; it thus belongs more in the realm of the philosopher and theorist than in the worlds examined by empirical researchers.

In his *Critique of Pure Reason* the eighteenth-century philosopher Kant speculated that space is a *synthetic a priori*, that is, an innate precondition of human knowing that makes it possible to understand the empirical world. Space is not another thing *in* the world, but a reality created by the interaction of human reason *with* that world. Thus it reflects properties of both the observed and the observer. Only three other concepts—time, morality, and aesthetics—have similar extra-empirical foundations, according to Kant. Although Kant's focus on Euclidean geometry (the only geometry known at the time) was later shown to be mistaken, his arguments on the extra-empirical nature of space in general continue to be at the center of scholarly discussions of the concept (Entrikin 1977). Whether they follow Kant or not, geographers must come to grips with the idea of space and with a variety of related concepts such as location, place, and region. These ordinary language terms may appear unproblematic and self-explanatory, but each corresponding concept is rooted in a somewhat different understanding of space, even though that understanding is usually tacit.

Geographers are not concerned with space for its own sake, only for what it may mean for the phenomena they study. No other empirical discipline allows space to play such a central role in its approach to the world and in its own self-definition. For physicists and other natural scientists the problem of space has long been solved, and the concept—as embedded in numerous formalisms—is for the most part taken for granted. Most social scientists on the other hand, and economists in particular, tend to ignore it; they have been accused (primarily by geographers) of seeing the world unfold on the head of a pin. For geographers, by contrast, a concern for space in its different implications and manifestations is an unbreakable common thread underlying an extreme diversity of interests that range from the humanities to social and physical science.

The primary goal of this chapter is to articulate a number of different conceptions of space underlying the work of geographers and others. Notions of space progress from more to less formal, from more to less well understood, and also from poorer to richer in human interest and meaning. The modern mathematician, the spacecraft designer, the industrialist trying to decide where to build a new factory, the urban commuter, the youth roaming the ethnic ghetto, the baby learning to reach for objects over its crib, and the mystic contemplating the perfection of a sphere all deal with space in its different manifestations. Indeed, we may speak of a hierarchy of spaces: mathematical, physical, socioeconomic, behavioral, experiential.

There is no single best conceptual scheme for discussing space. Sack (1980), who published an entire book on the different conceptions of space, chose as the basis for his framework a two-fold distinction between subjective and objective on one hand and between space and substance on the other. In Sack's scheme, conceptions of space that maintain these distinctions are called *so-*

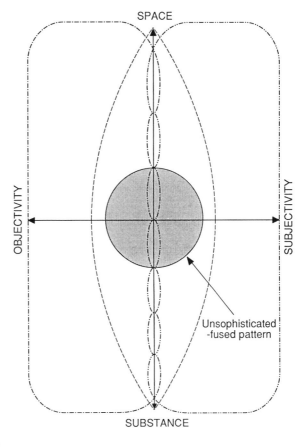

Fig. 10.1. Sack's Framework of Space Conceptions. SOURCE: Sack 1980:25.

Physical science domain   Artistic domain   Social science domain

phisticated-fragmented, whereas those that merge them are called *unsophisticated-fused* (fig. 10.1). The spaces of science and social science are of the first kind; those of everyday life, and of myth and magic, are of the latter. Many insights about the meaning of space emerge from such oppositions. Other insights arise from a view that different notions of space form an ordered sequence, as outlined in this chapter.

MATHEMATICAL SPACE

"It is said of the Socratic philosopher Aristippus, so Vitruvius wrote in the preface to the sixth book of his *De Architectura*, that being shipwrecked and cast on the shores of Rhodes and seeing there geometrical figures on the sand, he cried out to his companions, 'Let us be of good hope, for indeed I see the traces of men.'" (Glacken 1967, epigraph). For the ancient Greeks, and for much of the Western intellectual tradition that followed their lead, geometry was the signature of human intelligence. In a world that remained inexplicable and unpredictable for so long, the certainty of the geometrical truths and the conceptual elegance of their derivation were contemplated with almost religious reverence.

Geometry has its empirical roots in the need to reestablish field boundaries following the annual floods of the Nile in ancient Egypt. *Geometry*, in Greek, means measurement of the earth, just as *geography* means description of the earth. Literally, then, geometry was the oldest and purest form of what modern geographers think they have only recently invented—quantitative geography. It was, and remains, the formal science of space and spatial relations. David Harvey (1969), one of the foremost scholars to discuss space in the context of geography, called geometry "the language of spatial form." In recent decades pure geometry has moved in directions that few geographers care to follow, but the mathematical clarity and power it brought to the description of space remains a standard by which other approaches are measured.

For over two thousand years the Euclidean geometry of the ancient Greeks was the only one known. Now there are many more geometries, and inventing new ones is within the reach of today's generation of doctoral students in mathematics. But the geometry of two-dimensional and three-dimensional Euclidean space and its extension to the curved surface of the earth remain the foundations of geographical description. This is the geometry represented in most ordinary maps and the one underlying most mathematical models of geographic processes. But geographers have explored and made creative use of several other kinds of formal spaces as well. I mention here only two: *discrete* and *fractal* spaces.

Imagine a space made up of tiny grains that cannot be decomposed any further. These grains or cells make up a space that is *discrete*, unlike the smoothly continuous, infinitely divisible space that Euclidean geometry describes. A granular space runs counter to intuitive understanding of the world around us, and yet discrete spaces have become popular in geography and several other sciences, in part because the mathematics of discrete spaces (like that of discrete time) is on the whole simpler to work with. It is also more compatible with the binary logic of the digital computers, which cannot represent continuous entities without error. Models of geographical phenomena based on discrete space make the most of computer capabilities, and in some cases they allow geographers to explore processes that are difficult to represent by other means.

Members of one class of discrete-space models of special interest to geographers are called *cellular automata*. They represent the development over time of spatial processes based on local interactions, that is, on rules or laws that operate at particular locations and their neighborhoods rather than across an entire space. For example, a forest fire spreads from burning trees to neighboring trees; it does not attack all trees simultaneously, nor does it jump capriciously from one point to other distant points. The same is true of the spread of an epidemic, of a rumor or a fad, of an urban neighborhood that starts decaying once a few houses become derelict, and of many other phenomena where things that happen at one place strongly influence what happens in surrounding places. In fact, geographers made use of the properties of discrete space to study spreading or diffusing phenomena long before cellular automata became fashionable (Morrill, Gaile, and Thrall 1988). Still, the introduction of cellular automata provides a more general and systematic understanding of how complex and unpredictable the results of such processes can be, and of the ways small differences in initial local circumstances can lead to widely differing outcomes (see Couclelis 1988 for a discussion of cellular automata in geography). The irregular but far from random patterns (fig. 10.2) resulting from application of a cellular automaton rule operating at the scale of the individual cell remind one of the patterns of destruction left in the wake of a fire, of the form of an urban area, or of the distribution of species in an ecological community. For geographers, they are vivid demonstrations of how the small and large scales interact, and of how closely space, time, and substance are intertwined in spatial processes.

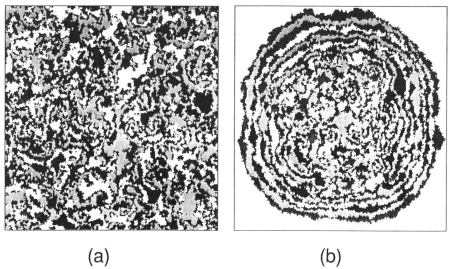

(a)   (b)

Fig. 10.2. Cellular Automata Patterns: *a*, from uniform randomness; *b*, from a small patch of randomness. SOURCE: Toffoli and Margolus 1987:92. Copyright © 1987 by MIT Press.

Fractal spaces became widely known only in the past decade, and their exploration has been closely linked with the analytical and display capabilities of modern computers (Mandelbrot 1982). Fractal geometry knows no straight lines, smooth shapes, or tidy, regular volumes. It is the geometry that best describes the tortuous, irregular forms usually found in nature: clouds, rocks, rivers, mountain-scapes, tree canopies, and bushes. Many human-induced forms such as urban areas seem to have fractal contours as well. Whereas conventional geometry sees the world in 1, 2, 3, . . . n dimensions, fractal geometry fills the noninteger dimensions in between. A convoluted shoreline may have a dimension of 1.4, rough terrain one of 2.25 or 2.6—somewhere between the perfectly flat plain of dimension 2 and a 3-dimensional parallel-epiped, cone, or sphere. Fractal geometry allows the reproduction of natural shapes on computers with unequaled realism. It also encourages geographers to think about scale in novel ways, as fractal shapes will keep revealing more and more detail as one looks closer and closer. Fractal geometry is a new tool for geographers, and applications to actual problems outside the area of graphic displays are only beginning to appear in the literature. What are the deeper implications of this tool for geographic research? The future will show whether fractals are a passing fad or a new departure (Batty and Longley 1986; Goodchild and Mark 1987).

PHYSICAL SPACE

Space station, space capsule, space travel, outer space, intergalactic space. Space is a term readily associated with the physical world, especially with the unthinkable vastness of what lies beyond. The scientific and technological conquests of the last few decades have made the unthinkable thinkable, and we have become comfortable with discussions of space probes that send signals from interplanetary space, and with images of humans in space suits floating in space.

A significant clue to how we understand space lies in the preposition *in*, which indicates containment. We conceptualize space as a container of objects: things are in space just as oranges are in a box or fish are in water. That intuitive view is formalized in the notion of *absolute space* in Newtonian mechanics: space is a neutral background against which the positions of objects can be pinpointed and their motions described. The classical scientific view of space is compatible with human experience of the everyday world, which is also the world of the geographer. But there are other kinds of physical space and other kinds of geographical space inspired by these alternative conceptualizations.

Hand in hand with the development of non-Euclidean geometries, modern physics introduced the notion of *relativistic space*. This is a space (actually, a space-time) whose structure both influences the distribution and motion of matter and is governed by it. Even stranger spaces are said to exist at the level of elementary particles, some of which have several tightly curled dimensions, others of which are structured more like Swiss cheese. In the physical world such unconventional notions of space become relevant only at either the astronomical or the infinitesimal scale; they do not apply at the scale of neighborhoods, regions, and oceans. Yet, the idea of a space with structure and properties that are intimately tied to process proved extremely attractive to geographers. Cartographic expressions of the relative-space idea soon appeared in the form of map transforms and cartograms. In these, distances between points on a map are made proportional to some measure that expresses the phenomenon under study, rather than to actual geographical distance. For example, cities can be placed on a map in such a way that the distances between them on paper are proportional to travel time or travel cost, which in many cases are more relevant measures of distance than actual mileage. In this way, a relative space can be mapped that represents more accurately than Euclidean space some of the key constraints governing interactions among places. In one of the earliest and best-known illustrations of this technique (fig. 10.3), Hägerstrand suggested that distances away from a place are experienced logarithmically rather than linearly, with longer movements costing (both materially and psychologically) proportionately less than shorter ones. It is indeed a common experience that a thirty-minute commute does not seem twice as long as a fifteen-minute commute.

More abstract expressions of the idea of relative space in geography are represented in the numerous mathematical models of socioeconomic processes developed since the 1960s. Indeed, the notion of a space constituted by spatial relations and processes proved particularly attractive to human geographers because the most relevant spatial notions from a social science perspective—spatial relations, spatial organization, spatial process, spatial dynamics, restructuring, and change—resist being conceptualized as objects in a container. Physical geographers by contrast have little practical reason to depart from the absolute conception of space that has served so well in the description of the physical world at geographical scales. Still, approaches developed in the physical sciences stressing spatiotemporal dynamics and used in physical as well as in human geography have given wide currency to the notion that space and process are interwoven in all branches of the discipline. The result of these conceptual and technical developments is a more abstract, flexible, sophisticated view of geographical space, one that is for the most part closer to relative than to absolute conceptions.

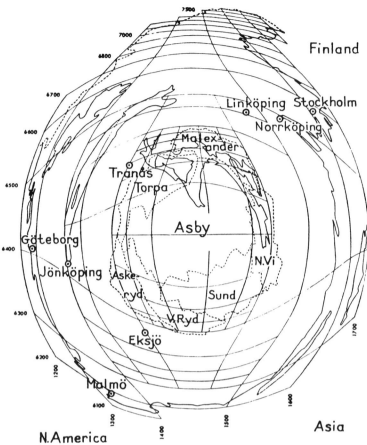

Fig. 10.3. A Celebrated Relative-Space Cartogram: Hägerstrand's Logarithmic Map of Distances from Asby, Sweden. SOURCE: Tobler 1963:65; by courtesy of Torsten Hägerstrand and the American Geographical Society.

SOCIOECONOMIC SPACE

Imagine a completely flat and featureless landscape with a town or two and a few copper and iron mines here and there. Workers live in the towns or perhaps in a few villages also in the area. The towns, the villages, and the mines are so compact that they are virtually points in the landscape. The transportation network is so homogeneous and dense that travel is equally easy in all directions, and you can always take a straight line between two points. An industrialist examines the area, trying to decide where to locate a steel factory. Being an economically rational person, the industrialist knows that the less money spent on transportation, the greater the profits will be. The best location will thus be the one where the combined transportation costs of raw materials from their sources to the factory and of the finished products from the factory to the markets are the lowest possible. If the workers have no means of transportation, the factory may have to be at one or the other of the towns where they live. In this case, cheaper labor in one village may offset a slightly lower total transportation cost at another.

That unlikely landscape could be anywhere, or rather, nowhere at all. It is an instance of an abstract *economic space*, defined by the spatial relations between consumers, producers, labor, and raw materials. It is a relative space, the properties of which (for example, the existence of points where total transport costs are minimized) arise through two features: the location of other critical points relative to each other, and certain economic conditions and principles assumed to hold in that space. The value of thinking about the world as an economic space is that such a view allows us to develop general theoretical statements about the effect of geographic factors such as distance on economic activity. What can be said of one economic space is true of all economic spaces of a similar kind at all times. Although the empirical world is always infinitely more complex, the basic interdependencies between spatial and economic factors highlighted in economic space are valuable guides for thinking about real issues in real places, and even for trying to predict or guide economic development in a region. Economic geography, developed around the investigation of the properties of economic space, is a thriving subfield in modern human geography (Lloyd and Dicken 1977).

The concept of economic space can be extended to other kinds of problems not explicitly concerned with economic factors but that also focus on the relative location of human activities. In such cases geographers speak more generally of *socioeconomic space*. As with economic space, the insights gleaned from an analysis of socioeconomic space are usually consistent with intuition. For example, all other things being equal, there will be more exchanges among places that are near each other than among places that are far apart, and more among large cities than among small villages, and the farther a commodity has to be transported, the higher its cost will be to the producer or the consumer. By developing analytical models of socioeconomic space it is possible to figure out not only the general thrust of these intuitively obvious relationships, but also their relative importance in quantitative terms (Haynes and Fotheringham 1984).

Spatial analysis is the general approach that uses mathematics and statistics to derive the quantitative properties of spaces of interest to geographers, and of socioeconomic space in particular. Spatial analysis is closely associated with the quantitative revolution in geography that took place in the sixties, though it is by no means restricted to geography. Regional science and urban economics have both developed spatial analytic methods in their investigations of regional or urban growth and restructuring, for example, how the industrial structure of a region may change as a result of local labor becoming more expensive, or of changing conditions in the supply of raw materials. Archaeologists have borrowed spatial analytic models and techniques to speculate on the locations of lost settlements of ancient cultures and connections among them.

Within spatial analysis, *location theory* looks at socioeconomic space from the special perspective of trying to determine optimal locations for specific services, facilities, or functions. In socioeconomic space, *optimal* means lowest-cost, least-time, least-effort, least-risk—generally speaking, least of something undesirable. Applications are numerous: find the location for a school that will keep the total time children spend on school buses as low as possible, or find the location for an ambulance station that will ensure that patients or accident victims in an urban neighborhood are reached and transported to a hospital as quickly as possible. The hypothetical steel factory case that opened this section on socioeconomic space, taken from a classical problem in industrial location theory (Weber 1929), is another example of seeking an optimal location within economic space.

Socioeconomic space, and in particular the kind that location theory examines, highlights a characteristic property of relative spaces that sets them apart from Newtonian absolute space: the various points in them are not undifferentiated and neutral; they are intrinsically better or worse for some purpose because of their position relative to some other meaningful points. The value that a place has in socioeconomic space by virtue of its relative position is captured in the notion of *situation*, in contrast to its value as a *site*, which consists of whatever relevant characteristics or attributes are to be found at that place— vegetation, slope, land use, buildings, and so forth. The site/situation distinction in socioeconomic space is a concrete geographical expression of the difference between absolute and relative space, and it highlights the significance of that difference for geography.

Socioeconomic space, the relative space defined by social and economic activities and relations, is also of interest to geographers who do not espouse the quantitative methodology of spatial analysis. A number of alternative approaches based on fundamentally different premises—realist epistemologies, Marxian theory, or the theory of structuration—have contributed insightful analyses of the interplay between social relations and spatial structure. For example, several recent studies have examined how the distribution of employment in urban regions, coupled with transportation conditions, places constraints on the lives of disadvantaged population groups such as working women or low-income minorities. Another series of studies has tried to determine how capitalist modes of production have led to a decentralization of the garment industry (which depends heavily on low-paid female labor) in Los Angeles. In historical context, others have documented how the replacement of open fields by enclosures in Northern and Western Europe led to a restructuring of social relations among the farming population (Pred 1986). From these perspectives, space is interesting only when viewed as a *social production*: something constituted, reproduced, and changed by social relations, and in turn constraining the unfolding of such relations. Abstract spatial properties, as defined through spatial analytic approaches to socioeconomic space, are, according to this view, of limited relevance to social processes.

This is the view taken by geographers inspired by critical social theory, and Marxist theory in particular. Others, arguing from the standpoint of the theory of structuration, object to what they see as an implied sociospatial determinism in such reasoning and have stressed the contributions of conscious human agents to the shaping of social (and therefore also spatial) relations. This latter emphasis parallels that highlighting the economic decision maker in traditional spatial analysis (Gregory and Urry 1985).

Because of profound philosophical and methodological differences, geographers on either side of spatial analysis often think that they have very little to share with each other. Yet at issue on both sides is a relative space, the properties of which are determined by social and economic relations and processes. It is thus appropriate to include both perspectives in a section on socioeconomic space. To highlight the still considerable distinctions between the two, let us call *social space* the conception of space underlying the work of geographers espousing the critical social theory viewpoint.

BEHAVIORAL SPACE

Everyone knows the feeling: rush to meet a loved one at the airport, and the trip seems to last forever; drive to a dreaded dentist's appointment, and you are there in no time. Drive away from a city center, and things appear fairly spaced out; drive back on the same road, and distances seem to contract. Distances as experienced by people are not the same as distances on the map. Perceptions of how near or how far things are from each other are affected by personal levels of knowledge of an area, by psychological effects of habit, anticipation, fear, stress, or boredom, and by a host of other subjective or even biological factors. These perceptions, in turn, affect human behavior in space. People may patronize a neighborhood store for years in the mistaken belief that it is closer to home than some other. They may fail to use a reasonably accessible public library because for some reason it feels too far away. They may avoid a perfectly efficient route to work that is perceived to be dangerous or stressful. The space people experience and in which they make daily decisions differs from the objectively definable, theoretical spaces that fall under the rubrics of mathematical, physical, and socioeconomic space.

The key insight contributed by the approach known as behavioral geography is that people respond to environments largely as they perceive and understand them (Moore and Golledge 1976). Mathematically optimal locations and routes are not necessarily the ones people will, or for that matter, should adopt. Behavioral geography recognizes two facts: first, individual behavior and decision making in space is based on knowledge that is incomplete and distorted; second, complexities of human psychology lead to behavior and decisions that may not be optimal in a theoretical sense, but that are considered best at the time by the individuals who make them.

According to behavioral geography, individuals function in a subjective world—a world in the head. Numerous empirical studies have explored behavioral space, documenting its properties with experiments involving subjects from different population subgroups: young and old, female and male, disabled and able, well-educated and illiterate. Major insights into the structure and properties of behavioral space gleaned from such studies are found in the extensive literature on *cognitive maps*. Cognitive maps attempt to represent graphically an individual's understanding of the spatial structure of the environment (Downs and Stea 1973; Gould and White 1974). In a series of studies that launched the specialty in the 1960s, Lynch (1960) explored people's perception of urban space by asking subjects to sketch from memory maps of the city in which they lived. Behind the wide variety of graphic abilities and styles, Lynch found that five kinds of spatial elements were almost always mentioned: landmarks, paths, districts, nodes, and barriers. In more recent years, studies of cognitive maps have become more sophisticated. Computer-aided approaches were developed that allow the cartographic representation of *cognitive configurations*, that is, presumed representations of the spatial structure of a city, neighborhood, or building in the minds of experimental subjects. These methods do not rely on subjects' map-sketching abilities but use instead more directly available and robust information, thus bypassing the problem of distinguishing the picture from the spatial knowledge contained in it. In addition, the geometrical properties of configurations obtained in this fashion can be thoroughly investigated with analytic tools. For example, samples of individual cognitive configurations were obtained from residents of Columbus, Ohio, along with an average configuration derived from the responses of all the subjects in the experiment (fig. 10.4). What would be a square grid on an ordinary map of Columbus appears to be stretched, twisted, folded or torn in various ways in the minds of the experimental subjects. Discussions of cognitive maps have addressed such issues as how these distortions arise and their relation to overt behavior (Downs 1981; Golledge and Stimson 1987).

There are other, more direct approaches to the study of behavioral space. One popular method is to observe the spatial choices of a group of people— say, where they go shopping, where they search for a house to buy, or where they migrate—and then try to find a relationship between that spatial behavior

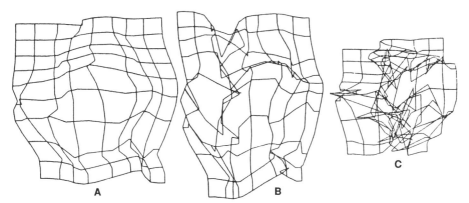

Fig. 10.4. Distortions in Cognitive Maps: How Two Residents of Columbus, Ohio (*b,c*) and the Average Subject in That Experiment (*a*) May Perceive the Spatial Structure of Their City as Referenced by a Square Grid. SOURCE: Golledge and Stimson 1987:80.

on the one hand, and socioeconomic and other personal characteristics on the other. Issues of behavioral space that may arise within such a framework are the ways in which income, education, race, and gender affect the extent of the urban space within which a person's daily activities unfold, or whether increased length of residence in a community produces less distorted perceptions of local distances (Golledge and Rushton 1976).

Another approach to the study of behavioral space is that of *time geography,* where individual movements are observed not only across space but also through time. In typical time geography experiments, participating subjects agree to keep daily records of when they go where, and how long they stay at each place. By the end of the day, each person has described her or his path in daily time-space: left home at 7:40, at work at 8:10, left for a dentist's appointment at 10:15, got there at 10:30, and so forth. Each day, the space-time paths of different individuals come together in shared activities, then separate, then regroup elsewhere in different combinations, and then separate again, giving rise to what one geographer has called the daily choreography of existence (fig. 10.5). The basic insight upon which time geography is built is that there is a direct trade-off between behavioral space and behavioral time, because time is needed to overcome space (distance). The more time spent by an individual moving from place to place, the less time left to spend at any one place engaging in productive or personally fulfilling activities; conversely, the more time one is obliged to spend at a particular place (say, home or work), the less that person's access will be to other places. This simple observation has profound implications for the relationship between the spatial distribution of daily activities on the one hand, and the constraints imposed on a person's life by societal demands of work and family care on the other (Pred 1981).

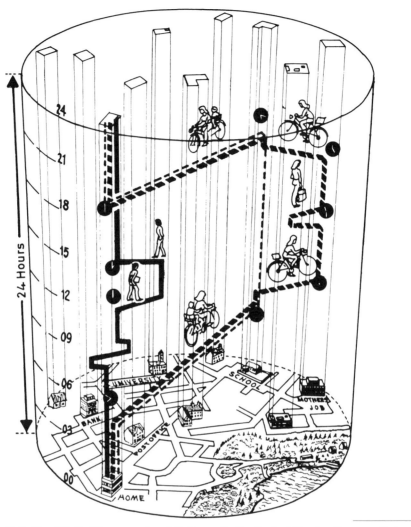

Fig. 10.5. The "Daily Choreography of Existence" of a Typical Family. Source: Parkes and Thrift 1980:252.

EXPERIENTIAL SPACE

Whether one sees them as creations of the human intellect or as discoveries of things that really exist, mathematical, physical, socioeconomic, and behavioral space are all constructs belonging to natural or social science. They are the result of theoretical thinking and of the application of sophisticated mathematical and statistical techniques created for particular rational purposes—for the scientific understanding of the natural and the social world. Even behavioral space, in its deliberate attempt to be objective in measuring the subjective, belongs in the category of sophisticated-fragmented spaces of Sack's (1980) scheme (see fig. 10.1).

Experiential space, on the other hand, is the space human beings actually experience before it is passed through the filters of scientific analysis. It embraces all the intuitive, unanalyzed, unexamined, or unarticulated forms of spatial understanding, including the practical, commonsense understanding of space in everyday life, the imperfect but growing understanding of the infant and the small child, that of the disabled, that of the alien culture, the tribe that time forgot. Experiential spaces also include the contemplative kinds of spatial experience inherent in the apperception of sacred and mythical spaces, as well as the aesthetic experience of symmetry, proportion, balance, and so on that is central to the creation and appreciation of art. Sack (1980) called these latter conceptions of space *unsophisticated-fused* because they seem to collapse the objective and the subjective, space and substance. In reality, it is only from the scientific, analytic perspective that these conceptions may appear unsophisticated. They are far from simple if one recognizes the refinements of intuitive spatial knowledge required by even the most ordinary of everyday tasks, or the subtlety and complexity of meaning projected on the sacred geographies of even the most primitive of cultures.

Take everyday tasks, for example. The spatial skill involved in grasping a cup of coffee, bringing it to one's mouth, and drinking out of it without spilling is barely matched by the most sophisticated robotic devices of the day. Anything more complex than that—be it the tennis-playing skill of a mediocre athlete or the navigational prowess of an illiterate islander—exceeds human analytic understanding and the ability to construct mechanical or formal models of the task. Nor is it human spatial skills alone that fill us with wonder. It is only in the last few years that one of the world's most advanced engineering laboratories at the Massachusetts Institute of Technology was able to concoct a device that can balance itself on its six spindly legs and walk like an insect.

Bodily mastery of spatial skills is thus one form of knowledge of experiential space that all living beings share. It is a preconceptual, gut knowledge of space, a basic intuition developed over eons of evolutionary adaptation to a world shaped by the constraints of physics, biology, and in the case of humans, culture as well. Gravity gives humans the sense of vertical and horizontal, up and down; the asymmetrical build of their bodies, that of back and forth; inertia and friction, their sense of motion; their abilities to receive and process light and sound, their sense of spatial layout; and some argue that being born and raised in a world of rectangular buildings, objects, and street patterns gives people a Euclidean conceptualization of space. Kant was right after all: space is a synthetic a priori, an innate precondition of conceptual knowledge, even though its Euclidean appearance may be to some degree a cultural artifact.

That part of experiential space that we know with our bodies rather than with our minds is called *sensorimotor space*. An aspect of sensorimotor space important enough to merit separate consideration is *perceptual space*, the space deriving from sight, hearing, and the other senses. The workings of the sense organs are easier to grasp with analytical apparatus than are the more primitive and obscure bodily senses of balance or movement. Still, decades of studying human vision have not come close to answering the simple question of how we comprehend what we see. Perceptual space remains more mysterious than the most arcane multidimensional space of mathematical geometry.

While sensorimotor and perceptual space make up the practical space of *skills*, on the other side of the analytic divide we find the symbolic space of *meanings*. Unanalyzed but not unsophisticated, unarticulated but not unexplicated, mythical and sacred spaces transform geography into a projection of the cosmos. Spirit becomes place, God becomes Eden, the Dreamtime becomes territory in the song lines of the aborigines; Mt. Olympus and Mt. Fuji have always been holy; Earth herself is the Great Mother Gaia. Myths are spun around these transformations, projecting timeless realities of one kind (spiritual) onto timeless realities of another kind (geographic). In the reverse transformation, the homeland (geography) becomes sacred (spirit). One's hometown is like no other place in the world; the home is where the heart is.

We have come a long way from the space of mathematics and physics. By now, space—space enriched with human experience and meaning—has become *place*. This is indeed how Tuan (1977) and other humanistic geographers view the distinction: place is space infused with human meaning.

It is easy to imagine a primitive geography bereft of the notions of mathematical, physical, socioeconomic, or behavioral space. But geography would be unthinkable without experiential space—without the intuitive notions of up and down, near and far, contiguous and disjoint, here and there; without the sense of vision in particular, to which we owe the knowledge of distant horizons; and without the meanings associated with the sense of place that is present everywhere where humans are, but condensed and sublimated in mythical and sacred places. Thanks to humanistic geography, the part of geography that considers itself a scholarly pursuit rather than a science, these most elusive and subtle of spaces are not lost to the discipline. For many cognitive scientists and linguists these days, experiential space is considered fundamental enough to underlie all human thinking and language (Lakoff 1987). Perhaps other traditions in geography will find experiential space increasingly relevant to what they study.

CLOSING THE CIRCLE

We have come full circle. Our sequence of spaces began with a pure space of formal symbols; it ended with another space of symbols—of affective and spiritual realities, private experiences, and collective memories. The former kinds of space are perfectly objective and can be fully described and communicated by the most rational means known; the latter are intensely subjective and can be shown but not defined, talked about but not described, acted in but not rationalized, shared but not communicated. Mathematical spaces are pure form, devoid of human meaning; experiential spaces have no form but are replete with human meaning. The symmetry appears complete.

Geography as spatial science spans the entire range of spaces: no wonder tension is felt within the discipline, but also symmetry. Analytic geographers, Marxist geographers, humanistic geographers—human and physical geographers, for that matter—sometimes think that they have little in common. And yet they all seek understanding side by side, along the same spectrum of space conceptions. Even the variety of spatial terms they use, a variety that can be confusing even to trained geographers, has purpose in the context of appropriate conceptions of space.

How the different spatial terminologies may mesh with each other is suggested in table 10.1. The mapping is only tentative; experiential space in particular defies orderly dimensional classification. Place is like location is like point in one sense, but it is not an element of zero dimensions; in fact, the whole world is place. Experiential space defines itself with its own semantics. Viewing the variety of spaces as a linear sequence provides the insight of gradually increasing substantive content against gradually decreasing formal structure. As with other complex concepts, a different understanding emerges each time space is viewed from a different angle. As a parting thought experiment, consider mathematical, physical, socioeconomic, behavioral, and experiential space not as a linear sequence, but as a nested hierarchy (fig. 10.6). The notion of gradual progression is maintained, but now behavioral, socioeconomic, physical, and mathematical space appear as increasingly constrained domains contained within the experiential. Does this make more sense? Perhaps the answer will become clearer in some future time.

Table 10.1. Four Spaces and Their Terminology

Mathematical	Socioeconomic	Behavioral	Experiential
Point	Location	Landmark	Place
Line	Route	Path	Way
Area	Region	District	Territory
Plane	Plain	Environment	Domain
Configuration	Distribution	Spatial layout	World

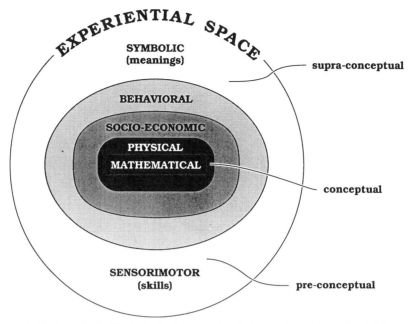

Fig. 10.6. Mathematical, Physical, Socioeconomic, Behavioral, and Experiential Space as a Nested Hierarchy.

REFERENCES

BATTY, M., and P. A. LONGLEY. 1986. The fractal simulation of urban structure. *Environment and Planning A* 18:1143–1179.

COUCLELIS, HELEN. 1988. Of mice and men: What rodent populations can teach us about complex spatial dynamics. *Environment and Planning A* 20:99–109.

DOWNS, ROGER M. 1981. Cognitive mapping: A thematic analysis. In *Behavioral problems in geography revisited*, ed. Kevin R. Cox and Reginald G. Golledge. New York: Methuen.

DOWNS, ROGER M., and DAVID STEA, eds. 1973. *Image and environment: Cognitive mapping and spatial behavior*. Chicago: Aldine.

ENTRIKIN, NICHOLAS. 1977. Geography's spatial perspective and the philosophy of Ernst Cassirer. *Canadian Geographer* 21:209–222.

GLACKEN, CLARENCE J. 1967. *Traces on the Rhodian shore*. Berkeley: University of California Press.

GOLLEDGE, REGINALD G., and GERARD RUSHTON, eds. 1976. *Spatial choice and spatial behavior*. Columbus: Ohio State University Press.

GOLLEDGE, REGINALD G., and R. J. STIMSON. 1987. *Analytical behavioural geography*. London: Croom Helm.

GOODCHILD, MICHAEL F., and DAVID M. MARK. 1987. The fractal nature of geographic phenomena. *Annals of the Association of American Geographers* 77:265–278.

GOULD, PETER R., and RODNEY WHITE. 1974. *Mental maps*. Harmondsworth, Eng.: Penguin Books.

GREGORY, DEREK, and J. URRY, eds. 1985. *Social relations and spatial structures*. London: Macmillan.

HARVEY, DAVID. 1969. *Explanation in geography*. London: Arnold.

HAYNES, KINGSLEY E., and A. STUART FOTHERINGHAM. 1984. *Gravity and spatial interaction models*. Scientific Geography Series 2. Newbury Park, Calif.: Beverly Hills.

LAKOFF, GEORGE. 1987. *Women, fire, and dangerous things: What categories reveal about the mind*. Chicago: University of Chicago Press.

LLOYD, PETER E., and PETER DICKEN. 1977. *Location in space: A theoretical approach to economic geography*. New York: Harper & Row.

LYNCH, KEVIN. 1960. *The image of the city*. Cambridge, Mass.: MIT Press.

MANDELBROT, BENOIT B. 1982. *The fractal geometry of nature*. San Francisco: Freeman.

MOORE, G. T., and REGINALD G. GOLLEDGE. 1976. *Environmental knowing: Theories, research, and methods*. Stroudsburg, Pa.: Dowden, Hutchinson, and Ross.

MORRILL, RICHARD, GARY L. GAILE, and GRANT IAN THRALL. 1988. *Spatial diffusion*. Scientific Geography Series 10. Newbury Park, Calif.: Sage.

PARKES, D., and NIGEL THRIFT. 1980. *Times, Spaces, and Places*. New York: John Wiley.

PRED, ALLEN, ed. 1981. *Space and time in geography: Essays dedicated to Torsten Hägerstrand*. Lund Studies in Geography, B, 48. Lund, Swe.: GWK Gleerup.

———. 1986. *Place, practice, and structure: Social and spatial transformation in southern Sweden 1750–1850*. Cambridge: Polity.

SACK, ROBERT D. 1980. *Conceptions of space in social thought: A geographic perspective*. Minneapolis: University of Minnesota Press.

TOBLER, WALDO R. 1963. Geographic area and map projections. *Geographical Review* 53:59–78.

TOFFOLI T., and N. MARGOLUS. 1987. *Cellular automata machines*. Cambridge, Mass.: MIT Press.

TUAN, YI-FU. 1977. *Space and place: The perspective of experience*. Minneapolis: University of Minnesota Press.

WEBER, ALFRED. 1929. *Theory of the location of industries*. Chicago: University of Chicago Press.

Movements, Cycles, and Systems

William L. Graf and Patricia Gober

The essence of geographic inquiry is a sensitivity to the worlds of place and space. For geographers the world is a complex mixture of human and physical phenomena that interact to produce distinctive locales, arrangements, and patterns. At the local scale these geographic signatures are the characters of places; at larger scales these signatures are connections and flows among places. Systems and patterns change over time. Geographers begin with static descriptions, but their explanations inevitably incorporate dynamic processes characterized by change. This chapter explores the ways geographers investigate geographic processes and patterns. The concepts in this chapter are relevant to all geographic systems; they are not limited to physical phenomena or to human behavior. They are a common heritage that explains why virtually all geographers share the same basic view of the world, irrespective of either their specialized research interests or the topics they investigate.

GENERAL SYSTEMS

Geographers use general systems thinking to simplify their worlds and to impose order upon earth features. A general system is a collection of elements and the connections binding them together (Hugget 1980:1). Elements of a system are recognizable entities with measurable attributes. Common geographic examples (and their measurable attributes) include cities (population), political or economic regions (wealth), drainage basins (water or sediment yield), and air masses (temperature or moisture). Connections in a system are the pathways by which energy, mass, people, goods, or information pass from one element to another. Taken together, the elements and their connections exist in a state that may change through time and that produces a mappable geographic pattern.

Geographers use systems concepts to model patterns and processes. In pattern analysis, places are elements and interactions are the flows that bind places together. In the analysis of processes, a system consists of one place (a city, state, nation, glacier, lake, or island ecosystem); elements are the salient features of that place. A city, for example, is a collection of people, businesses, and governments. Interactions are relationships among these features, such as government policies that affect business activities or the dependence of a business upon a population threshold for survival.

Systems can be open, with boundaries that are porous to external influences, or closed, with impervious boundaries. Most geographic systems are open. The economic system of a city, for example, responds to corporate decisions made in other cities, national fiscal policies, and vagaries of the global economy. The earth is an open physical system. It receives solar energy inputs that determine the geography of its environments. In order to simplify the analysis of a system, geographers may ignore external influences and assume that a system is closed. The cost of that assumption, however, may be ignorance of external forces critical to operations and states.

When a system is perturbed by external forces, it adjusts by feedback processes. Negative feedback occurs when adjustments return the system to a previous stable state. Positive feedback occurs when adjustments cause the system to depart from a previous stable state, resulting in permanent modification of the old state and establishment of a new state.

Entropy measures the uniformity of system elements. In mechanical terms, entropy measures the energy a system has expended. When a spring-driven watch is wound, it has little entropy; when the watch runs down and stops, its entropy is at a maximum. Entropy in geographic systems measures spatial uniformity. Systems that have little entropy are highly organized and therefore highly variable from place to place. Geographic systems sometimes operate to reduce spatial variability because extreme variations can be inherently unstable.

General systems theory also underlies the study of human geographic processes, sometimes explicitly, as in Mabogunje's (1970) model of rural-to-urban migration in Africa (fig. 11.1). A rural-to-urban migration system is open in that it exchanges information and people with a broader cultural environment. Communication, technology, modernization, and transportation reduce the isolation of potential migrants, whose behavior responds to a rural control subsystem. Control subsystems govern the operation of a general system and determine when and how to increase or decrease flows, much the way a thermostat controls room temperature. The rural control subsystem consists of family and community norms that encourage or discourage migration to cities. A corresponding urban control subsystem also regulates flows by providing employment and information about job vacancies. Negative feedback occurs

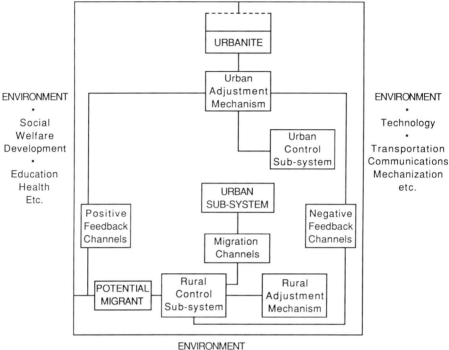

Fig. 11.1. Mabogunje's Systems Model for Rural to Urban Migration. SOURCE: Mabogunje 1970:3.

when city dwellers communicate the disadvantages of urban life to potential migrants, discouraging future migration. Positive feedback results when favorable information is transmitted to rural villages and the system moves away from a previously stable state through more rural-to-urban migration.

The role of rural-to-urban migration as a mechanism for population change is the subject of Rogers's (1985) systems-based model of migration in the Soviet Union and India. Dividing the national population into its urban and rural components, Rogers examined the respective roles of natural increase and rural-to-urban migration as causes of urban growth. Simulations showed that: (1) the principal effect of rural-to-urban migration is to establish levels of urbanization, whereas natural increase determines the rate of urban population growth; and (2) sharp increases in rates of rural-to-urban migration raise urban population growth rates in the short term but lower them in the long term. As more of a country's people come to live in urban areas where fertility (hence natural increase) is lower, urban population growth perforce declines.

At the metropolitan scale, Forrester's (1969) systems-based model of urban growth contains three major components: businesses, jobs, and homes. Simulations over a 250-year period suggest three phases in the life cycle of cities (fig. 11.2). The growth phase is dominated by positive feedbacks. Population growth increases population size, which in turn stimulates additional population growth for about 100 years. The second phase, lasting 50 to 75 years, is a period of stagnation caused by saturation of the fixed land area. Enterprises mature and eventually decline faster than new businesses evolve to take their places. In the third and final stage, over 75 to 100 years the system achieves equilibrium wherein the amount of housing passing from new to old categories equals the amount being constructed, in-migrants equal out-migrants, and new enterprises exactly replace dying ones.

General systems thinking also underlies much geographic research into the operations of natural processes. The work of Caine (1971) on alpine hydrologic and slope dynamics provides an example of a process-response system where adjustments result from changes in flows of mass and energy. When the U.S. Bureau of Reclamation began experimenting with artificially induced snowfall over mountains to augment water supplies, possible unwanted impacts included increased slope erosion. Caine's research produced a general systems model of slopes in the San Juan mountains of southwestern Colorado that was used to assess these impacts. His model focused on the creation, movement, and storage of sediment on ten-square-meter plots. Elements for each plot are accumulation of material, storage, weathering, and removal (fig. 11.3). Plots

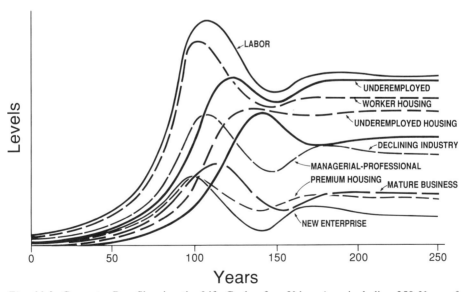

Fig. 11.2. Computer Run Showing the Life Cycle of an Urban Area including 250 Years of Internal Development, Maturity, and Stagnation. Components begin at low levels, overshoot sustainable levels of employment and housing, and eventually reach equilibrium. SOURCE: Forrester 1969:4. Copyright © 1969 by MIT Press.

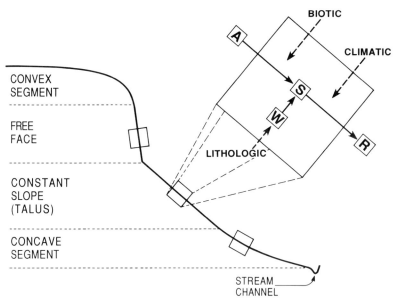

Fig. 11.3. A Simplified General Systems Model for the Down-Slope Movement of Surface Materials in Alpine Landscapes. *A,* accumulation from the next subsystem up slope; *W,* weathering; *S,* storage; *R,* removal to the next subsystem down slope. SOURCE: Modified from Caine 1971:326. Used by permission of the Regents of the University of Colorado.

are arranged along a hill slope profile so that output from each plot becomes input for the next one down slope. The external controls on the flows of debris from one element to another and from one plot to another are vegetation and climate, factors that can be influenced by human manipulation.

Although it is highly simplified, Caine's model clarifies the important variables for investigation and provides a path to understanding spatial variations in slope processes. Plots at different locations along the slope behave differently: storage elements vary in size, and transfer rates vary from one place to another. Human impacts take the form of adjusting the control variables that change transfer rates or storage elements along the slope. In subsequent research, Caine (1976) explored the nature of climatological controls by assessing the magnitude and frequency of precipitation episodes. He used the model as a descriptive framework and through further analysis explained the connections between erosion and changes in energy brought to the slopes by precipitation.

Caine's work typifies physical geographic applications of system concepts in that he evaluates a complex environment using a few simple elements connected by direct flows of mass and energy. His model led to the identification of the most important variable—precipitation energy—and provided a foundation for later analyses of a more explanatory nature. Geographers have yet to exploit the full potential of general systems thinking. Development of more sophisticated models that link form and process—models that combine description and explanation—would translate evidence about the past into tools for predicting the future (Terjung 1976; Strahler 1980).

MOVEMENTS AND FLOWS

Flows in human and physical systems are a focus of geographic inquiry because they involve movements across space that connect origins and destinations and because they redistribute water, sediment, energy, population, products, and economic activity. Flows keep geographic systems in balance. Changes in flow patterns often signal larger-scale changes in the environment that prod a system from one equilibrium state to another.

A geographic system is at equilibrium when forces for change are in balance; when in-migrants equal out-migrants, when the increased commuting costs of living on the periphery of a city equal the savings from the lower land costs there, when the amount of sediment entering a river segment equals the amount leaving, and when the quantity of short-wave solar energy absorbed by the earth equals that returned to space by long-wave radiation. Once a state of equilibrium has been reached, any change causes internal adjustments that seek to restore equilibrium. Disequilibrium results from external or internal disturbances that change the level or nature of systemic activity, such as when the shift to a postindustrial economy altered well-established interregional migration patterns, or when increased carbon dioxide alters the global radiation balance.

A strict condition of static equilibrium occurs when the system is at rest and there is no flow of matter or energy across system boundaries, that is, when the system is isolated from its environment (Huggett 1980). All geographic systems exchange people, goods, matter, and energy with their environments, and the balance of these flows maintains systems states. When processes operate but a balance exists, a system is in a steady or stationary state, or in equilibrium. Geographic systems can achieve equilibrium under varying rates of flows. Cities, for example, can maintain constant populations or experience similar growth rates under widely varying conditions of in- and out-migration. Population stability in neighborhoods can be achieved under high and low levels of mobility (Moore 1972). In-town apartments for young singles and newly married couples require high through-put (in- and out-migration) to maintain stable populations whereas single-family neighborhoods with low mobility rates may be equally stable in numbers. Steady-state concepts can be related to migration and used to examine changes from one time period to another. Certain levels of interregional population exchange are normal and expected outcomes of everyday life. Such flows maintain the stable distributions of population that reflect steady state-conditions. When migration favors one region over another and redistributes population across regions, non-steady-state conditions prevail.

To measure the presence of steady- versus non-steady-state conditions in interstate migration flows in the United States, Plane (1985) used a demographic efficiency ratio of net migration (in-migration minus out-migration) to total migration (in-migration plus out-migration). An efficient flow redistributes population and is a sign of non-steady-state conditions. An example is Arizona's exchange with Illinois between 1987 and 1988. Illinois sent 8,870 migrants to Arizona while only 4,171 persons moved from Arizona to Illinois (Internal Revenue Service 1988). Inefficient flows are those that shuffle people back and forth without significantly altering the current pattern of population. Arizona's interaction with California is extremely inefficient. Between 1987 and 1988, 27,666 Arizonans moved to California, while 29,423 Californians moved in the opposite direction. Inefficient migration flows are consistent with steady-state conditions. A summary measure of demographic efficiency for interstate migration flows in the United States is the percentage of flow that is nonreciprocal. That percentage declined from 21.1 for 1935–40 to 18.0 for 1955–60 and to 11.6 for 1965–70. Until 1970, migration was becoming less efficient. People moved back and forth without significantly effecting any net redistribution of population. As the pattern of population came more nearly into balance with the pattern of economic opportunity, migration was no longer needed to redistribute people. For the last census migration period (1975–80), demographic efficiency rose, a signal of the growing imbalance between the distribution of population and the changing locational needs of the postindustrial economy. Plane argued that the shift to postindustrial economic conditions caused the migration system to become more efficient and to resume its traditional role as a redistributor of population.

Embedded within these permanent residential changes are cyclical movements characteristic of Third World societies and seasonal movements of the elderly in the United States and elsewhere. Seasonal migration among the elderly substitutes for permanent changes in residence; therefore total migration among the aged is greater than census data rates lead us to believe (Behr and Gober 1982). Seasonal residence is also a substitute for rather than a precursor of permanent migration (Rowles 1983; McHugh 1990). Such short-term movements are required to keep longer-term patterns in equilibrium.

In contrast to those who view population movements as circulation systems, geographers view flows of energy and materials in the environment as components of cascading systems, wherein output from one element or subsystem is the input for another, producing chains of subunits with magnitudes and geographic locations (Chorley and Kennedy 1971: 5). Examples are water and sediment discharges through river systems, moisture and heat flows through boundary layer climates, and the movement of nutrients through biological niches. From a geographic perspective, the most important energy flow is global radiation. The earth's general radiation balance consists of two flows— one of short-wave radiation from the sun and one of long-wave radiation from the earth (fig. 11.4). Only half the short-wave radiation arriving at the top of

SHORT WAVE

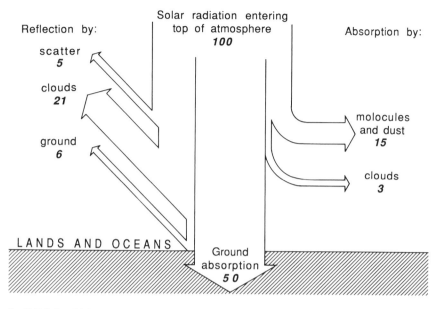

LONG WAVE

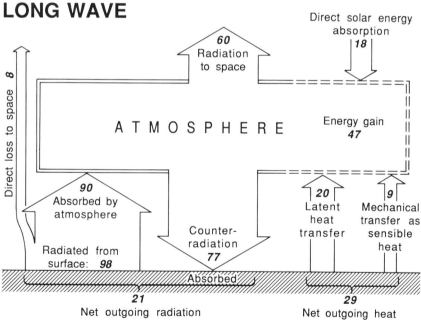

Fig. 11.4. Schematic Diagram Showing the Flow of Radiation within the Total Earth System. Numbers represent percentage of total input by solar radiation. Note that the earth/atmosphere system absorbs 68% in the short-wave budget and loses 68% in the long-wave budget, indicating an equilibrium system. SOURCE: Modified from Strahler and Strahler 1987:60. Copyright © 1987 by John Wiley & Sons. Used by permission.

the atmosphere ultimately flows to the earth's surface and is absorbed. Reflection from atmospheric and ground surfaces and absorption in the atmosphere
diverts the other half. Complex feedback flows, augmented by latent heat
transfers associated with the movement of moisture and with mechanical transfers of sensible heat through atmospheric convection, characterize the return
of heat to space via long-wave radiation. The view shown in figure 11.4 is a
nongeographic representation that describes the behavior of the entire system.

That budget becomes geographic when the transfer of energy into and out
of the system is not uniform over the surface of the earth. There is more solar
insolation than loss from earth radiation in equatorial regions and a deficit at
the poles, where more heat is lost than gained. The movements and flows
required to balance these geographic inequities drive the earth's climatic and
oceanic circulation systems. The greatest horizontal heat transfer occurs in the
midlatitudes, accentuating atmospheric turbulence in those regions (fig. 11.5).
Estimates of climatic changes resulting from human activities (such as carbon
dioxide infusions) are most likely to be successful if they take into account the

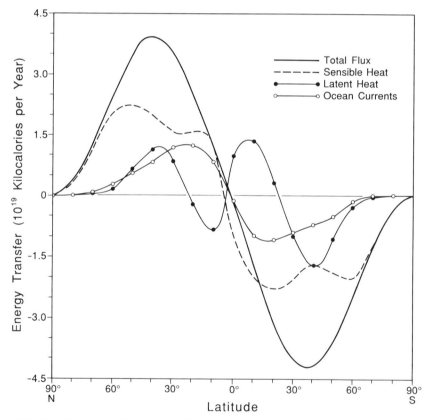

Fig. 11.5. The Geographic Distribution of Some of the Components of the Global Heat Balance.
Transport on the vertical scale is positive for northward movement, negative for southward
movement. Source: Sellers 1965:34. Copyright © 1965 by the University of Chicago Press.

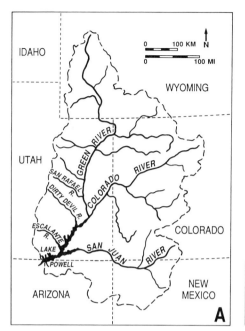
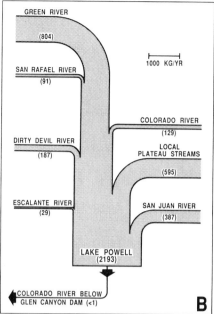

Fig. 11.6. Two Views of the Upper Colorado River Basin. *A,* the drainage subbasins; *B,* the flows of mercury in the river system. SOURCE: Graf 1985:553, 563.

total effects on the flows of energy. A specific example of geographic analysis of mass flows in an environmental system is Graf's (1985) investigation of mercury flows in the Upper Colorado River Basin. During the 1970s, biologists found high levels of mercury in fish in Lake Powell, a reservoir on the Colorado River in Arizona and Utah (Potter, Kidd, and Standiford 1975). They concluded that this potentially hazardous metal entered the food chain from lake-floor sediments. The geographic problem was to construct a regional mass budget for mercury that identified the metal's sources and defined its regional flows.

Analyses of mining areas, atmospheric fallout from coal-fired generating stations, naturally occurring bedrock outcrops, and ancient sediments that could not have been polluted by human activities showed that the mercury entering the lake came from natural sources. The movement of mercury from the source zones to the lake required evaluation of water flows (the source of transport energy), of the sediments to which the mercury is attached, and of metal from four basin areas (fig. 11.6A). The Green, Colorado, and San Juan rivers drain thousands of square kilometers and supply most of the lake's water. Local streams drain arid and semiarid terrain near the lake that account for less than 10 percent of the total drainage area, but because of their climates and highly erodible land surfaces, these streams supply 40 percent of the sediment entering the lake.

Chemical analysis of sediment samples collected from the three major rivers and from thirty-six local streams provided information on mercury concentration in sediments entering the lake from source areas. Combined with information from stream gauges on the total amount of sediment entering the lake from the various sources, these data provided a clear picture of the geographic characteristics of mercury transport for the region (fig. 11.6B). The Green River delivers high concentrations of mercury and moderate inputs of sediment to Lake Powell; local streams yield moderate concentrations of mercury but large amounts of sediment.

PATTERNS

Flows and geographic patterns are interdependent; flows are conditioned by patterns, and patterns are shaped and changed by flows. Intercity migration, for example, affects and is affected by the national network of cities. Similarly, vegetation distribution influences soil development through nutrient flows, and the resulting soil patterns in turn influence the growth of new plant communities. The relationship between patterns and flows depends upon whether a system is in equilibrium or disequilibrium. Under equilibrium conditions, flows arise from and perpetuate current patterns. Under disequilibrium, flows undermine the current order and redistribute human activities and natural phenomena.

The study of patterns in human systems often emphasizes the issue of agglomeration, the extent to which people and their activities cluster in space. Agglomeration at one scale may occur simultaneously with dispersal at another because geographic patterns are scale specific. The increased concentration of the population in metropolitan areas (urbanization) has been coincident with increased dispersal of people within those metropolitan areas (suburbanization). The question of agglomeration versus dispersal within settlement systems has been a major theme in urban and population geography (White et al. 1989). Historic trends have favored increasing agglomeration (Berry 1981), but the decline of many large metropolitan areas, the resurgence of small towns, and net movements from metropolitan to nonmetropolitan areas during the 1970s shifted attention to deconcentration. The term *counterurbanization* denotes the reversal of traditional patterns of settlement and movement (Berry 1976; Kontuly and Vogelsang 1988).

To investigate dispersal worldwide, Vining and Pallone (1982) examined internal migration patterns in twenty-two countries from 1950 to 1979. Using net migration rates into and out of core and peripheral regions as barometers of agglomeration versus dispersal, they identified three categories of countries: (1) those in northwestern Europe and North America where there was sustained net movement away from cores to peripheral regions (see fig. 11.7); (2) Japan,

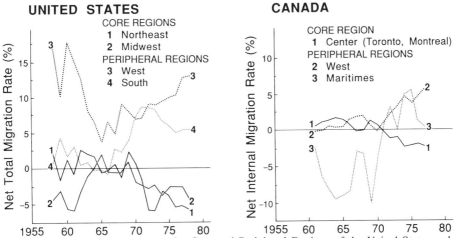

Fig. 11.7. Rates of Net Migration for Core and Peripheral Regions of the United States and Canada. Core regions shifted from positive to negative net migration between 1950 and 1980 while peripheral regions shifted from negative to positive or maintained positive net migration rates throughout the period. SOURCE: Vining, Pallone, and Plane 1981:345–346.

New Zealand, and countries on the edge of Western Europe where net migration to the core has slowed but not yet reached the point where significant net outflow of population to peripheries has occurred; and (3) economically less advanced countries of Eastern Europe, Taiwan, and South Korea where agglomeration continues to dominate. Vining and Pallone concluded that because the strongest dispersal tendencies appeared in the more advanced countries, dispersal is a function of advanced levels of economic development. At advanced stages of development, transportation and communication improvements enhance opportunities on the periphery, the disadvantages of congestion, pollution, and higher land costs outweigh the advantages of proximity in core regions, and people can afford to realize their preferences for smaller-scale, more relaxed living environments.

Just when geographers came to view a shift from core to periphery as the logical and inexorable outcome of more advanced levels of economic development, population and migration patterns in the United States in the early 1980s reversed. Metropolitan areas grew faster than nonmetropolitan areas, and rapid growth occurred in the nation's largest cities such as New York and Los Angeles. In addition, early patterns of dispersal were reinterpreted as artifacts of boundary changes and accounting inaccuracies. The pendulum swung back to a concern with agglomeration and its underlying causes.

In economic geography, the concern for aggregation is embodied in the study of the siting of productive activities in close proximity to each other. O'hUallachain (1989) investigated patterns of growth in employment and establishments for twenty-seven fast-growing service industries between 1977 and 1984. He found a strong relationship between rates of growth and city size, indicating the tendency for many services to cluster in large metropolitan areas. To sort out forces underlying patterns of service concentration, O'hUallachain examined the relative importance of urbanization economies versus localization economies as determinants of service growth. Urbanization economies are cost savings resulting from locations in large metropolitan areas and were measured by the sizes of local labor forces. Localization economies are the attractions of specialized industry clusters. They were measured by the employment size of a particular service industry in 1977. Localization economies were good predictors of employment growth, suggesting the emergence of service complexes where initial employment begets more employment. For a subset of service industries that included banking, management, public relations, and computer and data processing, urbanization economies predicted establishment growth more effectively.

Many geographic patterns are hierarchical and exhibit similar overall operations but significant differences in scale. In the atmosphere, for example, large and diffuse low-pressure systems affect subcontinental areas, whereas hurricanes several hundred kilometers in diameter are more intense but smaller in areal size. Tornadoes are even smaller and even more intense low-pressure systems. Though radically different in size, each of these systems exhibits

similar circulatory behavior. Other physical systems, particularly drainage basins, are hierarchical and nest within each other in repetitive patterns. They are
susceptible to analysis with fractals, a mathematical technique that describes
and measures repetitive patterns.

Patterns are represented by arrangements of points, lines, or areas. Researchers investigating the point pattern of karst development have explored
the degree of randomness in the distribution of sinkholes that result from solution and the collapse of underlying limestone. If the location of each sinkhole
is shown by a point on a map, the resulting distribution of points could be
highly regular, completely random, combined regular and random, or clustered. LaValle (1967) and McConnell and Horn (1972) superimposed grids on
the point distributions of karst sinkholes, counted the number of points in each
quadrat, and then compared the resulting frequency distribution of points-per-
quadrat to regular, random, and clustered spatial processes. They concluded
that sinkhole distribution was a combination of random and regular processes
that reflected the combined effects of exceedingly complex hydrologic processes and regularity imposed by joints in the limestone. Biologists analyzing
plant distributions and social geographers studying settlement patterns have
used similar techniques of point analysis.

Investigations into line patterns have included assessments of transportation
and river channel networks. Horton (1945), for example, believed that the
tree-like branching networks formed by river channels were unique products
of the erosion processes that formed the channels. Subsequent work by Shreve
(1967) demonstrated that the arrangement of network segments and the ways
they are connected are outcomes of a random topologic and statistical process
(fig. 11.8). Any network, whether of stream channels, airways in a lung, veins
on a leaf, or branches in a tree, has a common recognizable arrangement (Woldenberg et al. 1970). Networks in human and physical landscapes share certain
fundamental geographic properties and exhibit predictable arrangements of the
numbers and connections of tributaries with trunk segments of ever-increasing
size or magnitude.

In addition to points and lines, areas also define geographic patterns. Drainage basins assume particular shapes as they fill the available physical landscape
in response to runoff and erosion. Their shapes resemble the theoretical shapes
of economic market areas and may be modeled by hexagonal lattices. Air
masses dominate particular regions that may be defined by the frequency with
which specific air masses are observed there. Biogeographers often recognize
regions in the distribution of plants, animals, and complex ecological communities. Sometimes the distribution of the ecological community is controlled
by external forces such as climate, topography, or human influences. In many
cases, however, the internal properties of the community itself determine
which individual plants will be included in the distribution. Beatty (1984)
found that the types of forest canopy and the microtopography of the land
surface controlled the patterns of plants growing on forest floors. Biogeo-

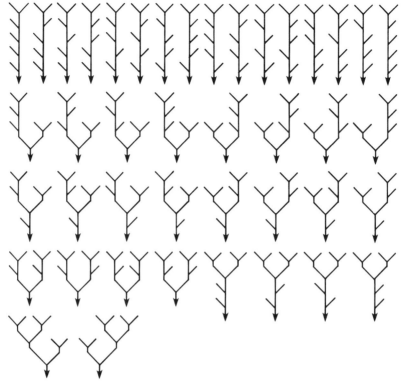

Fig. 11.8. Schematic Diagrams Showing the Forty-two Possible Topological Arrangements for Channel Networks with Six External Links. Within each row, each arrangement has an equal probability of occurrence. SOURCE: Shreve 1966:28.

graphic regions also reflect the operations of biological processes at different scales. Large-scale influences such as climate combine with small-scale features such as local topography to control the distribution of riparian vegetation along Colorado streams, for example. Bioregion definition must therefore be set in a complex systems framework operating at both macro- and microscales (Baker 1989).

SYSTEMATIC CHANGE

As flows of energy, mass, people, goods, and ideas move through their respective systems, equilibrium conditions eventually develop and system changes diminish. Occasional internal adjustments or external perturbations result in systematic change through time, often reflected in changes in the spatial patterns created by the system. Geographers find that these changes are cyclic and regular for some systems, leading to a certain degree of predictability. Other systems change on an irregular or episodic basis, and predicting those changes is more difficult.

A major system that experiences cyclic change is oceanic circulation near the western coast of South America. El Niño occurs when warm waters on the ocean surface produce heavy rainfalls and flooding in nearby coastal areas. Cold surface waters cause dry conditions. El Niño events occur on a regular basis in response to general Pacific Ocean water and atmospheric circulation patterns (Barnett 1984). As a result, floods on the northern coastal region of Peru occur on a cyclic basis that may become predictable (Waylen and Caviedes 1986).

While large-scale atmospheric and associated hydrologic systems change cyclically, many smaller surface systems adjust episodically, such as arroyo cutting and filling in the dry land southwestern United States. Streams in the humid upper midwestern United States have also undergone episodic changes since the end of the Pleistocene in response to changes in flood magnitude and frequency. Deposits along streams in Wisconsin's Driftless Area show evidence of a variety of floods that reflect irregular changes in the region's climate (fig. 11.9). The deposits also reveal irregular variations that result from human settlement (Knox 1987). Study of one drainage basin in the same area showed channel cutting also could be reversed by conservation practices (Trimble and Lund 1982).

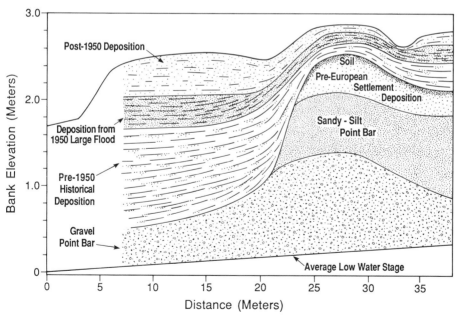

Fig. 11.9. Depositional Evidence of Episodic Changes in Processes in the Platte River, Southwest Wisconsin. Source: Knox 1987:167.

Whether cyclic or episodic, changes in system behavior have proven difficult to describe. Two competing philosophies emphasize gradual changes and catastrophic adjustments. Geomorphologists were long influenced by their roots in historical geology, where most workers adhered to·*uniformitarianism*, a concept that originally conveyed the idea that processes in the past were similar to processes operating now (Tinkler 1985 : 82). Modern usage expanded its meaning to include smoothly defined changes. Abundant evidence indicates that changes do sometimes occur abruptly, and catastrophic models and explanations are now common in geography.

Changes in system operation, whether abrupt or smooth, create new geographic patterns. Recent changes in the structure of American business, for example, have profoundly altered the geography of economic activity and population. The term *regional restructuring* has been used to refer to the geographic changes that accompany the rise in global capitalism (Susman and Schutz 1983). Large corporate enterprises headquartered in advanced countries establish branch plants in peripheral regions and less developed countries to gain new markets and to take advantage of cheap, nonunionized labor. A result is a decline in the number of blue-collar jobs, especially in the Manufacturing Belt of the American Midwest and Northeast. This decline is accelerated by the substitution of capital for labor and the deskilling of the work force resulting from the use of electronic information systems and computer programming. Accompanying the decline in blue collar manufacturing jobs is growth in the low-wage, low-skill consumer sector. The resulting labor force is increasingly bipolarized and bottom heavy (Ettlinger 1988). What little manufacturing employment remains is flexibly specialized; it consists of intermittent, part-time jobs (Scott 1985; Storper and Christopherson 1987). The term *post-Fordism* describes the shift away from vertical integration, where one firm controls all stages of the production process, to subcontracting by small local producers who employ individuals on a short-term or part-time basis.

Nowhere are the urban and regional patterns resulting from these shifts in American business more dramatic than in Los Angeles, where 1.3 million new jobs were added between 1970 and 1980. Significant changes in the sectoral and geographic distribution of employment include: (1) a decline in manufacturing's relative share of total employment; (2) a shift from specialization on aircraft production to greater diversification; (3) significant employment losses in the industrial zone between downtown and the ports of San Pedro and Long Beach (fig. 11.10A); (4) the development of large complexes of technologically advanced industry and services fed by massive growth in aerospace and electronics employment in suburban Orange County and around Los Angeles International Airport (LAX in fig. 11.10B); and (5) the emergence of downtown Los Angeles as a control and managerial center for international capital. The Los Angeles area simultaneously embodies the selective deindustrialization, plant closures, and layoffs of the Frost Belt, the dynamic overall growth

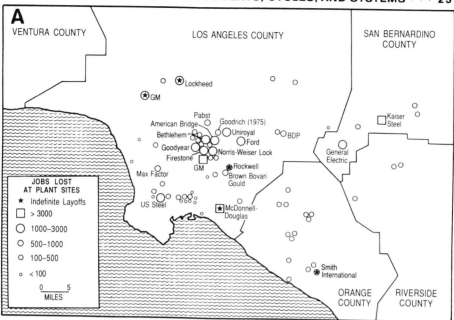

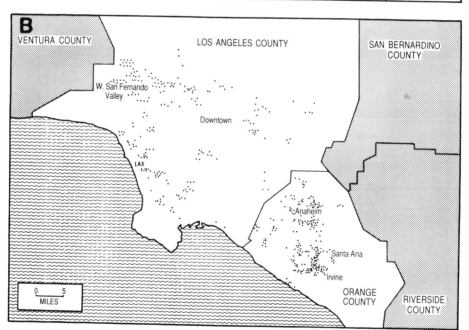

Fig. 11.10. A: Distribution of Plant Closings and Major Layoffs in Los Angeles, 1978–82. Most of the closings and layoffs occurred in the old manufacturing belt near the city center. *B:* Distribution of Electronics Component Plants in 1981. This rapidly growing sector of the Los Angeles economy was focused on Orange County to the south, in the area adjacent to LAX airport, and in the west San Fernando Valley. Areas experiencing plant closings and layoffs generally were not able to compensate through increases in electronics components. SOURCE: Soya, Morales, and Wolff 1983:213, 216.

and emphasis on high-tech activities typical of the Sunbelt, and a peripheralized labor force akin to that found in a Third World free enterprise zone (Soya, Morales, and Wolff 1983).

CONCLUSIONS

Human and physical systems are dynamic, constantly changing distributions of elements and connections, and a complete understanding of any terrestrial system requires an appreciation of its geography. In dealing with abstract problems and practical issues, geographers offer a tool kit of ideas that includes general systems concepts, movements, patterns, and change through time. The employment of these tools yields a clearer understanding of the major research problems facing the modern world. Recent debates concerning the likely consequences of increased carbon dioxide in the earth's atmosphere provide an example of the importance of including the geographic perspective. Inputs of carbon dioxide vary greatly on regional and global scales. Therefore understanding the input term for the global warming problem is inherently geographic. All global atmospheric circulation scenarios that incorporate continued increases of carbon dioxide produce dramatic regional variations in the climatic and surface processes likely to ensue if global warming occurs. Temperatures will surely increase in some regions, but they will decline in others; understanding the geography of those consequences of the process and problem is part of its solution. For the carbon dioxide issue, as for most major society/environment problems in the late twentieth century, resolution lies in prudent combinations of basic social and physical principles with a strongly developed geographic perspective.

REFERENCES

BAKER, W. L. 1989. Macro- and micro-scale influences on riparian vegetation in western Colorado. *Annals of the Association of American Geographers* 79:65–78.

BARNETT, T. M. 1984. Prediction of El Niño, 1982–1983. *Monthly Weather Review* 112:1403–1407.

BEATTY, S. W. 1984. Influence of microtopography and canopy species on spatial patterns of forest understory plants. *Ecology* 65:1406–1419.

BEHR, M., and PATRICIA GOBER. 1982. When a residence is not a house: Examining residence-based migration definitions. *Professional Geographer* 34:178–184.

BERRY, BRIAN J. L. 1976. *Urbanization and counterurbanization*. Beverly Hills, Calif.: Sage.

———. 1981. *Comparative urbanization*. New York: St. Martin's.

CAINE, NELSON. 1971. A conceptual model for alpine slope process study. *Arctic and Alpine Research* 3:319–330.

———. 1976. Summer rainstorms in an alpine environment and their influence on soil erosion, San Juan mountains, Colorado. *Arctic and Alpine Research* 8:183–196.

CHORLEY, RICHARD J., and B. A. KENNEDY. 1971. *Physical geography: A systems approach*. London: Prentice-Hall.

ETTLINGER, NANCY. 1988. American fertility and industrial restructuring: A possible link. *Growth and Change* 18:75–93.

FORRESTER, J. W. 1969. *Urban dynamics*. Cambridge, Mass.: MIT Press.

GRAF, WILLIAM L. 1985. Mercury transport in stream sediments of the Colorado plateau. *Annals of the Association of American Geographers* 75:552–565.

HORTON, R. E. 1945. Erosional development of streams and their drainage basins: Hydrophysical approach to quantitative morphology. *Geological Society of America Bulletin*, 56:275–370.

HUGGET, R. 1980. *Systems analysis in geography*. Oxford: Clarendon.

INTERNAL REVENUE SERVICE. 1988. Inter-state migration matrix, 1987–88. Unpublished. Washington, D.C.

KNOX, JAMES C. 1987. Stratigraphic evidence of large floods in the Upper Mississippi Valley. In *Catastrophic flooding*, ed. L. Mayer and D. Nash, 155–180. Boston: Allen & Unwin.

KONTULY, T., and R. VOGELSANG. 1988. Intensification of counterurbanization. *Professional Geographer* 40:42–53.

LAVALLE, PLACIDO D. 1967. Geographic processes and the analysis of karst depression distributions within limestone regions (abstract). *Annals of the Association of American Geographers* 57:794.

MABOGUNJE, AKIN L. 1970. Systems approach to a theory of rural-urban migration. *Geographical Analysis* 2:1–18.

MCCONNELL, H., and J. M. HORN. 1972. Probabilities of surface karst. In *Spatial analysis in geomorphology*, ed. Richard J. Chorley, 111–134. New York: Harper & Row.

MCHUGH, K. E. 1990. Seasonal migration as a substitute for, or precursor to, permanent migration. *Research on Aging* 12:229–45.

MOORE, ERIC G. 1972. *Residential mobility in the city*. Commission on College Geography Resource Paper 13. Washington, D.C.: Association of American Geographers.

O'HUALLACHAIN, B. 1989. Agglomeration of services in American metropolitan areas. *Growth and Change* 20:34–49.

PLANE, DAVID A. 1985. A systemic demographic efficiency analysis of U.S. interstate population exchange, 1935–1980. *Economic Geography* 60:294–312.

POTTER, L. D., D. E. KIDD, and D. R. STANDIFORD. 1975. Mercury levels in Lake Powell. *Environmental Science and Technology* 9:41–46.

ROGERS, ANDREI. 1985. *Regional population projection models*. Beverly Hills, Calif.: Sage.

ROWLES, G. D. 1983. Between worlds: A relocation dilemma for the Appalachian elderly. *International Journal of Aging and Human Development* 17:301–14.

SCOTT, ALLEN J. 1985. Location processes, urbanization, and territorial development: An explanatory essay. *Environment and Planning A* 17:479–501.

SELLERS, W. D. 1965. *Physical climatology*. Chicago: University of Chicago Press.

SHREVE, R. L. 1966. Statistical law of stream numbers. *Journal of Geology* 74:17–37.

———. 1967. Infinite topologically random channel networks. *Journal of Geology* 75:178–186.

SOJA, E., R. MORALES, and G. WOLFF. 1983. Urban restructuring: An analysis of social and spatial change in Los Angeles. *Economic Geography* 59:195–230.

STORPER, MICHAEL, and SUSAN CHRISTOPHERSON. 1987. Flexible specialization and regional industrial agglomeration: The case of the U.S. motion picture industry. *Annals of the Association of American Geographers* 77:104–117.

STRAHLER, ARTHUR N. 1980. Systems theory in physical geography. *Physical Geography* 1:1–27.

STRAHLER, ARTHUR N., and ALAN H. STRAHLER. 1987. *Modern physical geography*. New York: John Wiley and Sons.

SUSMAN, P., and E. SCHUTZ. 1983. Monopoly and competitive firm relations and regional development in global capitalism. *Economic Geography* 59:161–177.

TERJUNG, WERNER H. 1976. Climatology for geographers. *Annals of the Association of American Geographers* 66:199–222.

TINKLER, K. J. 1985. *A short history of geomorphology*. Totowa, N.J.: Barnes and Noble.

TRIMBLE, STANLEY W., and S. W. LUND. 1982. Soil conservation and the reduction of erosion and sedimentation in the Coon Creek basin. U.S. Geological Survey Professional Paper 1234. Washington D.C.: GPO.

VINING, D. K., and R. PALLONE. 1982. Migration between core and peripheral regions: A description and tentative explanation of the patterns in 22 countries. *Geoforum* 13:339–410.

VINING, D. R., R. PALLONE, and D. PLANE. 1981. Recent migration patterns in the developed world. *Environment and Planning A* 13:243–250.

WAYLEN, PETER R., and CAESAR N. CAVIEDES. 1986. El Niño and annual floods on the North Peruvian littoral. *Journal of Hydrology* 89:141–156.

WHITE, S. E., L. A. BROWN, W.A.V. CLARK, P. GOBER, R. JONES, K. E. McHUGH, and R. L. MORRILL. 1989. Population geography. In *Geography in America*, ed. Gary L. Gaile and Cort J. Willmott, 258–289. Columbus, Ohio: Merrill.

WOLDENBERG, M. J., G. CUMMING, K. HORSFIELD, K. PROWSE, and S. SINGHAL. 1970. *Law and order in the human lung*. Harvard Papers in Theoretical Geography 41. Cambridge, Mass.: Harvard Laboratory for Computer Graphics.

The Local-Global Continuum

William B. Meyer, Derek Gregory,
B. L. Turner II, and Patricia F. McDowell

The geographical imagination is most fully exercised when it wanders across a range of scales, teasing out the connective tissue that binds different levels of the local-to-global continuum together. Geography is not unique among disciplines in observing and working at various scales. As a pursuit in which space is central, however, geography has assumed priority in addressing a core issue: the role of spatial scales in inquiry and understanding, the issues embedded within the local-to-global or micro-to-macro research nexus.

The primary objective of this chapter is to review the geographical research agendas that have lately magnified questions of scale. Past forays into the subject suggest that a rigorous and detailed mapping of those agendas lies beyond our reach at present and may always remain so, because the issues are constantly redefined. But greater awareness of their outlines and dimensions will suggest promising directions for the immediate future.

We focus throughout this chapter on geographic research rather than on research by geographers, although our examples are biased toward geographic research conducted by geographers. The distinction is not trivial. Geographers commonly engage in research in technical areas or on subjects that are distant from their ultimate geographic goals; they may study pollen analysis to gain insights into past climates or document morbidity rates to inform research in population geography. Few among us would argue that a geographer's work at the scale of palynology or of individual-level data collection is per se geographic. On the other hand, the examples selected for this chapter should not be interpreted as the limits, domain, or core of geographic research. They are examples, nothing more.

SPATIAL SCALES IN HUMAN GEOGRAPHY

The difficulties of linking research across spatial scales have long been recognized. The regional tradition that dominated Anglo-American human geography at mid-twentieth century displayed a largely intuitive sensitivity to the scale problems encountered in describing and characterizing areas—an awareness made explicit by Bird (1956). Whereas empirical research was conducted principally at the mesoscale, considerable microscale work was also done, albeit amid controversy as to its proper role in geography. Microstudies were defended either as a way of epitomizing the patterns and processes of a region through examination of smaller areas deemed to be typical of it, or as ways of highlighting local variability apt to be obscured in regional generalizations.

A more sharply focused but correspondingly more abstract engagement with issues of scale emerged after mid-century in spatial analysis and in geographic applications of systems theory. The quantitative revolution in human geography clarified the nature of several technical problems of scale in empirical research based on areal data that had previously been addressed outside the discipline, including the ecological fallacy and the effects of boundaries on data units. One team of economic geographers noted:

> In geographic investigation it is apparent that conclusions derived from studies made at one scale should not be expected to apply to problems whose data are expressed at other scales. Every change in scale will bring about the statement of a new problem, and there is no basis for assuming that associations existing at one spatial scale will also exist at another. (McCarty, Hoak, and Knos 1956:16)

A concrete illustration was Haggett's (1964) demonstration that the amount and distribution of forest cover in southeastern Brazil correlated with different factors as the scale of analysis moved from regional to local levels. From a purely empirical perspective, explanations of forest patterns could change at different scales of analysis.

Current interest in spatial-scale issues centers on substantive and conceptual matters. There is more than one local-to-global continuum in human geography, for *global* and *local* are dual in meaning; besides their spatial denotations, they refer to conceptual levels. The *Oxford English Dictionary* (2d ed.) defines *global* not only as embracing the scale of the world, but as "pertaining to or embracing the totality of a number of items, categories, etc.; comprehensive, all-inclusive, unified, total." When Geertz (1983) writes of "local knowledge," on the other hand, he refers as much to its particularity, discreteness, and contextuality as to its spatial localization.

Many issues related to scale in process and explanation that were first raised in geomorphology (Schumm and Lichty 1965) now find parallels in human geography (Watson 1978; Taylor 1981; Philbrick 1982; Palm 1986; Smith and Ward 1987; Bird 1989:19–43). Contemporary human geographers are drawn increasingly to diverse scales of study because of the wide range of subjects they address, and also because of their use of explanatory modes in which sensitivity to scale effects is explicit, modes that themselves imply spatial meaning. None are more powerful than those that focus on *globalization.*

In its strongest version, recognition that scale matters would foster the argument that processes and phenomena are not connected at all as scale of analysis changes, and attempts to develop heuristics to guide efforts to cross scales or to match concept with analysis would be fruitless. The other extreme—the view that scale matters not at all because apparent problems are illusory and easily resolved—finds expression in holism and in naïve aggregation. Holism entails a belief that macro-states determine micro-states and that regional or local attributes matter little compared to forces operating at the global scale. Naïve aggregation relies on summations of local forces to explain broader ones. Though both of these positions are represented in the geographic literature, neither is satisfactory. Less simplistic perspectives are needed to unravel the real issues of scale disjuncture.

The issues outlined here are illustrated by reviews of work drawn from the space-society and nature-society branches of geography. We make no attempt to cover all the subfields in these two broad branches of human geography but draw upon specific themes that amplify scale problems. The global-to-local theme has been played out in discussions of modernity and postmodernity in historical and economic geography. The local-to-global trajectory has also been traced by land-use research in cultural-ecological and resource geographies. For the sake of contrast, the two extremes of spatial and conceptual scale—the local and the global—are emphasized throughout.

Society, Space, and the Global-to-Local Continuum

Countless commentators have drawn attention to the persistent globalization of political, economic, and cultural life in the late twentieth century. Although the images conjured vary—from the global village Marshall McLuhan (1964) saw as emblematic of modernity to the vast stretches of hyperspace that Jameson (1988) views as symptomatic of postmodernity—they share a heightened sensitivity to the interdependence that underlies contemporary societies. Webs of interaction are spun across greater spans of time and space; sometimes erasing and always transforming existing landscapes, they put in place new spatial structures that frame and shape the conduct of social life in new and often unexpected ways.

Modernity and Globalization. The need to think globally has never been greater, but it should not obscure the fact that globalization is not new. Following Taylor's (1981, 1982) lead, many geographers were inspired by Wallerstein's account (1979, 1984) of the genesis of the modern world system. Wallerstein locates the great divide between the traditional and modern worlds in the late sixteenth century. Until then, "the history of the world was the history of the temporal *coexistence* of three modes of production" (1984:164). They supported fragile and evanescent minisystems, unstable and transitory world economies "which were always transformed into empires: China, Persia, Rome" (Wallerstein 1974:16), and spectacular and encompassing world empires. In the sixteenth century, "for the first time, a world economy did not disintegrate but survived, and became the world capitalist system we know today" (Wallerstein 1984:164). Although it originated in Europe, the world economy was predicated upon ceaseless expansion and by the end of the nineteenth century included "virtually the whole inhabited earth" (Wallerstein 1984:165). What is particularly novel about Wallerstein's argument is that he conceives of the modern world system as a single system whose spatial structure is of central importance to the systematic surplus transfers that ensure its reproduction; "today," Wallerstein declares, "there is only *one* social system and therefore only *one* mode of production extant—the capitalist world-economy" (1984:165), (fig. 12.1).

One persistent objection to Wallerstein's thesis is the preeminence of a single scale of analysis (Agnew 1982). This preeminence has a number of implications. First, Wallerstein's theoretical apparatus and substantive analysis have a peculiarly Eurocentric cast, whereas Europe was "an upstart peripheral to an ongoing operation," an operation that during the thirteenth century stretched from northwestern Europe to China in a chain of interlocking spheres of exchange (fig. 12.2). This premodern world system was nonhierarchical and polycentric. Had its spheres and circuits not existed, Abu-Lughod insists, "when Europe gradually 'reached out,' it would have grasped empty space rather than riches" (1989:12). Also, even Europe's *outreach* cannot be explicated in uniquely European terms. In showing how the core subjugated the periphery, Wallerstein effectively confines the dynamism of the world system to the core and fails to recognize the dynamics of the periphery. Action be-

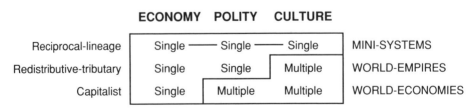

Fig. 12.1. The Relations of Economy, Polity, and Culture in Wallerstein's Schema.

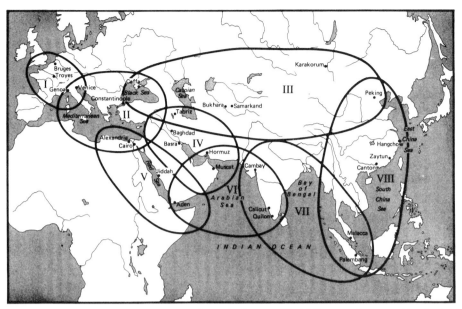

Fig. 12.2. The Eight Circuits of the Thirteenth-Century World System. SOURCE: Abu-Lughod 1989.

comes the preserve of Europeans and their descendants, to whom all other cultures merely or even passively respond (Wolf 1982).

A second objection, and another way of saying much the same thing, is that Wallerstein's framework makes it difficult to envisage regional geography as anything other than a screen on which the logic of the world system is projected. Yet from the colonial period onward, the incorporation of the United States into the world economy had profound consequences for different regions: "Indeed, regions have themselves often been defined by their relationships to the world-economy as much or more than by their relationships to each other" (Agnew 1987:20; see also Meinig 1986). This observation continues to be true, but Agnew also implies that from the closing decades of the nineteenth century, the rise to dominance of the United States within the world economy has fundamentally changed that economy's logic. In much the same way, Storper claims that "capital [has been] forever altered by the [subsequent] rise of Japan, not simply revealed 'in another instance'" (1987:420). The point of these reservations about Wallerstein's thesis is that in certain critical circumstances, "*the local makes the global*" (Storper 1987:420).

A third problem with Wallerstein's thesis is its reductionism. It would be foolish to deny the extraordinary power of the capitalist world economy, as recent events in Eastern Europe have made only too obvious, but it is wrong to fold the multiple dimensions of modernity into the single axis of capitalism. A host of scholars of different persuasions would now agree that no adequate historical geography of globalization can be written at an economic level alone, especially when one abandons the privileged perspective prevalent at Waller-stein's core. "Critical regional analysis of peripheral societies will stagnate [unless] we throw off the fetters of econocentrism" and develop a critical *political* theory (Sadler 1989:288). The horizons of such a project would of necessity be wide, and its contours cannot be confined to the state and to geopolitical alignments. The localization of experience is intense in many of these peripheral societies, and the struggles over meaning that inhere in par-ticular places are struggles over *cultural* modalities. As such, they reach not only into the abstract calculus of commodity production, at once totalizing and globalizing, but also into the intensely concrete and capillary microphysics of local cultures and local knowledge. Peasants in subaltern societies may well be "people on whom the wave of modernity has just broken," but geographers will make little sense of their struggles if they persist in measuring the power of the wave as a uniquely economic force (Watts 1988:31).

These qualifications suggest that globalization is a compound process, a se-ries of overlapping and interpenetrating changes that admit of no single, simple dynamic. In itself, perhaps, this does not say much. Certainly, it says nothing about the practical problems involved in conducting multilocale and multilevel inquiries, nor does it say anything about the contemporary crisis of represen-tation that such inquiries must confront, wherein "the problematization of the spatial" looms large (Marcus and Fischer 1986; Marcus 1990). But it does at least suggest why social (and other human and cultural) theories that remain uninformed by geographical imagination will continue to represent modernity as (for example) the colonization of a monistic lifeworld by a monistic system. These singular simplifications have little real purchase on the topography of modernity, but the debate over postmodernism and postmodernity has pursued these questions most keenly; it offers one way in which the scales question might be thoroughly addressed.

Postmodernity and Globalization. It is difficult to place an abrupt caesura between modernity and postmodernity because of the profound schism in the late twentieth-century world between the locus of experience and the loci of power and production.

> What tends to disappear is the meaning of places for people. Each place, each city, will receive its social meaning from its location in the hierarchy of a network whose control and rhythm will escape from each place and, even more, from the people in each place. . . . The new space of a world capitalist system . . . is a space of variable geometry, formed by locations hierarchically ordered in a continu-ously changing network of flows. . . . Space is dissolved into flows . . . cities into shadows (Castells 1983:314).

Such a disjuncture is a commonplace of modernity, where "the truth of experience no longer coincides with the place in which it takes place" (Jameson 1988:349). Similarly, myths come to surround consumption when its experience is divorced from the realities of production (Sack 1990). But postmodernity intensifies this disjuncture to such a degree that it becomes extraordinarily difficult to represent its configuration in modernist terms. To Jameson, for example, the spatial peculiarities of postmodernism are

> symptoms of a new and historically original dilemma, one that involves our insertion as individual subjects into a multidimensional set of radically discontinuous realities, whose frames range from the still surviving spaces of bourgeois private life all the way to the unimaginable decentering of global capital itself. Not even Einsteinian relativity, or the multiple subjective worlds of the older modernists, is capable of giving any kind of adequate figuration to this process. (1988:351).

Jameson's is not a counsel of despair. His analyses of postmodernism and postmodernity have sought to bring contemporary fragmentations within the comprehensive gaze of historical materialism. Parallel attempts by geographers to come to terms with the crisis of representation owe much to Jameson's reflections, and they bear directly on interpenetrations of the local and global.

Edward Soja's essays on the *internationalization* of Los Angeles provide some of the most illuminating and economical sketches of the ways in which the local and the global may be represented under the sign of postmodernism. Although the space-economy of the late twentieth-century city is situated within successive global and regional restructurings of capital, Soja rejects conventional models of modernization. Rather than trace the progressive march of modernity outward, through a series of metropolitan growth poles, Soja proposes a complex dialectic of globalization and localization in which, "more than ever before, the macro-political economy of the world is becoming contextualized and reproduced in the city" (1989:188).

> Ignored for so long as aberrant, idiosyncratic, or bizarrely exceptional, Los Angeles, in another paradoxical twist, has, more than any other place, become the paradigmatic window through which to see the last half of the twentieth century. I do not mean to suggest that the experience of Los Angeles will be duplicated elsewhere. But just the reverse may indeed be true, that the particular experiences of urban development and change occurring elsewhere in the world are being duplicated in Los Angeles. (Soja 1989:221)

This implosion renders conventional forms of spatial analysis inoperable, but the essentials of Soja's analysis are cast within the framework of David Harvey's thoroughly modern historico-geographical materialism (Gregory 1990).

One of Harvey's longstanding concerns has been to show that the production of space is an essential moment in the reproduction and transformation of the capitalist mode of production. He argues that one of the characteristic impulses of capitalism has been what Marx called "the annihilation of space by time," a propensity for capital rotation to accelerate through what geographers term *time-space convergence* (Janelle 1969) or *time-space compression* (Harvey 1989). Postmodernity results from a deep-seated crisis in the Fordist regime of capital accumulation that was established during the decades immediately following World War II. This crisis is currently being resolved through the precarious and painful installation of a new regime of post-Fordist flexible accumulation. The crisis is predominantly one of temporal and spatial form, so that the transition from one regime of accumulation to another has to be embedded in a dramatic transformation of the space-economy of advanced capitalism. Indeed, contemporary waves of time-space compression have become so powerful that flexibility connotes hypermobility. Spasmodic surges of financial capital now shape the fortunes of people and places throughout the world.

> The rapidity with which currency markets fluctuate across the world's spaces, the extraordinary power of money capital flow in what is now a global stock and financial market, the volatility of what the purchasing power of money might represent, define . . . a high point of that highly problematic intersection of money, time and space as interlocking elements of social power in the political economy of postmodernity. (Harvey 1989:298)

These instabilities are structurally implicit in the contemporary crisis of representation. As human experience of time and space becomes fragmented and radically destabilized, people become increasingly unsure how to represent the tense and turbulent world in which they find themselves. If this uncertainty seems to address Castells's concerns, however, it also feeds into a *revalorization* of place: "The collapse of spatial barriers does not mean that the significance of place is decreasing . . . as spatial barriers diminish so we become much more sensitized to what the world's spaces contain" (Harvey 1989:294). Contemporary accent on the local—which includes "the discursive preoccupation with distinction, fragmentation and uniqueness," all watchwords of the postmodern sensibility—derives from the dynamics of globalization. The valorization of place is to be theorized within "the contradictory dynamics of increasingly footloose and mobile capital seeking out profitable locations amidst a highly disjointed and fragmented mosaic of uneven [global] development in which competitive places try to secure a lucrative development niche" (Swyngedouw 1989:31). The essential paradox for Soja, Harvey, and Swyngedouw is that the heightened importance of globalization does not erase the significance of difference; it requires, rather, its recomposition.

These issues remain to be settled. Some critics have urged a still more contextual analysis of post-Fordism, one that is more sensitive to the cultural and social specificities (the extraeconomic geographies) inscribed within the very modalities of "flexible specialization" (Sayer 1990), one that parallels the objections to Wallerstein's framework. Others have urged greater attention to the coexistence of other modes of economic organization: to geographies of combined as well as uneven development (Gertler 1988). Still others have cautioned against overvalorization of place and the construction of "mythical geographies" that fail to grapple with the formidable political geography of the local-global continuum (Amin and Robins 1990). All these proposals will confront a host of technical as well as theoretical difficulties (Cox and Mair 1989). But taken together, they indicate, in different ways, that spatial or conceptual scale need not be a barrier to understanding. It often seems that the commonplace politics of scale constitute one of the most insistently ideological aspects of everyday life. Through these multifaceted explorations of the complex and changing geographies of the local-global nexus and the critical vision they provide, geographers will begin to glimpse at least a provisional anatomy on which to base future incisions, even if the skeletal frameworks remain hidden.

Nature, Society, and the Local-Global Continuum

The themes of modernity and postmodernity have thrust globalization to the center of space-society research. Concern with the persistence of premodernity and with the environment has kept the focus on locality strong among nature-society geographers, particularly those who study the rural Third World. Globalization themes have entered not only through the conceptual issues noted above but through practical concerns emerging from the research agenda on human dimensions of global environmental change (Turner 1989a; Kates et al. 1990).

Locality in Premodernity. Much of the local-to-global discourse has been grounded in the study of land use, especially in the have-not world. To address this subject is almost to confront a conceptual and terminological impasse: the behavior and rationales of actors—whether individuals, groups, or societies—are not fully indigenous and traditional (precapitalist), yet neither are they modern (capitalist). Rather, a set of unique hybrids of human conditions and decisions influences land use between these polar types (Brush and Turner 1987). For the most part, these cultural hybrids engage nature directly on a day-to-day basis, primarily through smallholder agriculture. It is through the agricultural engagement with nature that the socioeconomic facets of the relationship are articulated. Of particular significance is the dual nature of decision making that is grounded both in the rationalization of value captured in the market (commodity production) and in specific forms of production for consumption guided by indigenous motives and rationales (Brush and Turner 1987:33–34). What undergirds this work is the belief that the state of agriculture (and hence of land use) at any time and place is a product of farmer motives and rationales as the opportunities and constraints of the physical environment are assessed. Land use is obviously more than the product of human behavior, for it is intimately tied to the physical environment. Similar social, political, and economic conditions of decision making in dissimilar physical environments usually yield different land uses (Brookfield 1972). Environment mediates and conditions land use.

Contemporary geographic research on land use in the have-not world examines small areas from two perspectives. Microgeographies of land use focus on individual households and recognize the role of indigenous agricultural knowledge (particularly in microlevel manipulations of environments) and the importance of the structure of the household production unit (Rocheleau 1989). Land use is a function of the resources available to a household, and gender-based access to land or control of agriculture inputs in minispaces will affect overall land-use patterns. Regional assessments of land-use patterns, on the other hand, follow from comparative studies of these same themes at broader scales of aggregation; regions are bounded by socioeconomic and environmental contexts (Parry, Carter, and Konjin 1988).

The emphasis on studies of smaller areas has not followed solely from the pragmatics of research; it also reflects underlying views about the best ways to understand and explain nature-society relationships. At least three views favor microapproaches: (1) historians and humanists argue that complete understanding of land uses and their changes, the human "ways of belonging to an ecosystem," is best achieved "at the local level, where they become most visible" (Cronon 1983:12,14); (2) an emerging interest in melding various forms of social theory with nature-society relationships also brings analysis to the microlevel in order to capture the hermeneutics involved (Zimmerer 1990; Bebbington, in press); and (3) dissatisfaction with the metatheories of the nature-society relationship has led many practitioners to opt for midrange conceptualizations—those applicable to broad socioeconomic conditions mediated by local factors (Kates 1987; Turner 1989b). They have apparently accepted the argument that metatheory applied at macrospatial scales may lead to generalizations of questionable utility, while microapproaches alone may lead to no generalizations at all (Merton 1967).

These geographic approaches to nature and society employ a full palette to paint the richness of context, and they have added the variability of context to theories that hold much of context constant. They have typically stopped short of penetrating other important aspects of land-use understanding, especially the broader sociopolitical forces that structure conditions within which land-use decisions are made. Such forces typically originate outside the areas being studied and are linked to broad and even global elements of society that are not readily captured in theories of the microscale or the midrange.

Globalizing the Relationship. A global perspective pulls nature-society geographies from micro- and mesoscales to a macro view. Conceptually, Wallerstein's influence is again an appropriate point of departure, although precursors for the analysis of land use exist, such as the extension by geographers of von Thünen's land-use principles and patterns to transcontinental scales (Peet 1969). Much of the recent discourse has centered on "the intersection of world political economy and local processes of access to and struggle over resources" (Watts 1989:11). Colonialism, global capitalism, and underdevelopment have offered extremely limited and biased choices to smallhold farmers, usually in a context disadvantageous to them (Blaikie 1985; Wisner 1988). The influence of this work has exceeded what would be expected given the small number of geographers addressing questions of Third World agriculture, which suggests the degree to which this aspect of agriculture has been overlooked in the past. To be sure, local and regional inequities that influence land use have received considerable attention, but they are considered mere mediating influences on global forces such as the spread of capitalism and commodity production. Smallholders in the have-not world operate under structural constraints that deny them choice, leading to the dual behavior described above that drives land use and land-use change.

These macroperspectives shunt the principal questions of nature-society relationships away from individual or group activity within a socioeconomic structure to the evolution and nature of that structure itself, to the changing nature of global capitalism as it reconfigures ways to extract profit from smallholders. For nature-society geographies in general, this global emphasis runs the danger of so submerging the local beneath the global that contextual elements operative at the microscale—especially the physical environment—disappear from view. Globalism courts the excesses of holism just as much as the behavioral approach risks those of aggregation.

A second kind of global interest is also emerging—that of environmental change, with a major emphasis on global land-use transformations (Wolman and Fournier 1987; Kates et al. 1990). It focuses both on the human forces that lead to land-use changes that in turn ultimately affect the global environment, and on human responses to the environmental impacts that follow. Two kinds of global change are considered, differing in the scales of immediate operation (Turner et al. 1990). *Systemic change* affects a fluid physical system that operates globally. Examples include climate, the composition of the atmosphere, and sea level. Systemic change has worldwide impacts, even though the human sources of impact may be quite concentrated spatially and even though impacts will be felt differently in different places. *Cumulative change* is localized in its reach. It can be termed global when it attains a worldwide spatial distribution or a globally significant magnitude. Examples include soil loss or degradation, mineral resource depletion, and water withdrawal and pollution.

Critical to understanding both kinds of change are the roles given to human macroforces. The subject has not been tackled by the geographic community, although the candidate forces (population change, technological change, economic organization, and political structure) are well known and have been explored through a number of nature-society studies undertaken at local to mesoscales (Hecht and Cockburn 1989). An emerging issue is the effect that globally systemic human forces (e.g., the international oil market) will have on global environmental change. Most discussion to date has been speculative. Empirical demonstration of global-scale relationships—for example, between major forms of national political structure and the degree of deforestation or amount of carbon dioxide released—is absent beyond truisms such as the links between the rise of world population or per capita consumption and increasing land transformations. There is good reason to believe that such relationships, if found, will be tenuous at best and of little use in addressing the problem. Thus far, the significance of the macroforces that underpin others or form the basis from which mediating or conditioning forces can be added remains in the same speculative limbo as that described for macroconceptual themes of globalization.

Here Clark (1987) offers valuable insights about some of the broader ways to tackle problems of scale in environmental change. Identifying the space-time scales at which particular sets of phenomena processes operate and "mapping" them (fig. 12.3) illustrates the scale problems in the linkage of climatic, ecological, and social phenomena processes. Such exercises may help resolve scale issues in geography, particularly if the phenomena and processes are mapped against the scales of explanatory themes.

Scale need not be a barrier to understanding, but it has functioned as one in nature-society geographies. How much of this barrier is inherent in the subject and how much in the biases of its students remains to be seen. The need for a more complete understanding of the nature-society relationships that link the local to the global has been recognized but little explored. Attempts have been made to address land degradation in the have-not world from a perspective that purports to bridge the several local-to-global approaches, but because "land degradation has occurred in such a wide variety of social and ecological circumstances, it is clearly futile to search for a uni-causal model of explanation" (Blaikie and Brookfield 1987:4). Macro- and microforces in similar socioeconomic and environmental circumstances play different roles. An ambitious undertaking indeed, the Blaikie and Brookfield volume illuminates the conceptual disjunctures and inconsistencies that follow from bottom-up (micro to macro) or top-down (macro to micro) approaches; it stresses the need to nest different

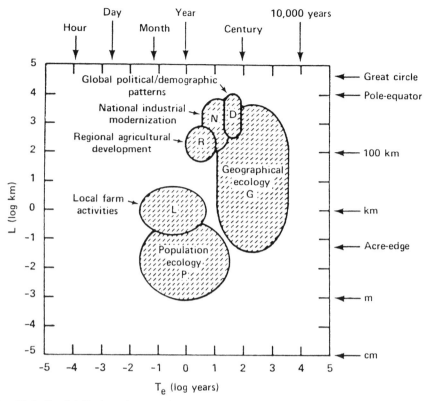

Fig. 12.3. Spatial Scales of Interacting Nature-Society Phenomena and Processes. SOURCE: Clark 1987. Used by permission of Kluwer Academic Publishers.

explanations at different scales, and it returns us to the assessments made in the mid–twentieth century (McCarty, Hook, and Knos 1956:16) with an enlarged and enriched appreciation of the conceptual, substantive, and technical challenges of scale disjunctures.

PHYSICAL GEOGRAPHY AND THE SCALE CONTINUUM

When the temporal-spatial range of environmental objects is considered (fig. 12.4), three critical facts emerge. First, natural objects of interest to physical geographers occur over a range of spatial scales spanning fifteen orders of magnitude or more. Second, the basic objects of study occur at somewhat different spatial scales in each of the earth systems—atmospheric phenomena considered important tend to be larger than those of the lithosphere surface and the biosphere. A mesoscale study in synoptic climatology, for example, might involve analysis of a regional drought affecting 10^5 to 10^6 km^2,

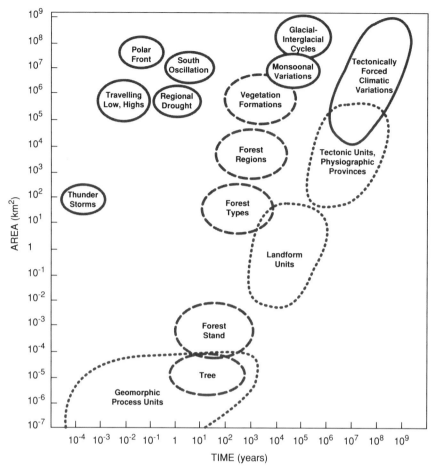

Fig. 12.4. Some Important Features of the Atmosphere, Biosphere, and Lithosphere Shown in Space-Time. SOURCE: Based on McDowell, Webb, and Bartlein 1990:144, 151, 155. Climatic units are shown by solid outlines, ecological units by dashed outlines, and geomorphic units by dotted lines. Ecological terms are functional rather than spatial in concept (Allen and Hoekstra 1990), but because only one ecological system (forest vegetation) is represented here, relative scale differences can be shown.

a mesoscale study in geomorphology might focus on a watershed of 10^1 to 10^3 km², and a mesoscale study in biogeography might focus on a forest stand of 10^{-1} km². Third, within the three earth systems—atmosphere, biosphere, and lithosphere—spatial and temporal scales of variation are related; larger entities tend to vary over longer temporal scales. Whereas spatial scales tend in human geography to be linked with conceptual scales, in physical geography the choice of a spatial scale to some extent focuses attention on processes that vary over a corresponding temporal scale.

Because phenomena of interest occur over such a wide range of scales, physical geographers, more than human geographers, have always worked at multiple scales. At any one time, however, research tends to focus on a particular range of scales. As physical geography has developed, this range has shifted from large-scale to small-scale to large-scale. Scale shifts in each of the subfields of physical geography are individualistic, driven by technical developments as well as by intrinsic and extrinsic theoretical developments.

Scales in the Practice of Physical Geography

In geomorphology as in human geography, the regional scale was dominant during the early part of the twentieth century (K. Gregory 1985). Regional-scale interests derived from the influence of William Morris Davis and his followers, who focused on questions of the long-term evolution of regional landscapes. Starting in the 1950s, explanation in geomorphology shifted away from the Davisian approach toward understanding the mechanics and dynamics of specific, small-scale geomorphic processes (K. Gregory 1985). Although a shift in time scales (from historical explanations to contemporary process explanations) is usually emphasized, a shift in spatial scales clearly occurred simultaneously. What caused this shift in spatial scale of research? One factor was the need to make direct field measurements of form and process rates. As a result, measurement capabilities constrained the spatial scale to specific sites or localities; the temporal scale was held to correspondingly short spans. Only contemporary processes can be directly measured. One researcher suggested that the attractions of quantitative field measurement and analysis may have propelled this scale shift more than did the rationale behind the use of those techniques (Gardner 1983). Another factor promoting scale reduction is that as spatial scale increases, complexity increases. The search for basic physical understanding, in fact, often leads in a progressively reductionist direction (Richards 1990).

Geomorphology as practiced today is eclectic and is not limited exclusively to microscale process studies. Process geomorphology includes both microscale studies of process units (e.g., stream channel features, slopes, and beach profiles) and mesoscale studies of landform units (e.g., small-to-large drainage basins and littoral cells). Another important theme at the mesoscale is drainage network analysis. In Quaternary geomorphology—the study of environmental influences over the last two million years—regional-scale studies are relatively common (Marston 1989). At present, there is relatively little work by American geomorphologists in the new field of megageomorphology (Gardner and Scoging 1983; Bridges 1990).

In the early twentieth century ecology and biogeography were dominated by the work of Clements on plant succession and plant formations (Viles 1988). While plant succession occurs on a time scale much shorter than that of Davisian landscape evolution, plant formations occur at a spatial scale similar to that of Davisian landscapes (see fig. 12.4). By the middle of the twentieth century, biogeography in U.S. geography departments was relatively inactive but focused on human-biota interactions and vegetation mapping. In the late twentieth century, activity has increased, with a trend toward more small-scale studies (Veblen 1989). Currently, some important themes in biogeography include vegetation dynamics and vegetation-environment relations, ecosystem structure and function, Quaternary paleoecology, and human-biota interactions. Studies at the microscale to the small end of the mesoscale appear from Veblen's review (1989) to be numerically dominant in the field as a whole. Like geomorphology, much of modern biogeography involves direct field sampling by the researcher, thereby limiting the spatial scale of projects that can be attempted.

Early climatology research was descriptive. It focused on climatic classification of sites and spatial patterns of climate at the scale of regions (K. Gregory 1985). Through time, focus shifted from description to explanation of large-scale patterns in response to both technical and conceptual developments. Today, hemispheric-scale to global-scale questions are much more important, although microclimatology also remains an active field. The development of the radiosonde produced consistent, large-scale data sets, and developments in the mathematical theory of atmospheric circulation allowed weather and climate at a station to be explained in terms of the dynamics of primary and secondary circulation features. These theories and observations form the basis of modern synoptic and dynamic climatology. At a variety of small spatial scales, geographers have also focused on the role of landscape and human activities in the energy and water balances at the earth's surface (Miller 1977; Oliver et al. 1989).

From Microscale to Macroscale

As physical geographers have become more confident of their concepts and techniques at the microscale (site scale), two forces are drawing attention back to the meso- to macroscales. First, there is a growing understanding of the interaction of factors across different scales. Questions defined and answered at one spatial scale may omit important relationships and controls that operate at other scales, and longer-term history may influence processes that are defined and studied at short time scales. In ecology, for example, species diversity at a site may be controlled not only by local-scale factors such as competition and environment but also by large-scale factors such as regional diversity (Ricklefs 1987). Regional diversity is largely the result of history on Quaternary or longer time scales, so controls are multiscale in time as well as in space. Indeed, there may be networks of interacting controls at different scales, and the relative importance of micro- and macroscale controls themselves may vary spatially and temporally (Baker 1989). Rather than focusing solely on microscale controls, microscale process studies should be preceded by analyses that consider controlling factors operating over a range of spatial scales (Phillips 1986).

Second, explanations of landscape-scale phenomena are held by most geomorphologists (Church et al. 1985) and by many biogeographers (Veblen 1989) to be the basic goals of research. In geomorphology, however, it is widely recognized that this aim is not presently being achieved. Many geomorphologists have called for integrating microscale, short-term process studies into explanations of larger-scale, longer-term regional landscape development (Gardner 1983; Bridges 1990; Richards 1990), but this integration is not easy (Mark 1980; Thorn 1982). It has been attempted, for example, in Quaternary stratigraphic studies and through space-for-time substitutions. The former in some cases do not adequately resolve temporal details, and the latter are theoretically flawed in most applications (Church and Mark 1980).

Major theoretical barriers may impede the integration of microscale understanding into landscape-scale explanations. Schumm and Lichty (1965) point out that the causal roles of various geomorphic variables are scale-dependent: characteristics that function as dependent variables at one time scale may be independent variables at shorter time scales. Schumm and Lichty's analysis in the time domain has corollaries in the spatial domain (Baker 1989), and the analysis can be extended from geomorphology to climatology and ecology (McDowell, Webb, and Bartlein 1990). Explanations based on observations at one scale cannot casually be transferred to another scale, and scale must be quantified as rigorously as possible. On the other hand, the time-dependence of variable status is not "a part of the inherent nature of the system. Rather, the effect is an artifact of statistical sampling . . . [or] is caused by a spatiotemporal stratification of sampling" (Montgomery 1989:56). Long-term history is the result of systems operating under the same physical laws as those that underlie small-scale, contemporary explanations of geomorphic processes. Despite such assertions, scale transference has not been accomplished extensively to date.

And despite the lack of a clear theoretical road map for extending microscale understanding to the macroscale, geomorphologists increasingly are attempting it. Three recent examples illustrate the variety of approaches they use. In a study of stream system response to climatic change at the 10^3- to 10^4-year time scale, Knox (1983) evaluated competing qualitative theories using three kinds of criteria: consistency with physical processes known from smaller-scale studies, comparison of their predictions to empirical data, and consistency in the spatial and temporal scales of behavior of the controls and the response. Currey (1990) suggested that records of Pleistocene closed-basin lakes be related to paleoclimatic and other paleoenvironmental changes by using a series of conceptual and semiquantitative models of lake response to various physical controls. Those models, in combination with stratigraphic data, can be used to pose hypotheses of past processes for testing against further stratigraphic and paleoenvironmental evidence. Morris and Olyphant (1990) adopted a combined physical modeling and empirical approach to explain spatial development of alpine lithofacies (moraines, rock glaciers, talus) at the regional scale. Based on local-scale process studies, they developed and parameterized a process model simulating talus and rock glacier development at a site as a function of time and varying microclimate. They then used the model to simulate development of lithofacies over a hypothetical three-thousand–year sequence of climatic conditions. A next step would be to test their model's performance against empirical data from one or more sites. These examples show a variety of approaches to integrating analysis at different scales, from conceptual schemes to quantitative models of physical processes. Empirical historical data are critical at all scales because these data provide the means for testing process models, and process studies are in turn critical in that they provide the means of formulating the models. Implicit in the work of Knox and Currey is the possibility of using historical data bases for regions or continents to evaluate the effects of controls such as climate that operate at large spatial scales.

Global Change and Scale Integration

A new impetus for scale integration in environmental science has emerged in recent years. The threat of human-induced modification of climate has resulted in the development of the new interdisciplinary field of global change and highlighted the importance of scale in geographic research. The natural science aspects of global change research focus on understanding interconnections between large-scale features of the atmosphere, hydrosphere (especially oceans and ice), land surface, and biosphere, and on predicting effects of large-scale variations in these systems. Since the late 1970s, remarkable progress has been made in understanding these large-scale processes (Prell and Kutzbach 1987; Broecker and Denton 1989; Intergovernmental Panel 1990). The greatest challenge in predicting effects of variations is to integrate across scales in the opposite direction from that described above. That is, increased understanding of climate variations on large spatial scales and over long time spans must be able to predict regional and local impacts.

In addition to its focus on the problem of near-future climatic change, global-change research has also influenced research on environmental change on long time scales. A major tool of global-change research has been global circulation models (GCMs) of the atmosphere and atmosphere-oceans. Empirical data on Quaternary paleoenvironmental conditions at specific times are used for model verification (COHMAP Members 1988). In addition, GCM simulations of past climate at specific times have been used to formulate independent hypotheses of climatic processes that can be tested using empirical geomorphic, ecologic, and hydrologic data (Kutzbach and Wright 1985; Bartlein and others, in press). GCM simulations can provide more specific resolution of seasonal climate and precipitation vs. temperature changes than was usually possible using paleoenvironmental data alone. In addition, this approach eliminates the problem of circularity in testing paleoecological reconstructions with paleoecological data. As a result, it will be possible to advance understanding of the long term behavior of biotic systems (Bartlein and Prentice 1989), geomorphic systems, and hydrologic systems.

The use of GCMs in global change assessments provides another example of the importance of linking different scales. On the one hand, the further development of GCMs will require global-scale data on microscale processes and on properties such as vertical profiles of leaf area and root density (Sellers and Dorman 1987; Wilson et al. 1987) that currently are known for only a few sites. On the other hand, translating GCM simulations of future climate into regional and local projections will require the development of process models of vegetation, soils, and hydrology (Prentice and Solomon 1991) to supplement those presently available for microclimatology.

Physical geography, like human geography, has suffered from scale disjunctures. The problems persist, but researchers are using a variety of approaches in trying to overcome it. The new field of global change presents an opportunity and a challenge to integrate research across scales; efforts in this area will have payoffs for meso- and microscale understanding as well.

THINKING GLOBALLY AND ACTING LOCALLY

The environmentalist injunction to think globally and act locally is as daunting in practice as it is appealing in principle. Yet if disjunctures in scale pose problems, they also offer opportunities. The persistent and growing attention given to the issues posed by the spatial-scale continuum testifies to their importance. We have raised some of these issues and resolved none. Note, however, that while scale issues have at times acted as barriers to understanding by isolating geographers with particular spatial, temporal, and conceptual scales from one another, they need not do so. Each new phase of interest leaves geographers conceptually richer if not disciplinarily wiser.

Geographers are unlikely to abandon their interest in the micro-to-macro continuum of spatial scales. Nevertheless, the persistence of the regional/mesoscale focus and the consistency with which geographers return to it after forays into other scales reflect more than tradition—the return to the mesoscale is recognition that it offers a meeting ground for the local and global poles of empirical research. Through that recognition, geographers will preserve the insights gained at the extremes of the scale spectrum and, ideally, move more nimbly up and down the continuum. The risk they run is that of finding a comfortable resting place in the mesoscale, and remaining so grounded there that they never grasp the levels of understanding found at either end. The challenge to geography is to devise a means for embedding the local and the global in the regional.

NOTE

McDowell gratefully acknowledges the support of the Swedish Natural Science Research Council (NFR) residential fellowship during the preparation of this paper, and helpful comments from P. J. Bartlein.

REFERENCES

ABU-LUGHOD, JANET. 1989. *Before European hegemony: The world system A.D. 1250–1350.* Cambridge: Cambridge University Press.

AGNEW, JOHN. 1982. Sociologizing the geographical imagination: Spatial concepts in the world-system perspective. *Political Geography Quarterly* 1:159–166.

―――. 1987. *The United States in the world-economy.* Cambridge: Cambridge University Press.

ALLEN, T.F.H., and T. W. HOEKSTRA. 1990. The confusion between scale-defined levels and conventional levels of organization in ecology. *Journal of Vegetation Science* 1:5–12.

AMIN, A. and K. ROBINS. 1990. The re-emergence of flexible economies? The mythical geography of flexible accumulation. *Environment and Planning D: Society and Space* 8:7–34.

BAKER, W. L. 1989. Macro- and micro-scale influences on riparian vegetation in western Colorado. *Annals of the Association of American Geographers* 79:65–78.

BARTLEIN, PATRICK J., P. M. ANDERSON, M. E. EDWARDS, and P. F. McDOWELL. n.d. A framework for interpreting paleoclimatic variations in eastern Beringia. *Quaternary International.* In press.

BARTLEIN, PATRICK J. and I. C. PRENTICE. 1989. Orbital variations, climate, and paleoecology. *Trends in Ecology and Evolution* 4:195–199.

BEBBINGTON, A. J. n.d. Indigenous agricultural knowledge: human interests and critical enquiry. *Agriculture and Human Values.* In press.

BIRD, J. 1956. Scale in regional study illustrated by brief comparisons between the western peninsulas of England and France. *Geography* 41:25–38.

―――. 1989. *The changing worlds of geography: A critical guide to concepts and methods.* Oxford: Clarendon.

BLAIKIE, PIERS. 1985. *The political economy of soil erosion in developing countries.* London: Longman.

BLAIKIE, PIERS, and HAROLD C. BROOKFIELD. 1987. *Land degradation and society.* London: Methuen.

BRIDGES, E. M. 1990. *World geomorphology*. Cambridge: Cambridge University Press.

BROECKER, W. S., and G. H. DENTON. 1989. The role of ocean-atmosphere reorganizations in glacial cycles. *Geochimica et Cosmochimica Acta* 53:2465–2501.

BROOKFIELD, HAROLD C. 1972. Intensification and disintensification in Pacific agriculture: A theoretical approach. *Pacific Viewpoint* 13:30–48.

BRUSH, S. B., and B. L. TURNER II. 1987. The nature of farming systems and views of their change. In *Comparative farming systems,* ed. B. L. Turner II and S. B. Brush, 11–48. New York: Guilford.

CASTELLS, M. 1983. *The city and the grassroots*. Berkeley: University of California Press.

CHURCH, M., B. GOMEZ, E. J. HICKIN, and O. SLAYMAKER. 1985. Geomorphological sociology. *Earth Surface Processes and Landforms* 10:539–540.

CHURCH, M., and D. M. MARK. 1980. On size and scale in geomorphology. *Progress in Physical Geography* 4:342–390.

CLARK, W. C. 1987. Scale relationships in the interaction of climate, ecosystems, and societies. In *Forecasting in the social and natural sciences,* ed. K. C. Land and S. H. Schneider, 337–378. Dordrecht: D. Reidel.

COHMAP MEMBERS. 1988. Climatic changes of the last 18,000 years: Observations and model simulations. *Science* 241:1043–1052.

COX, KEVIN R., and A. MAIR. 1989. Levels of abstraction in locality studies. *Antipode* 21:121–132.

CRONON, WILLIAM. 1983. *Changes in the land: Indians, colonists, and the ecology of New England*. New York: Hill and Wang.

CURREY, D. R. 1990. Quaternary paleolakes in the evolution of semidesert basins, with special emphasis on Lake Bonneville and the Great Basin, USA. *Paleogeography, Paleoclimatology, Paleoecology* 76:189–214.

GARDNER, R. 1983. Introduction. In *Mega-geomorphology,* ed. R. Gardner and H. Scoging, x–xiii. Oxford: Clarendon.

GARDNER, R., and H. SCOGING, eds. 1983. *Mega-geomorphology*. Oxford: Clarendon.

GEERTZ, CLIFFORD. 1983. *Local knowledge: further essays in interpretive anthropology*. New York: Basic Books.

GERTLER, M. 1988. The limits to flexibility: Comments on the post-Fordist version of production and its geography. *Transactions of the Institute of British Geographers* 13:419–432.

GREGORY, DEREK. 1990. Chinatown, part three? Soja and the missing spaces of social theory. *Strategies: A Journal of Theory, Culture, and Politics* 3:40–104.

———. n.d. *Geographical imaginations*. Oxford: Basil Blackwell. In press.

GREGORY, K. J. 1985. *The nature of physical geography*. London: Edward Arnold.

HAGGETT, PETER. 1964. Regional and local components in the distribution of forested areas in southeast Brazil: A multivariate approach. *Geographical Journal* 130:365–380.

HARVEY, DAVID. 1989. *The condition of postmodernity*. Baltimore: Johns Hopkins University Press.

HECHT, SUSANNA, and A. COCKBURN. 1989. *The fate of the forest*. London: Verso.

INTERGOVERNMENTAL PANEL ON CLIMATE CHANGE. 1990. *Climate change*. Cambridge: Cambridge University Press.

JAMESON, F. 1988. Cognitive mapping. In *Marxism and the interpretation of culture*, ed. C. Nelson and L. Grossberg, 347–357. Urbana: University of Illinois Press.

JANELLE, DONALD. 1969. Spatial reorganization: A model and a concept. *Annals of the Association of American Geographers* 59:348–364.

KATES, ROBERT W. 1987. The human environment: The road not taken, the road still beckoning. *Annals of the Association of American Geographers* 77:525–534.

KATES, ROBERT W., WILLIAM C. CLARK, V. NORBERG-BOHM, and BILL L. TURNER II. 1990. *Human sources of global change: A report on priority research and initiatives for 1990–1995*. Discussion paper G-90-08. Cambridge, Mass.: John F. Kennedy School of Government Global Environmental Policy Project, Harvard University.

KNOX, JAMES C. 1983. Responses of river systems to Holocene climates. In *The Holocene*. Vol. 2 of *Late Quaternary environments of the United States*, ed. H. E. Wright, Jr., and S. C. Porter, 26–41. Minneapolis: University of Minnesota Press.

KUTZBACH, J. E., and H. E. WRIGHT, JR. 1985. Simulation of the climate of 18,000 years BP: Results for the North American/North Atlantic/European sector and comparison with the geologic record of North America. *Quaternary Science Reviews* 4:147–187.

McCARTY, H. H., J. C. HOOK, and D. S. KNOS. 1956. The measurement of association in industrial geography. *University of Iowa Department of Geography Reports* 1:1–143.

McDOWELL, PATRICIA F., T. W. WEBB III, and PATRICK J. BARTLEIN. 1990. Long-term environmental change. In *The Earth as transformed by human action*, ed. B. L. Turner II, W. C. Clark, R. W. Kates, J. F. Richards, J. T. Mathews, and W. B. Meyer, 143–162. Cambridge, Mass.: Cambridge University Press.

McLUHAN, MARSHALL. 1964. *Understanding media: The extensions of man*. New York: McGraw-Hill.

MARCUS, G. 1990. The redesign of ethnography after the critique of its rhetoric. Paper presented at conference, "The Notion of Knowing in the Social Sciences," University of Southern California.

MARCUS, G., and M. FISCHER. 1986. *Anthropology as cultural critique: An experimental moment in the human sciences*. Chicago: University of Chicago Press.

MARK, DAVID M. 1980. On scales of investigation in geomorphology. *Canadian Geographer* 24:81–82.

MARSTON, R. A. 1989. Geomorphology. In *Geography in America*, ed. Gary L. Gaile and Cort J. Willmott, 70–94. Columbus, Ohio: Merrill.

MEINIG, DONALD W. 1986. *Atlantic America, 1492–1800*. Vol. 1 of *The shaping of*

America: A geographical perspective on 500 years of history. New Haven: Yale University Press.

MERTON, R. K. 1967. On sociological theories of the middle range. In *On theoretical sociology: Five essays, old and new*, 39–72. New York: Free Press.

MILLER, D. H. 1977. *Water at the surface of the Earth*. New York: Academic.

MONTGOMERY, K. 1989. Concepts of equilibrium and evolution in geomorphology: The model of branch systems. *Progress in Physical Geography* 13:47–66.

MORRIS, S. E., and G. A. OLYPHANT. 1990. Alpine lithofacies variation: Working towards a physically-based model. *Geomorphology* 3:73–90.

OLIVER, J. E., R. G. BARRY, W. A. R. BRINKMANN, J. N. RAYNER. 1989. Climatology. In *Geography in America*, ed. Gary L. Gaile and Cort J. Willmott, 47–69. Columbus, Ohio: Merrill.

Palm, Risa. 1986. Coming home. *Annals of the Association of American Geographers* 76:469–479.

PARRY, M. L., T. B. CARTER, and N. T. KONJIN. 1988. *The impact of climatic variations on agriculture*. 2 vols. Dordrecht: Kluwer.

PEET, J. R. 1969. The spatial expansion of commercial agriculture in the nineteenth century: A von Thünen interpretation. *Economic Geography* 45:283–301.

PHILBRICK, ALLEN K. 1982. Hierarchical nodality in geographical time-space. *Economic Geography* 58:1–19.

PHILLIPS, J. D. 1986. Spatial analysis of shoreline erosion, Delaware Bay, New Jersey. *Annals of the Association of American Geographers* 76:50–62.

PRELL, W., and J. KUTZBACH. 1987. Monsoon variability over the past 150,000 years. *Journal of Geophysical Research* 92:8411–8425.

PRENTICE, I. C., and A. M. SOLOMON. 1991. Vegetation models and global change. In *Global Changes of the Past*, ed. R. S. Bradley, 365–383. Boulder, Colo.: University Center for Atmospheric Research, Office of Interdisciplinary Earth Studies.

RICHARDS, K. 1990. "Real" geomorphology. *Earth Surface Processes and Landforms* 15:195–197.

RICKLEFS, R. E. 1987. Community diversity: Relative roles of local and regional processes. *Science* 235:167–171.

ROCHELEAU, D. E. 1989. Yours, mine and ours: The gender division of work, resources, and rewards in agroforestry systems. *Proceedings, 2nd Kenya National Seminar on Agroforestry*. Nairobi: ICRAF.

SACK, ROBERT D. 1990. The realm of meaning: The inadequacy of human-nature theory and the view of mass consumption. In *The Earth as transformed by human action*, ed. B. L. Turner II, W. C. Clark, R. W. Kates, J. F. Richards, J. T. Mathews, and W. B. Meyer, 659–671. Cambridge, Mass.: Cambridge University Press.

SADLER, D. 1989. Peripheral capitalism and the regional problematic. In *New models in geography*, ed. R. Peet and N. Thrift, 267–294. Boston: Unwin Hyman.

SAYER, ANDREW. 1990. Post-Fordism in question. *International Journal of Urban and Regional Research* 14:666–695.

SCHUMM, S. A., and R. W. LICHTY. 1965. Time, space, and causality in geomorphology. *American Journal of Science* 263:110–19.

SELLERS, D. J., and J. L. DORMAN. 1987. Testing the simple biosphere model (SiB) using point Micrometeorological and biophysical data. *Journal of Climatology and Applied Meteorology* 26:622–651.

SMITH, N., and D. WARD. 1987. The restructuring of geographical scale: Coalescence and fragmentation of the northern core region. *Economic Geography* 63:160–182.

SOJA, EDWARD W. 1989. *Postmodern geographies: The reassertion of space in critical social theory*. London: Verso.

STORPER, MICHAEL. 1987. The post-enlightenment challenge to Marxist urban studies. *Environment and Planning D: Society and Space* 5:418–426.

SWYNGEDOUW, ERIC. 1989. The heart of the place: The resurrection of locality in an age of hyperspace. *Geografiska Annaler* 71B:31–42.

TAYLOR, PETER J. 1981. Geographical scales in the world systems approach. *Review* 5:3–11.

———. 1982. A materialist framework for political geography. *Transactions of the Institute of British Geographers*, n.s. 7:15–34.

THORN, C. E. 1982. *Space and time in geomorphology*. London: Allen and Unwin.

TURNER, B. L. II. 1989a. The human causes of global environmental change. In *Global change and our common future: Papers from a forum*, ed. R. S. DeFries and T. F. Malone, 90–99. Washington, D.C.: National Academy.

———. 1989b. The specialist-synthesis approach to the revival of geography: The case of cultural ecology. *Annals of the Association of American Geographers* 79:88–100.

TURNER, B. L. II, R. E. KASPERSON, W. B. MEYER, K. M. DOW, D. GOLDING, J. X. KASPERSON, R. C. MITCHELL, and S. J. RATICK. 1990. Two types of global environmental change: Definitional and spatial-scale issues in their human dimensions. *Global Environmental Change* 1:14–22.

VEBLEN, THOMAS T. 1989. Biogeography. In *Geography in America*, ed. Gary L. Gaile and Cort J. Willmott, 28–46. Columbus, Ohio: Merrill.

VILES, H. 1988. Introduction. In *Biogeomorphology*, ed. H. Viles, 1–8. Oxford: Basil Blackwell.

WALLERSTEIN, I. 1974. *Capitalist agriculture and the origins of the European world-economy in the sixteenth century*. Vol. 1 of *The modern world system*. New York: Academic.

———. 1979. *The Capitalist World-Economy*. Cambridge: Cambridge University Press.

———. 1984. *The politics of the world economy*. Cambridge: Cambridge University Press.

WATSON, M. K. 1978. The scale problem in human geography. *Geografiska Annaler* 60B:36–47.

WATTS, MICHAEL. 1988. Struggles over land, struggles over meaning: Some thoughts on naming, peasant resistance, and the politics of place. In *A ground for common search*, ed. R. Golledge, H. Couclelis, and P. Gould, 31–50. Goleta, Calif.: Santa Barbara Geographical Press.

———. 1989. The agrarian crisis in Africa: Debating the crisis. *Progress in Human Geography* 13:1–41.

WILSON, M. F., A. HENDERSON-SELLERS, R. E. DICKINSON, and P. J. KENNEDY. 1987. Sensitivity of the biosphere-atmosphere transfer scheme to the inclusion of variable soil characteristics. *Journal of Climatology and Applied Meteorology* 26:341–362.

WISNER, B. 1988. *Power and need in Africa*. London: Earthscan.

WOLF, E. 1982. *Europe and the people without history*. Berkeley: University of California Press.

WOLMAN, M. G., and F. G. A. FOURNIER, eds. 1987. *Land transformation in agriculture*. Chichester: John Wiley and Sons.

ZIMMERER, K. S. 1990. Common field agriculture in the Central Andes: Struggles over production, space, and ecology in the 16th and 20th centuries. Paper presented at the annual meeting of the Association of American Geographers, Toronto, Ontario.

Scale in Space and Time

John C. Hudson

· · · ——————————————— ○ ——————————————— · · ·

This chapter explains how geographers define and use scale in the construction of thematic maps. The choice of map scale is guided by the nature of the problem under investigation, but maps are further constrained by the amount of information that can be presented in a limited space. Generalizing a map involves simplifying the information to be presented while simultaneously enhancing the main ideas the map is designed to show. Scale reductions nearly always prompt the need for an effective generalization of map information. Technology has recently given geographers much more information to map and more precise instruments for making maps, and these developments demand even more sophisticated approaches to the art and science of map generalization.

All fields of study that make empirical observations must consider scale. When the scope of any study is set, decisions are made about what is to be studied in detail and what is to be left in the background. Scale is not a constant in most cases: it is stretched here or compressed there to emphasize what is important at the expense of what is peripheral. Whether space or time is stretched or compressed, the language used to express scale relationships is much the same. Geographers address scale in these conventional ways, but they incorporate scale in a much more specific manner in the construction of maps.

Although it is convenient to think of a map as a miniaturized picture of some portion of the earth's surface (some maps are actual photographs overprinted with additional features and labels), a geographer's map is much more than a pictorial display. It is a thesis, an implicit argument about what is important about some place. A map portrays relationships between various features that seem to be related in a causal manner. The visibility of these features, suggested in phrases such as "bird's-eye view" or "as seen from space," actually has little to do with mapping. In fact, many maps show things such as language use, income, or population mobility that cannot be directly seen at all, but that are mapped in much the same way as more readily visible features. Such *thematic* maps are constructed for the purpose of showing some set of information in its relevant geographical context (see Chapter 4).

Like a thesis, a map originates in its creator's mind. A good map is constructed (as is a good argument), by admitting information selectively and then presenting it in sufficient detail (but no more) that its conclusions can be evaluated. Every decision made in the mapping process is informed by the mapmaker's knowledge of scale—the scale of the things depicted as well as the scale of the map for effectively illustrating the argument.

SCALE AND MAPS

Map scale is an intuitive concept that everyone uses without having to think about it in a formal way. It is simply the ratio between the size of some feature on a map and its corresponding size on the Earth. Scale is most often expressed as a definition of corresponding lengths, such as "one inch on the map represents one mile on the Earth's surface." A convention in writing such statements is to express them as fractions, in this case, 1/63,360; or, in words, one inch on the map is equivalent to 63,360 inches on the Earth's surface. Because the fraction 1/63,360 is a smaller number than, for example, the fraction 1/31,680, we say that a 1/63,360 map has a smaller scale than a 1/31,680 map.

One of the first things a geography student learns, then, is that although scale is an intuitive concept, the terminology of *large* and *small* scale is counterintuitive because the relevant size comparison is in the denominator of the fraction, not in the numerator. In modern practice, scale is usually written 1:63,360 or 1:31,680, but the meanings of large and small scale accord with historic practice. One mile is represented by one inch on the 1:63,360 map, whereas one-half mile is represented by one inch on the 1:31,680 map. Thus, any object on a relatively smaller-scale map will appear smaller than the same object on a relatively larger-scale map, and in this sense the usage of small scale and large scale agrees with our intuition.

Scales written in the form 1:63,360 (or 1/63,360) are called *representative fractions*. An advantage of this conventional definition of scale is that the representative fraction is unit-free; whether mapmakers and readers measure in inches, centimeters, or any other units, the representative fraction holds true. The reason for measuring in linear units, such as inches or centimeters, rather than in areal or volumetric units, or even a mixture of all three, is based on underlying mathematical properties. Although linear scale is never consistent over a whole map, there is always at least one point or at least one line where the linear scale is equal in all directions. Areal scales are generally reserved for the class of map projections on which area is consistent over the whole map. Volumetric measurements are rarely made from maps.

Every representative fraction is a ratio, equivalent to what mathematicians call a rational number—a number that can be expressed as the ratio of two integers (whole numbers). And since every rational number can be mapped as a point on a line, any arithmetic operation performed on a map scale (reducing scale or enlarging scale) results in another rational number. Thus, if the scale of a map is reduced from 1:31,680 to 1:63,360, every straight-line segment on the initial map will be half as long as it was (31,680/63,6360 = 1/2). But reducing the size of a line segment by half reduces the size of an area by 75 percent. If the definition of scale were in areal rather than linear units, the reduction in scale in this example would be $(31,680/63,360)^2 = 1/4$. Any segment of the reduced map's area will be only 25 percent as large as it was, in other words, a 75 percent reduction. Obviously, therefore, doubling the linear scale (63,360/31,680 = 2) yields a four-fold increase (2^2) in the map's size. And just as area is the second power of length, volume is the third power. Doubling the representative fraction of a globe would result in a globe eight times the size of the original (although its surface, because it is an area rather than a volume, would be only four times as large as it was).

It follows that reducing map scale by fixed amounts results in an even greater reduction in the amount of area available for depicting the information the map was designed to portray. Nearly all mapping must confront this dilemma, and in practice a variety of scales are used in geographic research.

For example, a convenient map size of 5 by 7 inches that might be found in a textbook can show the following areas at different scales:

- a house at a scale of 1:100;
- a city block at a scale of 1:1000;
- an urban neighborhood at a scale of 1:10,000;
- a small city at a scale of 1:100,000;
- a large metropolitan area at a scale of 1:1,000,000;
- several states at a scale of 1:10,000,000;
- most of a hemisphere at a scale of 1:100,000,000; or
- the entire world with plenty of room to spare at a scale of 1:1,000,000,000.

These examples, ranging from largest ($1:10^2$) to smallest ($1:10^9$) scale, span eight orders of magnitude and, as a practical matter, cover the spectrum of scales at which geographers are likely to use maps.

It might seem that this sizeable range of scales results from the problem of portraying two-dimensional areas at a variety of sizes and is therefore a natural outcome of the fact that geography deals with areal relationships at many scales. But the problem is neither unique to geography nor does it result solely from a concern with areas. For example, geologists use time scales that have even a broader range. Whereas the earth is about 4.7 billion years old (4.7×10^9), a sand spit along a coastline might have a life span of only a few decades, while earthquake waves travel in units of time measured in minutes (10^{-5}) by comparison. Scientists concerned with nuclear processes measure time in units ranging from 10^{-12} to 10^{-16}. Thus many sciences use a range of both temporal and spatial scales in their investigations.

Viewed comparatively, geography focuses on a small range of scales. If all the kinds of scientific studies done at scales ranging from the subatomic to the intergalactic are included, geography occupies a narrow middle ground, because its subject matter is focused on place-to-place relationships involving the human use of the Earth. Geographers construct global models of climate, but they are also interested in how humans transform their immediate surroundings. The range in orders of magnitude of geographic scales thus becomes clear. The scales of maps geographers make are those that are relevant for understanding people-environment relationships. Many problems involving the human use of the Earth require simultaneous consideration of many scales, hence we speak of a *local-global continuum*, referring to the many orders of scale magnitude that separate a local focus from a global one (see Chapter 12).

Geographers are also concerned with time scales. Geologic time, beginning with the origin of the Earth 4.7 billion years ago, can be represented as an outward spiral, with more space on the time map devoted to the better-known later periods than to the distant past (fig. 13.1). Geographers' time scales are strongly skewed toward the present. A geography of the western plains and mountains of the U.S. might mention the age of the Earth (4.7×10^9), give some details of the origin of the Rocky Mountains and the subsequent development of the Great Plains surface (6.5×10^7), and go on to describe the ice ages (10^6).

But the resolution of the geographer's focus would then sharpen to treat in some detail the end of glaciation, the arrival of humans, the emergence of the grasslands (about 1.1×10^4 years before present), and the eastward diffusion of maize-based agriculture from the Southwest (2×10^3). Greatest attention would focus on the period since permanent Euro-American settlement began in the region (1.4×10^2). Recent trends in geography are defined in terms of changes observable in the most recent decennial census (10^1 years ago). The geographer's time scale is more compact than a geologist's, broader than a historian's, and much broader than that of an economist who focuses on contemporary trends. Geology and archaeology, as well as history and economics, are thus temporally cognate fields for geography.

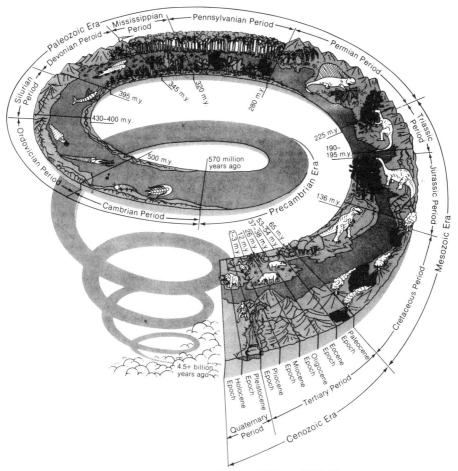

Fig. 13.1. Time and Earth History. SOURCE: Press and Siever 1974:79.

SCALE AND MAP READING

Although the range of scales common in geography is an obvious consequence of its subject matter, there remains an additional reason for the practical restriction of geographical map scales to the middle range of values. All information required to construct a map of the entire world—showing features of the Earth's surface such as relief, land use, population, and rivers—could be stored electronically in a space many orders of magnitude smaller than the map that would be needed to show that information. Furthermore, a cartographer might construct such a map, translate it into electronic form, store it in a computer, or produce a microphotograph of it. Neither case constitutes an exception to the rule that geographers study at the mid-range of scales for the simple reason that geographers' conceptions of maps are intimately tied to questions of whether they can be seen and read. It is in the practice of map reading that the true domain of geographical expertise often is found.

In an age of electronic image processing, some elaboration is required to justify why a field like geography, which is now so thoroughly suffused with computer-based technology, still adheres to the apparently quaint notion of map reading. An analogy might help: geographers read maps for the same reasons cardiologists read electrocardiograms. For both, patterns that appear to the untrained eye as specks of ink on paper or bits of light on a video screen are meaningful. There is no substitute for this direct, visual transfer of patterns from display form into a trained viewer's mind. Although anyone with minimum familiarity with road maps uses them as aids to getting around the countryside, a geographer looking at a road map designed to be read by the general public will see things others do not. The geographer sees patterns and their explanations and implications: a uniform, rectangular grid of roads on the map suggests the historic impact of the geometric land survey, a topography that is most likely flat and free of topographic barriers, and a settlement pattern that is pretty much the same from one area to another; where the rectangular road grid stops there is likely to be a decline in the quality of the land.

Geographers see things that are not explicitly depicted on maps because they know that the patterns they infer go together as a bundle of attributes defining a particular kind of landscape. To geographers, then, any map of an area instantly calls to mind other maps of the same or similar areas. Mental correlations are made. They, in turn, become hypotheses that might be tested either by direct observation or via comparisons with other map patterns. Geographers do not merely correlate patterns. Underlying the kinds of visual comparisons that can be made among maps is a knowledge of physical and social processes appropriate to the scale of a map.

An example of how scale guides geographic research is found in John R. Borchert's (1950) classic study of the North American grassland. Borchert sought to explain the wedge-shaped intrusion of grasses known as the Prairie Peninsula into the humid, forested Great Lakes region of Illinois and western Indiana. Because the Prairie Peninsula's outline does not correspond with isohyets (lines of equal precipitation) on maps of annual total precipitation, some researchers sought explanations of a more local nature that emphasized the role played by forest-destroying fires set by early human occupants of the region. But without some process operating at a regional scale, there was no reason to presume that this particular wedge-shaped area would have been especially affected by fire. Borchert devised a regional-scale interpretation by demonstrating that the Prairie Peninsula, although no drier on an annual basis than the areas bordering it to the north and south, had the same shape as did a zone that was especially apt to be influenced by the eastward flow of warm, dry air from the High Plains at the foot of the Rocky Mountains (fig. 13.2). Droughts in the Prairie Peninsula were correlated with droughts in the dry, western

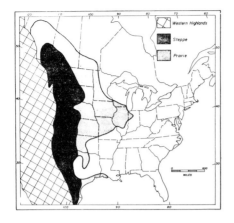
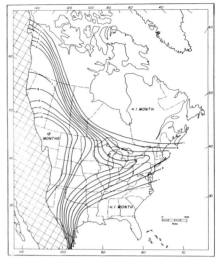

Fig. 13.2. John Borchert's Explanation of the Shape of the Prairie Peninsula (*left*) Was Based on a Model of Drought Frequency (*right*) on a Regional Scale. SOURCE: Borchert 1950:2, 26.

plains. And, since drought conditions favor fire, Borchert's explanation was able to accommodate the evidence concerning fire as a proximate cause of the west's prairie grassland.

Regional-scale atmospheric processes that produce such patterns over North America are components of a worldwide system. But neither a global-scale model nor a microscale model of climate over the Prairie Peninsula itself would have isolated and revealed the relationship Borchert discovered. Climatological phenomena of interest to geographers occur at many spatial and temporal scales, ranging from brief local storms through millennial trends in global temperature and precipitation. Generalizations formulated about trends at one scale are not necessarily appropriate at others, a well-known fact that must be considered whether a subject is being charted through time or mapped over areas (see chapter 12).

SCALE AND SPATIAL DEPENDENCE

Although relationships can be identified visually from comparing medium-scale maps such as those Borchert constructed, others are more complex and require simultaneous consideration of processes operating at several scales. Robert Haining's (1978) research on High Plains agriculture illustrates how scale can be made an explicit component of a study. Haining constructed a model that incorporated regional trends and local variations in observed patterns of wheat and corn yields in western Kansas and Nebraska, an area that exhibits strong year-to-year fluctuations in precipitation and crop yields. At a regional scale, Haining found that yields declined from east to west and from north to south, corresponding to drier and warmer environments, respectively. But these trends, already well known, did not account for all of the systematic variation in the data.

Haining incorporated two other scales at which dependencies between adjacent counties and variations within counties could be detected. First, counties with contiguous borders might be expected to exhibit similarities because of "identical, exogenous, uncorrelated variables" (1978:495), namely weather episodes such as droughts that could well occur on a twenty-five- to thirty-mile scale of variability, which is roughly the distance between centers of counties in the area. Second, some droughts are more extensive and relatively isolated, affecting only a few contiguous counties. A third component, intracounty variation, would reflect the effects of local severe storms that produce wind damage or hail. These geographical scales have direct temporal analogues, such as variations that are diurnal, seasonal, annual, decadal, and so on, to long-term cycles of changes.

Haining's model isolated the relative impacts of regional-, intercounty-, and local-scale effects on the total variability of crop yields by explicitly incorporating spatial autocorrelation—a statistical phenomenon well known to geographers. *Spatial autocorrelation* designates the frequently observed condition of dependence between adjacent locations; it is a pervasive feature of patterns over the Earth's surface. In practice, the existence of spatial autocorrelation means that if A and B are close together, what happens at A is related to what happens at B, and vice-versa. While recognition that adjacent places are likely to be similar is a fundamental truism of all geographic study, the logical complement of this proposition—that widely separated places are likely to be dissimilar—is neither obvious nor is it always true. The question, instead, is How far apart do places have to be in order to be considered independent? and the answer nearly always involves considerations of scale in the process being studied.

Anthropologists have been intrigued with this question for a century. Known as Galton's problem, from critical remarks made by Sir Francis Galton on a study that purported to show differences among several hundred ethnic units distributed around the world, the question is formulated in terms of the number of independent groups that can be identified in a continuously varying, worldwide array. If two groups have a specific behavior in common, does this constitute one instance of group behavior or two? If the groups were widely separated, the answer no doubt would be two. But if they are close together, it is likely that the two groups share a common influence. Such questions lead to discussions of how many truly different cultures there are on Earth. Behavioral traits can be classified as common or rare by asking How many cultures engage in behavior x? A geographer, however, would be unlikely to formulate the problem in exactly this way because the central question revolves around the definition of *group*, which has to have some notion of territorial limit in its specification. The problem becomes one of scale and regions (areas of relative homogeneity with respect to one or more factors). The title of a 1981 book by journalist Joel Garreau, *The Nine Nations of North America*, reflects the common perception that regions, like the anthropologists' groups, are precisely definable and hence countable. But there is no definitive way to delimit a region any more than there is a definitive way to delimit a human group.

Regions—of landforms, climates, settlement patterns, voting behavior, or any other subject—are discernible to the trained eye on geographical maps at scales ranging from $1:10^3$ to $1:10^8$. Furthermore, the number of possible regions found on a map is not a function of the map's scale. Given two maps at different scales, there is no reason to suppose that the smaller-scale map, which covers more of the Earth, will have more discernible regions just because it spans more territory. New details appear the more closely one looks, and hence diversity is not necessarily decreased when map scale is increased (or decreased). One can map cultures of the world, cultures of the United States, cultures of Missouri, cultures of the Ozarks, and so on, simply by continued refinement of the definition of what is being mapped.

SCALE AND MAP COMPLEXITY

In principle, then, complexity need not be a function of scale. If all place-to-place variations were *clinal*, that is, if they exhibited only continuous gradations in a given direction, then scale would be an irrelevant property of spatial distribution; a closer look at the data (in other words, more closely spaced observations plotted on a larger scale map) would produce no new insights. In such a case, measurements at equally spaced points along the cline would yield a function describing the rate of change, and a trend could be interpolated to make a larger-scale map or extrapolated to create one at a smaller scale. A variable that can be measured at one scale as well as at another is exemplified by the amount of radiation that the Earth receives from the sun (disregarding the effects of the Earth's atmosphere). The potential incoming solar radiation at any location on any day is a function only of the latitude of the location and of the day of the year. Gradations in both time and space are continuous around every spatial or temporal point. But geographers find relatively few variables that can be described so simply, and at the opposite extreme lie features of the Earth's surface that appear different no matter what scale is employed. The classic example of such a pattern is the case of water-land boundaries. An irregular coastline, such as that of the state of Maine, gets more irregular the more closely it is viewed. At a sufficiently large scale, the coast line becomes a broad zone that has a geography of its own, one that varies periodically with tide cycles. Yet, for many purposes, the coast of Maine can be drawn as a line on a map, and there is no problem of understanding what the line is meant to portray.

There are two issues involved. One problem is that the coastline of Maine actually gets longer the more accurately it is measured. There is no single number that can describe the length of a physical feature unless that number is qualified by a statement about the scale and method of measurement. All physical objects are measured by imposing some human standard of measurement. Highly irregular linear features, such as coastlines or meandering rivers, actually come closer to being areal features because of the amount of area that a line representing the feature occupies on a map. Linear features that have space-filling characteristics more like areas and two-dimensional surfaces so irregular that they are more like solids are known as *fractals* (see chapters 4 and 10). The other scale-related problem lies in the domain of conventional cartography. If a map is constructed from an aerial photograph, the cartographer generally will trace the outline of the feature being mapped with as much accuracy as feasible, but the resulting line will not be exactly the same as the feature. The coast is not a line, and even if it is treated as one, its width varies in both space and time. Some variability is short-term and periodic (waves, tides), but even the average shape of a coast changes, at least on the temporal scales relevant to geologists. So the generalization of a line, like other generalizations, is unlikely to be true to the extent that within it one can find all the properties of the original and no others.

Granting the problem of earth-to-map generalization, what about the apparently simpler issue of map-to-map generalization? Since reducing map scale often requires simplification (information removal) or generalization, (*selective* information removal), a desirable property of such a procedure would be the ability to ungeneralize the generalized map. If simplification merely muted details, they could be recovered by applying the inverse transformation (enhancement) to the processed map. Waldo Tobler (1969) has considered a class of transformations that can be characterized as having inverses. The height of some point on a mapped surface can be redefined as the weighted sum of its own elevation plus a discounted-for-distance function of its neighbors' elevations. Applying such a transformation to elevations, whether the heights represent terrain, population density, or any other numerically defined surface, has the effect of smoothing it by decreasing peaks and filling in valleys. The procedure is a two-dimensional analogue of time-series smoothing (such as numbers representing daily prices in a stock market) whereby short-term random shocks are filtered out by redefining each day's observation as a weighted function of itself and the adjacent days. The problem posed by Tobler is whether a profile that has been smoothed can be restored to its original form. It can, provided starting points (boundaries) are specified and provided certain conditions are met in the smoothing function. But the process must be deterministic, that is, it must be totally predictable and not subject to random fluctuations; otherwise, no unique reconstruction of the original is possible. These conditions might be met by the standards of a machine; they are less likely to be approximated when humans do the smoothing.

Although Tobler used a mathematical process to reversibly generalize, a physical example illustrates the difference between deterministic and human treatment of scale change. Suppose the coastline of Maine that appears on a 1 : 250,000 scale map is photographically reduced to a scale of 1 : 1,000,000. The photographic reduction is deterministic, and it can be undone by enlarging the new map four times $(250,000/1,000,00)^{-1}$, back to where it was. But since the reduced-scale coastline has all the detail it had before—now shown by a line only one-fourth as wide—a human cartographer would most likely smooth the line by eliminating some irregularities and, for the sake of visibility, draw the coast as a line almost as wide as it was before. A wiggly line may get straightened a bit when it is redrawn, but it may also acquire new wiggles that weren't there originally. The result is that the coastline has been generalized as well as simplified, and it can no longer be restored to what it was, given the inevitable random fluctuations introduced by redrawing it.

THE SCALE REDUCTION PROBLEM

Although geographers are apt to describe map symbols as representing point, line, or area features, these designations are merely conceptual. A zero-dimensional point (dot) representing a small town would be invisible unless it had a diameter; a one-dimensional line, signifying a road, would be invisible unless it had a width. Because symbols or letters on a map have to possess area in order to be seen, and because the conventional description of map scale is linear (a representative fraction), the scale reduction problem refers to the fact that area reductions are the second power of linear scale reductions, and the area available to show the same amount of information therefore decreases rapidly with decreasing map scale.

The purpose of either simplification or generalization is to reduce the information shown on the map because of the necessity to reduce scale; too much information on a small-scale map results in excessive complexity, whereas too little information on a large-scale map produces a map that is too empty. The dotted lines on fig. 13.3 show how information reduction accompanies scale reduction provided that information is kept constant relative to the amount of area on a map. If the scale of the map is to be reduced from S_1 to S_2, then simplification—reducing the amount of information the map depicts—is one way of coping with the problem; but it results in a large decrease in map infor-

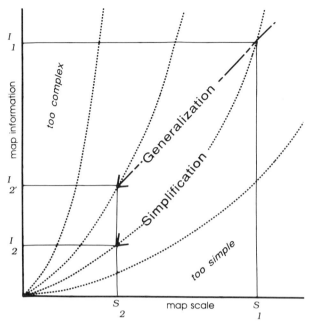

Fig. 13.3. Map Simplification and Map Generalization.

mation, from I_1 to I_2. Alternatively, if the definition of information is changed, such as by aggregating small, irregular units into larger, more regularly shaped units, information need not be decreased as much (only from I_1 to $I_{2'}$). Map generalization thus evades the scale reduction/information reduction problem by redefining map categories and symbols to retain only essential information.

SIMPLIFICATION VERSUS GENERALIZATION

Simplification refers to removing information from a map so that when scale is reduced the map is not too cluttered. Simplification may involve smoothing irregular features and removing extraneous information not necessary to the map's message. When the information to be mapped is stored in a computer, the computer may select only a part of the data for subsequent use. Whether the data are thinned by a person or by a computer, however, removal is undertaken for the purpose of reducing the amount of information to be mapped.

Simplification and generalization are undertaken with the intended message of the map in mind; the message, of course, is determined by the nature of the problem. The map of the Prairie Peninsula described earlier offers an example. Vegetation patterns are highly discontinuous, but that characteristic was not

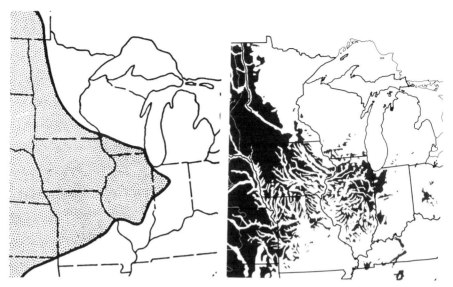

Fig. 13.4. Comparison of the Borchert Map of the Prairie Peninsula (*left*) with Certain Details of the Kuchler Vegetation Map (*right*). SOURCE: Borchert 1950:2; Kuchler 1964.

what Borchert wished to illustrate. His purpose was to show the general outline of grassland vegetation, especially its wedge-shaped eastward penetration.

Some insight into the nature of generalization is gained by comparing Borchert's map, printed at a scale of approximately 1:32,000,000, with one redrawn from the highly detailed wall-sized natural vegetation map of the United States published by A. William Küchler (1964) at a scale of 1:3,168,000, reduced here to match the Borchert map (fig. 13.4). The linear map scales differ by one order of magnitude and their area scales therefore by two orders. Comparison of the two maps at the same scale shows how much detail Borchert's map does not show. Although Borchert did not have the more detailed map (published fourteen years later) to work from, one may nonetheless see in Kuchler's map the relative accuracy of Borchert's generalization of the eastern edge of the prairie. The fact that grassy vegetation grows on uplands rather than on river bottoms is obvious on the Kuchler map but totally absent on Borchert's. The role of fire in creating grassy expanses on the flat uplands while leaving the valleys unscathed is an additional message of the Küchler map.

Why would two geographers map the same thing at areal scales such that one is more than one hundred times larger than the other? And how can we avoid concluding that Küchler's larger-scale map is preferable? The answer to both questions comes from understanding the purposes the maps served when they were constructed. The scale of Borchert's vegetation map was appropriate for comparison with the mesoscale climatic patterns he was studying. He was not invoking climate as an explanation of vegetation differences between uplands and valley bottoms, but rather as an explanation of the shape of the Prairie Peninsula. His highly generalized map suited his purpose. Küchler, on the other hand, had as his primary aim the depiction of vegetation patterns for their own sake; his 1964 map depicted 116 different vegetation types and used colors as well as pattern textures to distinguish among types. One would consult the Küchler map for the answers to many questions, but the Borchert map for the answers to only a few.

The difference between simplification and generalization thus becomes clear. The small-scale Borchert map is not merely a simplification of the details of a larger-scale map such as Küchler's. It is, instead, a highly selective generalization in which some details were suppressed for the sake of conveying a single message. To know how to generalize a map, then, is to understand the relationships one is trying to map. Random removal of information will not accomplish the task, and even orderly removal, such as deleting every nth piece of information from the map data, does not constitute an informed generalization; information has to be reduced, when going from large scale to small, at a rate proportional to the second power of the scale change. But all of the essential information can be retained—and even enhanced—by an appropriate choice of symbols, lines, and patterns.

SCALE AND GEOGRAPHICAL RESEARCH

It was not many years ago that only small-scale maps could be drawn with confidence for some areas of the world, especially areas that were sparsely populated, remote, and rugged. Ground-based mapping had achieved coverage of many countries, but extensive blank areas remained. A land-use map of the world that might have appeared in a standard atlas was accurate to the scale of $1:10^4$ for, say, Great Britain, but only $1:10^7$ for Siberia. Such a map would have been printed at a single scale, but its accuracy would have varied from place to place because details were lacking in some areas. Today, given the availability of satellite images, there need be few such place-to-place differences. Most surface areas are equally accessible to this technology, although for political reasons not all of the detail reaches the general public.

Accompanying the advent of space photography has been a shift in emphasis from speaking in terms of scale to speaking of image resolution. Resolution refers not to the image but rather to the device that records the image (such as a scanner) or to the medium on which it is recorded (such as film in a camera); it describes how finely features may be differentiated, given the construction of the device. Resolution requires attention to the largest-scale elements that such a device can record. The term *pixel* (from picture element) has been coined to define any two-dimensional area that is the smallest nondivisible element of an image. In other words, a high-resolution device will yield information in small pixels and is appropriate for constructing accurate, large-scale maps. The quest for greater accuracy on maps has naturally followed developments in high-resolution image-recording devices, including cameras, scanners, and digitizers.

At this point the art and the science of geography begin to diverge a bit. Map generalization is at least as much an art as it is a science. Many factors of judgment are involved in deciding how to reduce scale while preserving essentials. The art tends to be forgotten, however, when the only concern is greater accuracy, that is, higher resolution. The search for greater accuracy nearly always leads to an exponential increase in the amount of information. A map at a scale of $1:10^5$ requires one hundred times as much information as a map at $1:10^6$ in order to keep the accuracy per unit area constant. This comparison sounds staggering, until it is realized that the resulting map will also be 100 times as large. But there is little chance that atlases will become tomes 100 times larger than they now are just for the sake of showing all the new information that has been collected. Although the search for greater accuracy demands higher resolution and produces more information, that information must necessarily be generalized down to nearly the same scale as before simply to keep the product a manageable size.

From this observation it is natural to conclude that collecting more information, ever more accurately, produces diminishing returns for the effort expended. This is true in geography and in every other science that collects data: increasing the size of the sample adds new insights, but at a diminishing rate. Although increasing the size of a sample produces fewer new conclusions per observation as the sample gets larger, confidence in those conclusions continues to grow. And so it is with gathering high-resolution data to make more accurate maps: geographers and cartographers continue to generalize the information down to an appropriate scale, but their confidence about what the map reveals becomes stronger the larger the amount of information upon which it is based.

The information explosion that has accompanied the availability of high-resolution satellite-based images of the Earth has made small-scale maps much more accurate and has made large-scale maps of many areas possible. But it has not led to a general increase in the scales at which geographers study things. Physical processes do not change their natures just because scientists are able to monitor them more accurately or more frequently. What the new technology has provided instead is a clearer idea of the interrelatedness of the whole global-atmospheric system. The information gained from large-scale resolution, combined with the ability for frequent updates, has fostered global (small-scale) models that could only have been dreamed of a generation ago. Computer technology has advanced in response to the information-processing requirements that these models pose.

Rapid machine processing of millions of data points also has made it possible to use existing information to create new map products. An example is the shaded relief map of topography in the United States recently published by the U.S. Geological Survey (see fig. 6.9). What does the land surface of the United States look like when viewed from space? An actual photograph from space would seem to be the answer, but the terrain itself would be masked by clouds in some places and obscured by vegetation in others, and the sun's angle of illumination would vary continuously across the image. Instead, semiautomated processing of elevation data from 1:250,000-scale topographic maps allowed the map's authors to encode more than two billion elevations at a resolution of approximately one observation every 200 feet on the ground. The map information was then both simplified and generalized. First, elevation data were thinned to a more manageable file of twelve million observations appropriate to the scale of the final product. Then the data were generalized in several intriguing ways. One was to simulate the effect of low sun-angle illumination by letting the computer produce shaded relief to portray more effectively the ridges and valleys. The sun never shines at the same angle over the entire United States (when it is late afternoon in California it is dark in New England), nor does it ever strike at a horizontal angle that is particularly useful for representing surface form. But the consistency and choice of angle selected by the map's authors provide a meaningful visual image uninhibited by the complexities of reality.

Their map is a good illustration of how simplification and generalization can be combined to create a map that can be read and appreciated by specialists as well as the general public. The map reader gains an instant impression of the topography of the United States. The map is not a picture, even though it was deliberately created to look like one. It is a thematic map that, like all others, is a generalization—a selective portrayal of relevant information, with some details suppressed and others enhanced, that was created for a specific purpose.

SCALE AND VOLUME: THE DATA DIMENSION

So far, discussion has concentrated on scale as it relates to two-dimensional space. Had the depiction of landforms been a three-dimensional model, say in plaster or plastic, the model builder would have had to choose a vertical scale as well as a horizontal one and would undoubtedly have chosen a vertical scale larger than the horizontal. The highest point in the contiguous United States is less than three miles above sea level, whereas the east-west extent of the nation is more than three thousand miles. Thus, a model of the United States one foot wide would have a vertical variation about the thickness of a piece of paper. In such situations, geographers typically exaggerate the vertical scale in order to enhance surface contours.

Even when mapping data on two-dimensional maps rather than on three-dimensional models, vertical scaling is a vital consideration. The problem of categorizing data for mapping purposes is essentially one of scaling those data. Mapmakers decide whether to divide data into few classes or many, and whether to divide the data range evenly or to use some other means of clumping values into groups. Scaling the data dimension requires special attention to the range of values being depicted. Any categorization that does not divide the range evenly emphasizes some values at the expense of others. The Pike-Thelin map provides an example of scaling the data dimension. The authors converted their elevation data to logarithms, the effect being to create greater contrast in low-relief topography so that valleys in the Great Plains, for example, are more visible.

SCALE IN GEOGRAPHY

Scale is an important consideration in nearly all geographical studies. Careful selection of appropriate scales in space and in time is required to formulate and answer meaningful geographical questions. Because trends discernible at one scale are often invisible at another, geographers commonly use a range of scales to ensure that the conclusions drawn from a study are well matched to the economic, social, and physical processes known to underlie observed patterns. A map can effectively portray a geographer's thesis when an appropriate scale is chosen. A geographer's skills in constructing a map that shows only the relevant information resemble the crafting of a well-argued point in a debate, and when the map argument succeeds, a geographical thesis is substantiated.

REFERENCES

BORCHERT, JOHN R. 1950. The climate of the North American grassland. *Annals of the Association of American Geographers* 40:1–39.

GARREAU, JOEL. 1981. *The nine nations of North America*. Boston: Houghton Mifflin.

HAINING, ROBERT P. 1978. A spatial model for high plains agriculture. *Annals of the Association of American Geographers* 68:493–504.

KÜCHLER, A. WILLIAM. 1964. *Potential natural vegetation of the conterminous United States, 1:3,168,000*. Special Publication no. 36. New York: American Geographical Society.

PIKE, RICHARD J., and GAIL P. THELIN. 1989. Shaded relief map of U.S. topography from digital elevations. *Transactions, American Geophysical Union* 70:843,853.

PRESS, FRANK, and RAYMOND SIEVER. 1974. *Earth*. San Francisco: W. H. Freeman.

TOBLER, WALDO R. 1969. Geographical filters and their inverses. *Geographical Analysis* 1:234–253.

PART IV

WHY GEOGRAPHERS THINK THAT WAY

Paradigms for Inquiry?

John Pickles and Michael J. Watts

In recent years geographers have been drawn into the debates about social life in other social sciences, and they have rejected much of the disciplinary exceptionalism displayed earlier by areal differentiation, regional geography, and spatial analysis. For some this twentieth century fin de siècle seems to be exploding into a chaotic array of ambiguous and even contradictory trends, leaving a geography characterized by fragmentation and uncertainty, desperately in need of rational reconstruction and the imposition of new forms of unity. For others it reflects new, exciting concerns for multiplicity and multiple voices through which any attempts to recreate geography as a hegemonic project must be opposed. In this sense, contemporary human geography reflects what we might call a postparadigm condition in which disciplinary practice and concepts appear, for good and bad, to have broken loose from any notion of disciplinary closure and unitary coherence. More than at any time in the past, geography is comprised of competing, cross-cutting, flexible forms of knowledge production and assessment.

PARADIGMS AND MULTIPLICITY

The idea of *paradigms* in geography arises from Thomas Kuhn's *The Structure of Scientific Revolutions* (1962). Like Kuhn, geographers have used the term to mean exemplar, model, theoretical framework, political position, viewpoint, and system of protocols and rules. Common to these usages is the idea that systems of thought do not possess the stability they were once presumed to have. Not only do ideas change, but objects of scientific inquiry are transformed as they are situated within different frameworks of meaning and theory. Ideas are seen as historically and socially produced and contested.

Geography has always incorporated multiple perspectives: earth science, human science, landscape study, spatial systems, and environmental relations. A discipline that embraces oceans, mountains, agriculture, water problems, climate, atmospheric emissions, industrial location, capital accumulation, racial discrimination, graffiti, urban social movements, and national conflicts will encounter internal problems of definition, scale, purpose, and common ground. To many geographers Kuhn's ideas about paradigms seemed appropriate to the discipline's heterogeneity, and the ideas provided an organizing framework for geography's intellectual history from Darwinism and environmental determinism through possibilism, probabilism, areal differentiation, cultural ecology, landscape study, locational analysis, spatial science, systems analysis, humanism, Marxism, structuralism, social geography, and postmodernism.

PARADIGMS AND THE ABSENCE OF CRITIQUE

If the diffusion of the Kuhnian notion of paradigms had the beneficial consequences of challenging dogmatism and established orthodoxy, it did not provide any ontological or epistemological foundation from which to judge between competing paradigms or claims. Paradigms are incommensurable. Many geographers have therefore assumed that each geographical paradigm could operate as a closed analytical framework independent of theories and concepts used in other frameworks. Thus, although the notion of paradigm captures the essential character of all knowing (that is, it is perspectival and dependent upon context or the point of view), paradigmatic views of science suggest that ideas within a paradigm represent a partial totality, incommensurable from other paradigms—a world of self-contained and closed relativities. In this situation, choosing between one framework and another has as much to do with temperament, training, demographics, and power as it does with analytical power and rigorous argument. Furthermore, Derek Gregory contends,

> many of those who have—in my view, mistakenly—made use of Kuhn's notion of a paradigm have done so *prescriptively*. They have claimed the authority of 'positivism', 'structuralism', 'humanism' or whatever as a means of legislating for the proper conduct of geographical inquiry and of excluding work which lies beyond the competence of these various systems. (1989:69)

Two immediate dangers arise. First, the existence of a multiplicity of viewpoints becomes a justification for naïve relativism, in which concepts, methods, and theories are chosen pragmatically and tested only against others from within the same framework. The outcome is a fragmentation of research, a narrowing of perspectives, and a loss of rigor. Second, the resulting unwillingness to debate the merits of competing frameworks encourages reliance on values: assertion, training, and faith become sufficient conditions for selection. A new dogmatism is asserted in the name of science; and a narrow orthodoxy is established in the name of openness and multiplicity. As long as discussion of conceptual frameworks yields no more than a taxonomy of positions held to be equally valid ways of thinking about geography's worlds, confrontation with competing perspectives can be evaded.

POST-PARADIGMATIC GEOGRAPHY

Three emerging fields of inquiry (or analytical frames) exemplify postparadigmatic geographies: regional political economy, political ecology, and feminist geography. Each has emerged from a critical engagement with the knowledge-power and knowledge-interest nexus. Each addresses classical geographic themes: nature and society, town and country, space and time. But the way those themes are treated is distinctive: space is more than a container and nature is more than a context. These fields of inquiry focus on the social production of nature and space, and on the complex dialectical relations between them (Smith 1984; Soja 1989). In so doing, they are not developed separately from other social sciences, but in conjunction with them and informed by broader movements in social and literary theory (Thrift 1983; Gregory 1985).

While the three analytical frames do not constitute a unity, they do have several things in common that lend to each its postparadigmatic character. First, there is an openness to ideas within and across disciplines. Second, each analytic frame takes seriously the need to give an account of the nature of explanation and the role of criticism within its own development. Third, each analytic frame rests on an engagement with the causal role of structures and the need for an adequate theory of human agency. Each field explores the causal powers of persons conferred by virtue of their location within specific structures—what Callinicos (1988) calls "structured capacities." In sum, *contemporary human geography*, in the sense we have employed the term, strives to situate class, gender, culture, and social relations with respect to nature and space as necessary structures in the complex formation of particular geographies rather than as incidental "variables."

Can physical geography also be conceived of as in some way undergoing parallel changes in the face of the critique of paradigmatic thinking in the social sciences? There do indeed seem to be some parallels: physical geographers have been innovators in breaking down disciplinary boundaries and developing cross-disciplinary studies of Earth systems; geographers interested in nonlinear mathematics and chaos theory, for example, are beginning to challenge traditional forms of explanation; and physical geographers increasingly incorporate in their analyses the role of human agency in changing the environment.

Given these parallels, the absence of attempts to rethink physical geography in the context of postparadigmatic approaches is surprising. For the most part physical geographers engaged in cross-disciplinary research operate within a common paradigm model of "natural science." Contemporary human geography largely rejects this model as the model for all science and the mathematical manifold of nature as an appropriate metaphor for social life. In particular, it rejects the reduction of the sociospatial world to the parameters of a natural space, the adoption of biological metaphors for social process, and the unproblematized linking of scientific findings to prescriptive social agendas.

Political ecology and environmentalism, for example, argue forcefully that environmental problems cannot be treated as merely given but must be situated as symptoms of underlying social forces and struggles. Although contemporary physical geography recognizes the role of agency, the theoretical implications of this recognition have not been developed. Recent interest in nonlinear mathematics and chaos theory may reflect a rethinking of long-held concepts of nature and how we explain natural systems. However, since physical geography itself has been reticent to engage the new mathematics and new philosophies of contemporary science, this is little more than speculation. Insofar as they wish to speak about matters of environmental and social policy, however, physical geographers must begin to reject the partial closure of paradigmatic thinking and critically engage the arguments of postparadigmatic thinking that have so invigorated the contemporary human geographies to which we now turn.

REGIONAL POLITICAL ECONOMY AND INDUSTRIAL RESTRUCTURING IN SOUTH AFRICA

The development of contemporary human geography corresponds with a major restructuring of urban and regional systems in industrialized nations and with the rapid geographical extension of capitalism. Traditional economic geography failed to account for the growing internationalization of patterns of production and consumption, offshoring of American factories, deindustrialization of traditional centers of production, labor-market segmentation, new spatial divisions of labor, and rise of the north Asian newly industrializing countries (Peet 1989). Unable to deal with issues of social causality and the role of space in these complex social, political, and economic restructurings, orthodox industrial geographies neglected critical issues such as the relationship between structure and agency, the role of local struggles in the production of the built environment, and what David Harvey (1987b:374) called the total "penetration of capitalist social relations and of the commodity calculus into every niche and cranny of contemporary life."

Consequently, within political economy a new geography of restructuring arose that was informed by Marxist analysis. It deals explicitly with relationships between structure and agency, between theory and specificity, and among the roles of locality, class, race, and gender. "Industrial geography is now more definitely a branch of social theory directed at industrial forms of production where significant change poses crucial theoretical questions. In this new perspective, industry is not seen as the 'neutral' transformation of matter, but a nature-transforming activity structured by social relations" (Peet 1989:35).

The economic stagnation and stagflation that has typified Western economies, the expansion of capitalist modes of production into Third World areas, and the expansion of a commodity calculus to all areas since the 1960s have created new systems of production and trade, new forms of consumption, expansion of international markets, and increased investment in world financial markets. Industrial geography was traditionally dominated by a concern with mass production, but by the 1980s mass production was viewed as only one of several forms of industrial organization and scholars had shown that other forms, such as small-scale production by subcontractors, had been enduring correlates of mass production throughout the century (Storper and Christopherson 1987). The Third (northeast) Italy illustrated how the use of computers, flexible technologies, and new ways to link suppliers and sources, such as subcontracting and just-in-time scheduling, had created new institutional flexibilities that permitted new forms of locational flexibility that in turn tapped new labor pools and reorganized labor relations.

Rapid capital and information flows, the rise of multiplant companies, and the expansion of transnational corporations gave location renewed importance as a factor in production (Castells and Henderson 1987). States and municipalities were thrust into a global competition—place wars—where they vied with other municipalities and states to provide incentives for prospective employers. The combination of technical change and locational flexibility as factors in this contest among places to attract new industries led to changes in the "social organization of production and its consequences for regional differentiation" (Massey 1984:4). Major shifts ensued in national politics, social forms of capital, and gender relations, and geographers began to attend to the personal and social costs of deindustrialization and of other forms of economic restructuring.

Restructuring became most clearly associated with attempts to find cheap labor markets in areas peripheral to traditional centers of production and to renegotiate labor contracts and benefits in traditional industries and centers of production. That is, industrial restructuring at the point of production was intimately associated with the restructuring of social and institutional relationships between capital and labor. Such interpenetrations of industrial, regional, and social policy on the one hand, and the state and capital on the other, required that traditional categories of explanation be rethought. In the 1980s many geographers found some assistance for this rethinking in the work of the French regulation theorists, who periodized geographic change in terms of regimes of accumulation and corresponding modes of social regulation (Lipietz 1987).

In recent years geographers have also begun to give more attention to the impact of industrial restructuring and regional change in the Third World. In so doing they have written new geographies of social and regional change, in which issues of local economic development are increasingly theorized in the context of new patterns of globalization. Application of concepts derived from the literature of restructuring to the industrial and regional geography of South Africa, for example, has shown how processes operating elsewhere in the world are articulated in both traditional and reform versions of apartheid, and that South African regional systems of production have responded to these broader changes in the international economy. These geographies also recognize that class, race, and gender differentiation are integral elements of current restructuring and state policy.

Regional patterns of production, the nature of the labor process, and systems of industrial organization have been restructured in South Africa as the state has moved from territorial apartheid to new forms of regionalism.

> Existing spatial units have been identified as a source both of economic inefficiency and political stability. The purpose behind identifying new spatial forms would therefore be to disorganize oppositional groups that have challenged existing state boundaries and the systems of political representation predicated upon them, and to provide the spatial framework for renewal of capital accumulation (Cobbett et al. 1986:138).

But the South African state cannot manipulate capital accumulation merely by altering the regional structure of production to parallel administrative regions (McCarthy and Wellings 1988). State policies toward industry, regions, and apartheid laws are themselves influenced by changes in the economics and organization of production, the adoption of new forms of technology, systems of organizing production, and patterns of work.

Until recently the productive capacity of South Africa was based on cheap labor and continued investment of foreign capital. Both were guaranteed by apartheid's system of race privilege, labor reserves, and controls on the movement and power of black workers. Central to the continued provision of cheap labor and the maintenance of high rates of profitability were the black rural homelands where workers' families were confined and to which workers themselves were repatriated when their contracts or their "useful" working lives expired. By the 1970s, however, cheap migrant labor from the homelands was less effective in meeting the growing need for skilled labor in industrial, mining, and agricultural sectors, and by the 1980s rapid industrialization of the major metropolitan centers had produced sharply differentiated systems of production. Apartheid kept down costs in labor-intensive factories in lagging industrial regions such as East London, Port Elizabeth, and peripheral locations, but the growing industrial centers such as Durban and Johannesburg experienced increased difficulties in finding skilled workers and in maintaining productivity. Illegal and largely urban labor unions developed in the 1970s, and from this period to the mid-1980s, the positional power of black workers grew dramatically in some factories and in industries such as automobiles and chemicals.

Partly in response to growing black economic and political power in metropolitan areas, a small but increasing proportion of new factories located in South Africa's peripheral regions, where companies from metropolitan South Africa, Taiwan, Israel, and other newly industrializing countries can take advantage of generous state incentives and low wage rates.

Race, class, and gender relations have been reconstituted in this process of political and regional restructuring, and they now are being recombined in the emerging geographies of the "new South Africa." Companies seeking to overcome the economic crisis and to offset the new political power of the urban unions have adopted three strategies. One has been to develop new systems of production and correspondingly new forms of labor relations for urban workers: black unions have been permitted, wages and benefits have been improved, and greater levels of skill and productivity have been demanded. A second and related tactic is to develop flexible forms of production and organization to circumvent sanctions and disinvestment campaigns. Flexible arrangements also allow entrepreneurs to penetrate new labor markets, institute cost-efficient systems of production, and slow the decline in profit rates (Pickles 1991). A third strategy has been to restructure the geography of production itself by creating new social and spatial divisions of labor (Pickles 1988b).

The South African state plays a key role in effecting these strategies. From the mid-1970s, when pressure began to build for the business community and the state to accept black labor organizations, political leaders discussed a parallel strategy of geographical dispersal of industry that would simultaneously undermine any political and economic gains made by urban workers. The outcome was a spatially articulated industrial and labor policy. Capital attracted to the low-wage periphery serves the interests of the state and of sections of business by subverting organized, urban industrial labor. State incentives for industrial dispersal increased during the 1980s despite fiscal crises that reduced state expenditures for other programs. In South Africa's urban areas, new technologies and new labor relations are being forged in manufacturing and mining. In South Africa's peripheral areas and border towns such as East London, King Williamstown, Harrismith, and Ladysmith, the state provided factories at low cost, housing for managerial staff, transport rebates to offset the distance to major markets, preferences on government contracts of up to 10 percent, and subsidies of up to 95 percent of wage bill for all workers employed in the factory. Local development corporations staffed largely by Afrikaner technocrats provided additional assistance in the form of loans, marketing assistance, and bookkeeping.

State industrial policy and the changing economics of production have produced localized peripheral industrialization and substantial industrial complexes such as Isithebe and Phuthaditjhaba. But these industrial complexes are fragile: they continue to depend on state subsidies, they face severe international competition, and they consist of production stages that are unlikely to generate local linkages, spin-offs, or purchasing, and therefore they contribute little to the local economy.

This new geography of production links peripheral Fordist regimes of accumulation, a new labor-capital-state compact in urban areas, and new central-state incentives to relocate industry into the cheap labor reserves of the homelands. This spatially articulated industrial policy extends the range of production complexes in South Africa: increased levels of technical composition, skilled workers, and higher wages have developed in urban areas; in dispersed locations, the state has sanctioned cheap, labor-intensive production and exploitative and dangerous work conditions. South African industrial and regional restructuring also incorporates efforts to attract foreign capital. Western sanctions and disinvestment have forced South Africa to look to Taiwan, South Korea, and Hong Kong for economic and political support, thereby forging new international geographies and a reoriented geopolitical outlook (Pickles and Woods 1988, 1989).

POLITICAL ECOLOGY AND THE ENVIRONMENTAL CRISIS IN AMAZONIA

Two decades after the first Earth Day a worldwide resurgence of environmentalism has occurred. Despite the dismantling of U.S. environmental regulation since 1980, complex and wide-ranging green movements have appeared across North America, environmental parties have arisen in a dozen West European states, and popular ecology movements have been centrally involved in the struggle for democracy in Eastern Europe and the Soviet Union. Ecology and Third World development have been united in institutions as diverse as the World Bank and local grassroots organizations promoting sustainable development. The ecological enthusiasms of the 1990s are not a case of history repeating itself. They are responses to new and more destructive technologies and substances, to the devastating impact of decades of unregulated industrialization in newly industrialized countries such as Taiwan, South Korea, and Brazil, to state-sanctioned pollution in Eastern Europe and the Soviet Union, and to aggressive deregulation in the advanced capitalist states (O'Connor 1988).

Geographers have traditionally been concerned with human ecological questions. The 1955 Princeton Conference on Man's Role in Changing the Face of the Earth (Thomas 1956) was a watershed in the history of academic and popular studies of society-nature relations, bringing together Carl Sauer's geography, the intellectual histories of Clarence Glacken, and the incipient environmentalism that gave rise to the 1960s ecology movement. Three decades later, in 1987, Clark University hosted the Earth Transformed Conference, a self-conscious effort by geographers to update and reassess the global human environment (see Turner et. al 1990). Current geographic scholarship speaks powerfully to the contemporary crisis on the basis of a critical theory of the environment and is especially well equipped to assault Malthusian orthodoxies (Yapa 1979; Grossman 1984; Blaikie 1985; Hecht 1985, Richards 1985, Watts 1986; Blaikie and Brookfield 1987; Turner and Brush 1987; Wisner 1988; FitzSimmons 1989). Political ecology has emerged from a confluence of two broad intellectual traditions within contemporary human geography: cultural ecology and political economy. Cultural ecologists studying noncapitalist rural societies have delved into questions of adaptation and regulation in ecosystems of which humans are a part (Watts 1983; Butzer 1989). They focus on the role of culture and ritual in homeostatic regulation of the environment, relations that are often disrupted by external economic and cultural change (Nietschmann 1973; Butzer 1980; Denevan 1983).

Political economists build upon a long tradition of Marxist work to question the spatial unevenness of development on a world scale, the historically changing circuits of capital accumulation, and the character of dependent development in the periphery (Roxborough 1977). Dependency theory, modes of production analysis, unequal exchange, postimperialism theory, and so on—in short a huge and diverse literature (see Corbridge 1985 for a review)—ignited an explosion of peasant studies in the 1970s in which the central focus was differential control over and access to resources among household enterprises (i.e., peasants) inserted in various ways and at different times into an expanding world capitalism. While not explicitly concerned with the physical environment, political economy emphasizes the social relations of production—what

Eric Wolf in his influential *Europe and the People Without History* (1982:73) calls social labor—that provide the realms of possibility and constraint for managing environmental resources.

Political ecology has linked insights drawn from the study of underdevelopment and rural poverty (the "agrarian question") to the "environmental question." A critical theory of the environment is therefore the meeting ground of traditional geographic concerns. It weaves together the strengths of 1960s cultural ecology, such as sensitivity to patterns of indigenous knowledge and ethnobotany and focus on the resiliency and stability of ecosystems, and the powerful tools of Marxian political economy that examine structures of access and control. In this way geographers studying the environment have developed links with scholars from other social scientific disciplines who study the interface of environment and systems of exploitation, social differentiation, and class relations (see Redclift 1987; de Janvry and Garcia 1988; Martinez-Alier 1988).

The interdisciplinary nature of political ecology is personified by Piers Blaikie and Harold Brookfield (1987), who attempted to integrate physical and human with what they call Marxist and behavioral approaches to land degradation. In this perspective the center point of any nature-society study is the "land manager" in his/her historical, political, and economic context. Society and land-based resources are mutually causal: poverty can induce resource degradation that further deepens poverty. Local land managers balance spatial variations in their environments and links to "external structures" that include class relations, the state, and ultimately "almost every element in the world economy" (Blaikie and Brookfield 1987:68). Like B. L. Turner (1989), Blaikie and Brookfield share a concern with theorizing the environment at the intersection of the local and the global and in doing so focus on three key concepts. The first is the notion of political, economic, and ecological marginality, in which environmental degradation is the outcome of rational survival strategies by poor (i.e., marginal) households responding to changes in physical and political-economic contexts. The second is the idea of pressure of production on resources, where surplus extraction and exploitation among classes or individuals may impose excessive demands on the environment. And the third is the concept of *landesque* capital, where the investment in land beyond the life of the crop only takes place when other factors of production are present (see also Perrings 1987).

Political ecology has refuted the view that population pressure on resources causes poverty by affirming the need to understand the political economy of poverty as a major cause of ecological deterioration (Richards 1985; Bassett 1986; Zimmerer 1991). It also questions the compatibility of market or price systems with the uncertain and temporal nature of environmental systems, and hence the utility of much conventional resource economics. Political ecology demands a pluralistic approach rather than unicausal theories and analyses in which "a few strategic variables . . . relate together in a causal manner" (Blaikie and Brookfield 1987:48). Because one person's degradation is another's soil fertility, one must accept "plural perceptions, plural definitions, and plural rationalities" (Blaikie and Brookfield 1987:16). This quite extreme pluralism, which has some similarities with the move toward discourse analysis and postmodern geographies (see Watts 1990), does allow Blaikie and Brookfield to examine how discussions of ecological change necessarily involve contested facts and meanings (for example, the wildly different estimates of rates of erosion in Nepal), and to examine more critically the political content of ecological discourses (for example, conservationism).

The Brazilian Amazon, one of the most biotically diverse environments on earth, has always fascinated geographers. But the extraordinary rates of forest destruction over the past two decades have drawn a new generation of scholars to account for the horrifying speed, character, and consequences of the Brazilian rain-forest depletion. Susanna Hecht (1985) has shown how Malthusian ideas, the tragedy of the commons, and inappropriate technology arguments fail to explain the conversion of forest to pasture and the declining resilience of the livestock economy in eastern Amazonia since the military coup in 1964. She traces the Amazonian catastrophe to the opening of the Amazon to the 1960s military-business alliance that articulated a strong ideology of national security based on geographic integration. Enormous ranches were established on the basis of tax holidays and state subsidies, a program that produced land speculation and a scramble to prove land titles by clearing the forest. On the heels of the ranchers came an unregulated influx of land-poor peasants. The outcome was intense land conflict and local corruption coupled with further degradation of the forest at the hands of marginalized and impoverished peasants desperate for income. Environmental degradation in eastern Amazonia is a consequence of the role land plays in inflationary economies, the deliberate use of ranching to capture land and institutional rents from the state, and the role of the government subsidies in creating land markets and fueling excessive speculation. The productivity of the land was immaterial, and cautious land management was consequently irrelevant (Hecht 1985:680).

The state of Rondonia—another part of the Brazilian Amazon and roughly the size of western Germany—has experienced the most rapid deforestation ever seen in the humid tropics (Milliken, forthcoming). Following the completion of the Cuiaba–Porto Velho highway in 1964 Rondonia became a demographic frontier, absorbing a massive influx of peasants, sharecroppers, and rural workers dispossessed by land consolidation in the south. After 1981 the region received substantial support for a farming program designed to encourage sustainable smallholder production of perennial crops such as rubber, coffee, and cocoa. In practice, forest clearing increased, crop production lagged, and the predominant peasant land use has been cattle pasture, although only 40 percent of colonists owned cattle at the outset. Colonist attrition rates were also extremely high. Milliken shows that pasture production is situationally rational for smallholders; it establishes title to land, demands less capital and labor than perennial crops, is less subject to a deterioration in terms of trade (unlike cacao and rubber), and is readily sold in the event of calamity or personal crisis. Many Rondonia smallholders find themselves in a reproduction squeeze. They are vulnerable to downturns in prices, to which they respond with self-exploitation and by mining their Lilliputian resources base. They are frequently compelled to sell their land to local speculators, producing the high attrition rates for the Rondonia projects. Compelled by economic misery, they often resort to extractive gathering, mining, and rubber tapping in the remaining forests to survive, thereby drawing a subsidy from nature that may impose further burdens on the fragile forest ecosystems. The frontier is a zone of ecological deterioration, but it is also a zone of land conflict, economic marginalization and differentiation, and complex patterns of accumulation from which environmental degradation is inseparable.

The rapid demise of Brazilian forests has often been at the expense of indigenous peoples such as the Kayapo Indians in Para or the rubber tappers in Acre. Indeed, the expanded demands for foreign exchange revenue in the 1980s caused by Brazil's debt crisis intensified the pressures on Amazonian resources in the form of unregulated mining and forestry, leading to conflicts over access to resources. Whitesell (1987) describes the semi-indentured conditions under which most rubber tappers work, typically attached to a wealthy patron or *seringalista*, and how they have successfully organized to defend their rights, limit the incursions of ranching and timber interests, and promote alternative models of forest use (so-called extractive reserves). Whitesell's work is a model of political ecology, documenting the changing forms of access and control over resources and the new configurations of Green power, a sort of Liberation ecology. The rubber-tapper movement led by Chico Mendes before his assassination and the emerging coalitions and alliances between Amazonian environmental movements and political parties in preparation for the 1989 presidential elections illustrate well what Blaikie and Brookfield call the shifting dialectic of society and nature.

GENDER, FEMINIST THEORY, AND THE FAMILY FARM

Gender refers to "socially constructed and historically variable relationships, cultural meanings, and identities through which biological sex differences become socially significant" (Laslett and Brenner 1989:382). Feminist theory argues that gender, like class and race, is a basic building block of social organization. To this extent gender relations sustain a system of power capable of shaping the structural capacities of individuals. Gender, in other words, is not a structural given of biology but is a product of male and female actions that is institutionalized through families, schools, the workplace, and the state (Barrett 1980; Cockburn 1983; Beechey 1987; Scott 1987). Central to feminist debates are the ways in which a gendered division of labor specific to a culture has been constructed and maintained. How, in other words, are men's and women's work organized, how are cultural meanings attached to specific gender roles, and what is their relation to material conditions?

Feminism provides a strong critique of analytical frames that privilege class or the labor process because each is fundamentally gendered. Hart (1989) shows, for example, how masculinity is a critical but untheorized component of class and that the social processes underlying gender inequality have been a critical foundation for class formation in modern Britain. The link between gender and class as oppressive sets of relations forged a view of masculinity in the nineteenth and twentieth centuries as identified with the rights and duties of the breadwinner. Writing on the origin and changing character of such gendered divisions of labor, feminist research relies upon a specific reading of the concept of social reproduction that refers to the activities, attitudes and responsibilities attached to the maintenance of life on a daily and intergenerational basis. Unlike political economy, feminism sees the reproduction of life as a form of work as fundamental as other forms of work such as the production of things.

There is considerable debate over the analytical weight accorded to key feminist concepts, but patriarchy, gender regime, public/private realms, and the sexual division of labor are unquestionably central to its focus (Moore 1988). Ortner (1974) has made the claim that gender distinctions, and the subordinate status of women generally, are rooted in women's "closeness to nature." Women's physiology and reproductive functions appear closer to nature (while men are associated with the creative power of culture), and these same reproductive functions confine women to the domestic (private) realm. Ortner was thus able to link sexual ideologies and stereotypes with certain symbolic associations (public versus private, for example). These deeply embedded gender oppositions also represented patterns of structured inequality and subordination—in short, the culturally varied forms of patriarchy understood as "social systems and practices in which men dominate, oppress, and exploit women" (Walby 1989:214). Some theorists have argued that relations such as biological reproduction, heterosexuality, marriage, and the nuclear family are central to patriarchy (Foord and Gregson 1986). At a more abstract level, what Connell (1987) calls the "gender order," patriarchy intersects with other large structures (for example, capitalism), although this intersection is a subject of substantial debate. Hartmann (1979) and Delphy (1984), for example, argue that there is exploitation in housework (as in the wage relation), but it is by men over women. Patriarchy is also embodied in specific material practices such as work, sexuality, family, and the state, what Connell calls the "gender regime." Understanding the material, ideological, and symbolic character of these patriarchal structures is the critical starting point for an understanding of the sexual division of labor, and hence of any system of social reproduction.

Feminist geographers have employed gender perspectives in ways that present challenging interpretations of longstanding problems (Momsen and Townsend 1987). The rapid growth of women's participation in the labor force—in Britain, for example, from 29 percent in 1901 to 45 percent in 1988—has been associated with a substantial decline in male employment over the last two decades (Massey 1984). Women's employment, coupled with industrial restructuring, poses important questions for the rise of flexible labor markets, the growth of part-time employment among women, wage differentials and labor market segmentation, and the politics of the local state (Mackenzie 1988; Bowlby et al. 1989; Fincher 1989). Studies of the spatial, sectoral, and social character of women's participation in the work force has produced some especially important findings (McDowell 1983; Women and Gender 1984; Pratt and Hanson 1988). How women experience new work processes, urban decline, and social transformation has led to some preliminary explorations of language and gender, what Pred calls "gendered worlds" (1989). Women have been key actors in the rapid industrialization of the periphery since the 1960s, a transformation that has generated social tensions and struggles within families in which the wife emerges as the breadwinner (Christopherson 1983; Ong 1988). Questions pertaining to the sexual division of labor, household gender relations and wage work, property rights, and conjugality have wide-ranging implications for demographic and economic relations in all societies.

Since the seminal work of the Russian agrarian populist A. V. Chayanov (1986) at the turn of the century, theoretical and empirical work has refined our understanding of family farms in both advanced capitalist states and in the Third World (Shanin 1988). The distinctiveness of household enterprises resides in the centrality of the family, in domestic relations generally, in how labor is mobilized, and in the ways property is controlled. As a consequence, attempts to link the question of the survival of the family farm under capitalism with the complex internal relations of the family and the farm labor process have often used gender as a lens to bring these connections into sharper focus. This is not simply a question of examining the contribution of farm wives or of identifying the labor patterns of certain gender-specific divisions of labor. Rather it (1) expands the notion of farm labor beyond production narrowly defined to all activities, including the child care and subsistence that are necessary to household reproduction; (2) deconstructs the notion of the household as a unit of consumption and production, recognizing that the family is both socially and historically diverse and also the site of multiple and conflicting interests (Folbre 1986); and (3) situates the sexual division of labor within the family enterprise in terms of the power and ideological relations that are interpreted, legitimated, and contested in a variety of cultural forms (Moore 1988).

As Friedmann (1986) puts it, family farms are enterprises in which patriarchy and property are always at stake.

Judith Carney (1988) has explored the intersection of political economy and gender in her study of technological change and household structure among Mandinka peasants in the Gambia, West Africa. Mandinka society is characterized by multigenerational, patrilineal family structures in which there is a marked sexual division of labor by task and by crop. Since the middle of the nineteenth century, women have dominated the production of rice (a major staple), while men have controlled the principal cash crop (groundnuts) and secondarily contributed to other food requirements (millet and sorghum). Women are responsible for child care and most domestic functions (water collection, cooking, etc.) and generally gain access to land for individual use through the male head of household or occasionally by inheritance through the maternal line. As in many African societies, a fundamental building block for the domestic unit is the conjugal contract, the locally accepted terms on which goods, services, and property are exchanged between husband and wife. This is, however, part of a larger social architecture, a complex pattern of reciprocities between family members, which confers upon all junior members certain obligations to contribute to collective needs in return for individual access to resources, particularly land.

Carney's research documents the impact of a new technology—double-cropping rice on large-scale mechanized irrigation perimeters—on customary work routines and on the internal functioning of the household enterprise. The rice project appropriated (i.e., centralized land rights over) existing rice swamps, some of which were owned by women as individual property, and redistributed improved irrigated plots to the male heads of household. In sum, the rice scheme not only dispossessed certain women, who saw their economic autonomy eroded within the farming family, but also introduced new labor obligations, namely a dry season crop for which there were no customary precedents. The new rice work routines were arduous and labor intensive but also very profitable, potentially tripling local incomes. As Carney shows, two issues surfaced with great clarity, the first of which was the classification of the irrigated rice land as household property. If the irrigated plots were designated as "collective" (for domestic consumption) but under the jurisdiction of the patriarch, all juniors (i.e., wives and sons) were obliged to work on these fields even though the senior male controlled the disposition of the product. Second, in view of the expanded production regime, who would work, under what conditions, and for what return? Specifically, wives as skilled rice growers felt that some sort of material compensation was required for the claims on their labor power. In some cases this turned out to be a share of the crop, in other cases the allocation of land rights in peripheral swamps. In some cases, however, no resolution was negotiated, and women withdrew their labor and de facto abrogated the conjugal contract.

The state effort to increase peasant productivity animated struggles over work conditions within households that were fought out in a variety of idioms: conflicts over systems of classification of property (i.e., systems of meaning and forms of representation), over traditional labor obligations, and over conjugality. Because of a series of technological innovations, the sexual division of labor was transformed: men now cultivated rice, previously a women's crop, and they exercised patriarchal authority to capture the investible surpluses generated by the project. At the same time, irrigation technology introduced new social practices alongside customary domestic mores, in the course of which women came to understand their lifeworlds differently.

Despite great differences in society and technology, researchers interested in family farms in England have fruitfully explored similar questions about the ways in which labor is mobilized and resources controlled (Marsden, Munton, and Whatmore 1986; Whatmore, Munton, and Little 1987; Munton, Whatmore and Marsden 1988). They develop a relational typology of family farms on the basis of the extent to which property rights and farm labor process have been transformed into commodities and the extent to which family control of production has been usurped by external capitalist interests. *Family labor* farms and *family business* farms mark the ends of a spectrum reflecting a shift from family labor to family property as the organizing principal of the farm unit. The basic unit of analysis is the conjugal household, whose composition changes through time, and in which the farm labor process is examined in terms of various circuits of labor, specifically agricultural labor, domestic labor, and off-farm labor.

Amidst the growing differentiation of family farms in England, women's positions are structured by marriage and their conjugal status as wife, whereas men's locations are structured primarily by inheritance and their filial status (Whatmore 1988a, b). Like Carney, Whatmore is concerned with not only the sexual division of labor but also property rights and control over labor and income. In other contexts—the periurban green belt of London and the more rural southwest of England—field research has confirmed that the principal axis of the gender division of labor in family farms is between domestic and agricultural labor. Domestic household labor is almost exclusively women's work and is definitive of the notion of a farm wife. Nonetheless, farm wives also play critical administrative roles as bookkeepers, secretaries, and receptionists, and 70 percent of farm wives are also involved in manual labor on the farm. A higher proportion of women from family labor farms were involved in agricultural work, and on a regular rather than on a casual basis. Women typically worked on seasonal tasks and, more often than not, were confined to the immediate farmyard. They were least associated with the technology- and machinery-dominated sectors of production. In line with the findings of research on Third World peasantries, the nondomestic responsibilities of farm wives were in addition to household tasks, and differentiated gender roles were more closely associated with patriarchal power and the degree of commoditization of the farm than with stage in the life cycle (Whatmore 1988b:12).

Farm wives' interest in farmland was limited, and when women held a fiscal interest it was typically joint ownership. When women brought inherited property or capital to the conjugal farm it was typically associated with the family farm business, and it was usually a minority holding. Sharp distinctions exist between the family labor farms, in which wives have a legitimate working domain beyond the domestic realm, and the family business farms, in which the women's work domain is tightly domesticated. This division between "farm women" and "incorporated wives" was legitimated through ideologies of gender stereotyping and domestic consensus, such as agreement regarding the so-called natural responsibilities of farm women. Nonetheless, as we saw in Carney's study in Africa, these consensuses are fragile and subject to challenges that generate tensions that are accommodated and contested within what Whatmore calls a "known lifeworld" (1988b:22).

Though different in detail, these feminist geographies show that patriarchy is a relatively autonomous domain that intersects with local institutions of kinship, property, and conjugality. Both the Gambian and the English cases show that gender relationships are structures of power within which agents possess certain capacities. How these terrains of structure and agency are constituted in specific household enterprises in specific localities is central to the functioning and development of the agricultural sector.

THE POSTMODERN CONDITION IN CONTEMPORARY GEOGRAPHY

. . . what appears on one level as the latest fad, advertising pitch and hollow spectacle is part of a slowly emerging cultural transformation in Western societies, a change in sensibility for which the term post-modernism is actually, at least for now, wholly adequate. The nature and depth of that transformation are debatable, but transformation it is. I don't want to be misunderstood as claiming that there is a wholesale paradigm shift of the cultural, social, and economic orders; any such claim clearly would be overblown. But in an important sector of our culture there is a noticeable shift in sensibility, practices and discourse formations which distinguishes a post-modern set of assumptions, experiences and propositions from that of a preceding period. (Huyssens 1984:8).

Contemporary human geography faces a series of theoretical, methodological, and political issues that go to the heart of the current debate about philosophy. Geographers concerned with restructuring debate the homogenizing effects of capitalism versus the place-specific nature of struggles over resources. Within this broader debate, narrower issues arise such as the relationship between necessary and contingent relations, between structure and agency, and among different scales of analysis. Each raises questions about the objects geographers study (manufacturing plants, firms, production complexes), the significance of scale (how local, regional, national, and international scales interrelate), and the roles of causal structures (class, gender, race). The South African case illustrates the intricacies of race, class, and state interactions in restructuring the geography of economic activity and reform politics. The Amazonia example illustrates that discussion of ecological processes must be situated within the context of politics, ideology, and local resource management dilemmas. The discussion of agrarian systems illustrates the importance of integrating gendered geographies with studies of class and ethnicity.

Each case demonstrates that contemporary geography must address two difficult and sometimes conflicting needs: partial explanations must be situated within the context of broader theory, and geographers will enrich their understanding of societies and particular times and places by probing relationships such as class, agency, and patriarchy. To these considerations should be added questions regarding practical action: in the face of the scope and pace of the ongoing restructuring of the world's landscapes, what is the geographer to do (Pickles 1988a; Walker 1989)?

Many of these concerns have arisen in discussions of postmodernism. Postmodernists see themselves as conscious of the distinctiveness of the present as it undergoes rapid and radical restructurings. In this sense postmodernism is an overcoming of modernism or a separation from it. For Michael Dear (1986) postmodernism rejects the rationality of modernism with its master narratives of progress, development, humanity, and truth. That is, it is a rejection of the possibility of developing totalizing theory. Nigel Thrift (1983) defines such totalizing discourse as "the kind of holism which both sees society as an integrated system of 'universal' processes which function as a totality and as a system which (therefore?) can be apprehended from a total vantage point (such as the working class)" (404). Totalizing discourses impose unity where none exists, and postmodernists are skeptical of explanations that reduce causality in human geography to central principles such as economy, culture, and environment—or even space. In this view, societies themselves seem to be only partially integrated, and it seems particularly inappropriate for geographers to impose conceptual unity when (and where) its existence is suspect. As Michael Mann (1986) has pointed out, societies are much messier than our theories of them. Geography, like other disciplines, is becoming more pluralistic and less comfortable with calls for grand theory. In place of unified theory are varied frameworks of analysis that in one sense reflect the lesson of Kuhnian science that our scientific concepts do not mirror nature but are socially constructed. That our concepts and theories represent the way in which the world is constituted within our philosophical systems further demands from us that we do not conceive of those systems as closed incommensurable paradigms, but as open and competing systems of explanation.

Another way in which our worlds are constituted is through language. Knowledge is socially produced in the context of a plurality of language games. Since language is not able to get at reality, language itself is to be questioned. For some this implies that there is no prelinguistic reality: "language goes all the way down" (Rorty 1979), a claim that has led postmodernists such as Lyotard (1984:xxiv) to suggest the need for a pragmatics of language particles and language games in the context of a linguistic relativism. This is, in Dear's view, a turn to "relativism in human knowledge" (1988:271); in Rorty's terms it is a rejection of the "foundational epistemologies" that modernism accepts so easily.

Language itself becomes problematical. Representational notions of maps and words are discarded as foundationalist and essentialist, presupposing forms of objective reality beyond (yet accessible to) our conceptual categories. Such a rejection of representationalism raises questions about the tools of the geographer: the map and the written description (Harley 1989; Pickles 1992). It is in the work on language by Gunnar Olsson that the most systematic geographic investigation of the impacts of the loss of epistemological certainty is to be found. Olsson has been influenced successively by the early Wittgenstein, Saussure and structuralist linguistics, the surrealists, Derrida, and the later Wittgenstein. On this path through the analysis of language, Olsson has repeatedly returned to the contextual nature of language and meaning, and how such contextuality can itself imprison us in frequently invisible ways, the outcomes of which are often the uncritical reproduction of existing social and spatial relations (Olsson 1972, 1980).

Acknowledging the significance of context and local specificity in the formation of geographies raises the parallel question of how geographers are to deal with the fact that hegemonic or dominant discourses marginalize other forms of discourse and other voices. Postmodernism challenges the exclusionary practices within hegemonic discourse: it seeks on the one hand to place the marginalized back into the world, and on the other hand to challenge the authority of traditions. The challenge to the totalizing influence of the commodity calculus is, for Foucault, the reconstruction (archaeology) of hidden or marginal discourses. The challenge to the authoritarian voice comes from the deconstruction of the master narratives of texts and from deconstructive interference in which undisputed authority and demands for allegiance to the tradition are challenged.

Such claims raise many questions about the validity of generalizing claims that transcend or ignore spatiality and that fail to accord elevated status to local specificity. Allan Pred's (1986) work illustrates the specificity (and agency) of language use. The title of his book—*Place, Practice and Structure: Social and Spatial Transformation in Southern Sweden*—reflects the conjunction of issues with which the project deals. It is concerned with the formation and constitution of places and regions, of "successive variation, unfolding processes and historical sedimentation" in place (Pred 1986:195). This work is located at the intersection of two elements central to our discussion: the revitalization of regional geography in terms of the "constitution of regional social formations" (Gregory 1978; Pudup 1988) and the geographical anatomy of modes of production in which spatial structure becomes a medium through which social relations are produced and reproduced (Gregory and Urry 1985; Scott and Storper 1986). Pred's work makes explicit the larger poststructuralist critique in which spatiality is situated at the very heart of social theory, and it does so from an equally explicit structurationist perspective.

The question then becomes How do we write geography? The debate about the relationship between grand and local theory has given rise to vigorous argument about appropriate methodology and the role of structuralist analysis versus the study of local specificity (Harvey 1987a; Smith 1987a, b). The need to account for the role of structured agency in the formation of particular places has also stimulated two related trends in contemporary human geography: historical geography and the new regional geography. Related to each of these is the question of method and the importance of narrative (Sayer 1989).

As postmodernism has "spiralled its way out of architecture, aesthetics, and literary theory to confront the terrain of the human and social sciences as a whole" (Gregory 1987), geographers have begun to realize more clearly how their concepts, theoretical frameworks, methodologies, categories, and language arise out of a particular historical and spatial conjuncture—modernity. Behind this recognition is also acceptance of the need to rethink many of the approaches they use to deal with the world.

In this chapter we have tried to show how contemporary human geography, in taking seriously the postparadigmatic needs of openness (not closure) and criticism, has generated a series of scholarly endeavors that are linked to rich intellectual traditions inside and outside geography. These endeavors are fostering new critical approaches to the discipline that no longer merely challenge traditional theories, methods, and practices. Instead, they are restructuring traditional areas of geographical concern, such as space-society relations and society-nature relations, in ways that are generating new theories, transforming our ways of carrying out empirical research, and opening up new possibilities for geographical practice and pedagogy.

REFERENCES

Barrett, M. 1980. *Women's oppression today*. London: Verso.

Bassett, T. 1986. Fulani herd movements. *Geographical Review* 76:223–248.

BEAUREGARD, ROBERT A. 1988. In the absence of practice: The locality research debate. *Antipode* 20:52–59.

BEECHEY, V. 1987. *Unequal work*. London: Verso.

BLAIKIE, PIERS. 1985. *The political economy of soil erosion in developing countries*. London: Longman.

BLAIKIE, PIERS, and HAROLD C. BROOKFIELD. 1987. *Land degradation and society*. London: Methuen.

BOWLBY, SOPHIE, JANE LEWIS, LINDA MCDOWELL, and JO FOORD. 1989. The geography of gender. In *New models in geography*, ed. R. Peet and N. Thrift. New York: Allen & Unwin.

BUTZER, KARL. 1980. Civilization: Organisms or systems? *American Scientist* 68: 517–523.

———. 1989. Cultural ecology. In *Geography in America*, ed. Gary L. Gaile and Cort J. Willmott. Columbus, Ohio: Merrill.

CALLINICOS, A. 1988. *Making history*. Ithaca: Cornell University Press.

CARNEY, J. 1988. Conflicts over land and crops in an irrigated rice scheme. In *Women and land in agriculture*, ed. J. Davison, 59–78. Boulder, Colo.: Westview.

CASTELLS, M., and J. HENDERSON, eds. 1987. *Global restructuring and territorial development*. London: Sage.

CHAYANOV, A. 1986. *The theory of peasant economy*. Madison: University of Wisconsin Press.

CHRISTOPHERSON, SUSAN. 1983. Households and class formation. *Environment and Planning D: Society and Space* 1:322–338.

COBBETT, W., D. GLASER, D. HINDSON, and M. SWILLING. 1986. South Africa's regional political economy: A critical analysis of the reform strategy in the 1980s. *South Africa Review* 3:137–168. Reprinted in *Regional restructuring under apartheid: Urban and regional policies in contemporary South Africa*, ed. R. Tomlinson and M. Addleson. 1987. Johannesburg: Ravan Press.

COCHRANE, A. 1987. What a difference the place makes: The new structuralism of locality. *Antipode* 19:354–363.

COCKBURN, C. 1983. *Brothers*. London: Pluto.

CONNELL, B. 1987. *Gender and power*. London: Polity.

CORBRIDGE, S. 1985. *Capitalist world development*. Totowa, N.J.: Allanheld.

———. 1988. Deconstructing determinism: A reply to Michael Watts. *Antipode*. 20:239–259.

DEAR, MICHAEL. 1986. Postmodernism and planning. *Environment and Planning D: Society and Space* 4:367–384.

———. 1988. The postmodern challenge: Reconstructing human geography. *Transactions of the Institute of British Geographers*. n.s.13:262–274.

DE JANVRY, A., and R. GARCIA. 1988. Rural poverty and environmental degradation in Latin America. Typescript. University of California, Berkeley.

DELPHY, C. 1984. *Close to home*. London: Hutchinson.

DENEVAN, WILLIAM. 1983. Adaptation, variation and cultural geography. *Professional Geographer* 35:399–407.

FINCHER, R. 1989. Class and gender relations in the local market and the local state. In *The power of geography*, ed. Jennifer Wolch and Michael Dear, 93–117. London: Allen & and Unwin.

FITZSIMMONS, M. 1989. The matter of nature. *Antipode* 21:106–120.

FOLBRE, N. 1986. Hearts and spades. *World Development* 14:245–256.

FOORD, J., and N. GREGSON. 1986. Patriarchy: Towards a reconceptualisation. *Antipode* 18:186–211.

FRIEDMANN, H. 1986. Patriarchy and property. *Sociologia Ruralis* 26, 2:186–193.

GREGORY, DEREK J. 1978. *Ideology, Science and human geography*. London: Hutchinson.

———. 1985. Editorial: Thoughts on theory. *Environment and Planning D: Society and Space* 3:387–388.

———. 1987. Editorial: Postmodernism and the politics of social theory. *Environment and Planning D: Society and Space* 5:245–248.

———. 1989. Areal differentiation and post-modern human geography. In *Horizons in human geography*, ed. Derek J. Gregory and R. Walford, 67–96. Totowa NJ: Barnes and Noble.

GREGORY, DEREK J., and J. URRY, eds. 1985. *Social relations and spatial structures*. London: Macmillan.

GROSSMAN, LAWRENCE. 1984. *Peasants, subsistence ecology, and development in the highlands of Papua, New Guinea*. Princeton: Princeton University Press.

HARLEY, B. 1989. Deconstructing the map. *Cartographica* 26, 2:1–20.

HART, N. 1989. Gender and the rise and fall of class politics. *New Left Review* 175:19–47.

HARTMANN, H. 1979. Capitalism, patriarchy, and job segregation by sex. In *Capitalist patriarchy and the case for socialist feminism*, ed. Z. Eisenstein. New York: Monthly Review.

HARVEY, DAVID. 1987a. Flexible accumulation through urbanisation: Reflections on "post-modernism" in the American city. *Antipode* 19:260–286.

———. 1987b. Three myths in search of a reality in urban studies. *Environment and Planning D: Society and Space* 5:367–376.

HARVEY, DAVID, and ALLEN J. SCOTT. 1989. The practice of human geography: Theory and empirical specificity in the transition from Fordism to flexible accumulation. In *Re-modelling geography*, ed. W. MacMillan, 217–229. Oxford: Basil Blackwell.

HECHT, SUSANNA. 1985. Environment, development, and politics. *World Development* 13:663–684.

HUYSSENS, A. 1984. Mapping the post-modern. *New German Critique* 33:5–52.

KUHN, THOMAS S. 1962. *The structure of scientific revolutions*. Chicago: University of Chicago Press.

LASLETT, B., and J. BRENNER. 1989. Gender and social reproduction. *Annual Review of Sociology* 15:381–404.

LIPIETZ, A. 1987. *Miracles and mirages*. London: Verso.

LYOTARD, J. 1984. Answering the question: What is postmodernism? In *The postmodern condition: A report on knowledge*. Minneapolis: University of Minnesota Press.

MACKENZIE, S. 1988. Women's responses to economic restructuring. In *The politics of diversity*, ed. M. Barrett and R. Hamilton, 81–100. London: Verso.

MCCARTHY, J., and P. WELLINGS. 1988. The regional restructuring of politics in contemporary South Africa. Paper presented at the Association of Sociologists of South Africa Conference, University of Durban Westville, July.

MCDOWELL, L. 1983. Toward an understanding of the gender division of space. *Environment and Planning D: Society and Space* 1:59–72.

MANN, M. 1986. *A history of power from the beginning to AD 1760*. Vol. I in *The sources of social power*. Cambridge, Cambridge University Press.

MARSDEN, T., R. MUNTON, and S. WHATMORE. 1986. Towards a political economy of capitalist agriculture. *International Journal of Urban and Regional Research* 15:498–521.

MARTINEZ-ALIER, J. 1988. Poverty as a cause of environmental degradation. Typescript, World Bank, Washington, D.C.

MASSEY, DOREEN. 1984. *Spatial divisions of labour: Social structures and the geography of production*. London: Macmillan.

MASTERMAN, M. 1970. The nature of a paradigm. In *Criticism and the growth of knowledge*, ed I. Lakatos and A. Musgrave, 59–90. Cambridge: Cambridge University Press.

MILLIKEN, B. n.d. Tropical deforestation, land degradation, and society in Rondonia, Brazil. In *The social dynamics of deforestation in Latin America*, ed. Susanna Hecht and J. Nations. Ithaca: Cornell University Press. Forthcoming.

MOMSEN, J., and J. TOWNSEND, eds. 1987. *Geography of gender in the Third World*. London: Hutchinson.

MOORE, H. 1988. *Feminism and anthropology*. Cambridge: Polity.

MUNTON, R., S. WHATMORE, and T. MARSDEN. 1988. Reconsidering urban fringe agriculture. *Transactions of the Institute of British Geographers* 13:123–146.

NIETSCHMANN, BERNARD. 1973. *Between land and water*. New York: Academic.

O'CONNOR, J. 1988. Editorial. *Capitalism, Nature, Socialism* 1:1–5.

OLSSON, GUNNAR. 1972. Some notes on geography and social engineering. *Antipode* 4:1–22.

———. 1980. Hitting your head against the ceiling of language. In *Birds in egg/eggs in bird*, 4e–18e. London: Pion.

ONG, A. 1988. *Spirits of resistance and capitalist discipline*. Albany: State University of New York Press.

ORTNER, S. 1974. Is female to male as nature is to culture? In *Woman, culture, and society*, ed. M. Rosaldo and L. Lamphere, 76–88. Stanford, Calif.: Stanford University Press.

PEET, RICHARD. 1989. Conceptual problems in neo-Marxist industrial geography: A critique of themes from Scott and Storper's *Production, Work, Territory. Antipode* 21, 1:35–50.

PERRINGS, C. 1987. *Economy and environment.* Cambridge: Cambridge University Press.

PICKLES, JOHN. 1988a. Knowledge, theory and practice: The role of practical reason in geographical theory. In *A ground for common search*, ed. Reginald Golledge, Helen Couclelis, and Peter Gould. Goleta, CA: Santa Barbara Geographical Press.

———. 1988b. Recent changes in regional policy in South Africa. *Geography* 73:233–239.

———. 1992. The re-internationalization of apartheid: Information, Flexible Production, and Disinvestment. In *Collapsing space and time: Geographical aspects of communications and information*, ed. Stanley D. Brunn and Thomas R. Leinbach, 170–192. New York: Harper Collins.

PICKLES, JOHN, and J. WOODS. 1987. Undermining disinvestment: Recent changes in international investment in South Africa. Regional Research Paper 8718. Morgantown W. Va.: Regional Research Institute.

———. 1988. Reorientating South Africa's international links. *Capital and Class* 35:49–55.

———. 1989. Taiwanese investment in South Africa. *African Affairs*, Oct., 507–528.

PRATT, GERALDINE, and SUSAN HANSON. 1988. Gender, class and space. *Environment and Planning D: Society and Space* 6:15–36.

PRED, ALLEN. 1984. Place as historically contingent process. *Annals of the Association of American Geographers* 74:279–297.

———. 1986. *Place, practice and structure: Social and spatial transformation in southern Sweden, 1750–1850.* Cambridge: Polity.

———. 1989. In other wor(l)ds. *Antipode* 22, 2:33–52.

RAPPAPORT, R. 1967. *Pigs for the ancestors.* New Haven: Yale University Press.

REDCLIFT, M. 1987. *Sustainable development.* London: Methuen.

RICHARDS, P. 1985. *Indigenous agricultural revolution.* London: Hutchinson.

RORTY, R. 1979. *Philosophy and the mirror of nature: Fragmented and Integrated Observations in Gendered Spaces and Local Transformations.* Oxford: Basil Blackwell.

ROXBOROUGH, I. 1977. *Development and underdevelopment.* London: Routledge.

SAYER, ANDREW. 1989. The "New" regional geography and problems of narrative. *Environment and Planning D: Society and Space* 7:253–276.

SCOTT, ALLEN J. 1988. Flexible production systems and regional development: The rise of new industrial spaces in North America and Western Europe. *International Journal of Urban and Regional Research* 12:172–185.

SCOTT, ALLEN, and MICHAEL STORPER, eds. 1986. *Production, work, territory: The geographical anatomy of industrial capitalism*. London: Allen & Unwin.

SCOTT, J. 1987. On Language, gender and working class history. *International Labor and Working Class History* 31:1–36.

SHANIN, T., ed. 1988. *Peasants and peasant society*. Oxford: Basil Blackwell.

SMITH, N. 1984. *Uneven development*. Oxford: Basil Blackwell.

———. 1987a. Dangers of the empirical turn: the CURS initiative. *Antipode* 19:59–68.

———. 1987b. Rascal concepts, minimalizing discourse, and the politics of geography. *Environment and Planning D: Society and Space* 5:377–383.

SOJA, EDWARD W. 1989. *Postmodern geographies: The reassertion of space in critical social theory*. London: Verso.

STORPER, MICHAEL, and SUSAN CHRISTOPHERSON. 1987. Flexible specialization and regional industrial agglomeration: The case of the U.S. motion picture industry. *Annals of the Association of American Geographers* 77:104–117.

STORPER, MICHAEL, and RICHARD WALKER. 1989. *The capitalist imperative*. Oxford: Basil Blackwell.

THOMAS, WILLIAM, ed. 1957. *Man's role in changing the face of the earth*. Berkeley: University of California Press.

THRIFT, NIGEL J. 1983. Editorial: The politics of context. *Environment and Planning D: Society and Space* 1:371–376.

———. 1987. No perfect symmetry. *Society and Space* 5:400–407.

TURNER, B. L., II. 1989. The specialist-synthesis approach to the revival of geography: The case of cultural ecology. *Annals of the Association of American Geographers* 79:88–100.

TURNER, B. L., II, and S. BRUSH, eds. 1987. *Comparative farming techniques*. New York: Guilford.

TURNER, B. L., II., WILLIAM C. CLARK, ROBERT W. KATES, JOHN F. RICHARDS, JESSICA T. MATHEWS, and WILLIAM B. MEYER, eds. 1990. *The Earth as transformed by human action. Global and regional change in the biosphere over the past 300 years*. New York: Cambridge University Press.

WALBY, S. 1989. Theorising patriarchy. *Sociology* 23, 2:213–234.

WALKER, RICHARD. 1988. The geographical organization of production-systems. *Environment and Planning D: Society and Space* 6:377–408.

———. 1989. Geography from the Left. In *Geography in America*, ed. Gary L. Gaile and Cort J. Willmott, 619–650. Columbus, Ohio: Merrill.

WATTS, MICHAEL. 1983. On the poverty of theory: Natural hazards research in context. In *Interpretations of calamity*, ed. Kenneth Hewitt, 231–262. Boston: Allen & Unwin.

———. 1986. Drought, environment, and food security. In *Drought and hunger in Africa*, ed. M. Glantz, 171–212. New York: Cambridge University Press.

———. 1988. Struggles over land, struggles over meaning: Some thoughts on naming, peasant resistance, and the politics of place. In *A ground for common search*, ed. Reginald G. Golledge, Helen Couclelis, and Peter R. Gould, 31–50. Goleta, Calif.: Santa Barbara Geographical Press.

———. 1988. Deconstructing determinism: Marxisms, development theory and a comradely critique of capitalist world development. *Antipode* 20:142–168.

———. 1992. Sustainable development and struggles over nature. In *The political economy of sustainable development*, ed. F. Buttel and L. Thrupp. Ithaca: Cornell University Press.

WHATMORE, S. 1988a. Life-cycle or patriarchy? Paper delivered to the World Congress for Rural Sociology, Bologna, Italy.

———. 1988b. The other half of the family farm. Ph.D. diss., University of London.

WHATMORE, S., R. MUNTON, and J. LITTLE. 1987. Towards a typology of farm businesses in contemporary British agriculture. *Sociologia Ruralis* 1:21–37.

WHITESELL, E. 1987. Rubber extraction on the Jurua in Amazonas, Brazil. Master's thesis, University of California, Berkeley.

WISNER, BENJAMIN. 1988. *Power and need in Africa*. London: Earthscan.

WOLCH, JENNIFER R., and R. LAW. 1989. Social reproduction in a post-Fordist era. *Environment and Planning D: Society and Space* 7:249–252.

WOLF, ERIC. 1982. *Europe and the People without History*. Berkeley: University of California Press.

WOMEN AND GENDER STUDY GROUP. 1984. *Geography and gender*. London: Hutchinson.

YAPA, LAKSHMAN. 1979. Ecopolitical economy of the green revolution. *Professional Geographer* 31:371–376.

ZIMMERER, KARL. 1991. Wetland production and smallholder persistence: Agricultural change in a highland Peruvian region. *Annals of the Association of American Geographers* 81:443–463.

Humanism and Science

in Geography

Melvin G. Marcus

in collaboration with Judy M. Olson

and Ronald F. Abler

· · · ———————————— ◯ ———————————— · · ·

Geography is one of many disciplines experiencing the push-pull tensions of humanism and science, an age-old theme that has attained particular pertinence in the late twentieth-century world of exploding population and technology. For geography, it is probably a key issue for the 1990s.

The interconnections between humanism and science come in many forms. For example, almost any leap forward in science, in technology, or indeed in the health of nation-states is seen to be in conflict with the tenets of humanism. Yet, many pinnacles of humanistic thought and action have flourished during exactly such times. In the twentieth century, C. P. Snow's *Two Cultures and the Scientific Revolution* (1961) was one of the first and best expressions of the Western world's struggle to reconcile the diverging paths of scientists and humanists. Although Snow defined the latter rather narrowly as the literati and various academics, the implications of the two-culture syndrome he described extend across the larger spectrum of social and humanistic territory.

In geography the humanism-science tension has regularly reemerged through the counterpoints of changing paradigms. These were initially implicit in the works of venerated grandfathers of the discipline, from Alexander von Humboldt to Woeikoff and George Perkins Marsh to Paul Vidal de la Blache. Later, in America, humanistic threads emerged in the writings of Americans J. K. Wright, who explored the humanistic attributes of maps and geographical description, and Clarence Glacken, who traced the roots of humanism in the ebb and flow of civilization. Today, however, humanism is often cast in some postmodern, deconstructional struggle with traditional science, that is, positivism. But both humanism and science in geography go far beyond this transient (albeit important) intradisciplinary argument about how best to understand what are mostly urban-based political economies. In any case, there is confusion in the semantics of *humanism* and *science*, and it is useful to explore various expressions and definitions of these terms.

HUMANISM

Humanism has many faces, many meanings. Anne Buttimer, in her historical perspective, "Geography, Humanism, and Global Concern" (1990), has clarified much of the vocabulary of humanism. While "sketching the broad contours of the recent turn toward humanism in geography," she reminds us: "Humanism can scarcely be regarded as an autonomous field of knowledge inquiry. Rather it is a stance on life and world shared by people of diverse walks of life, including geographers" (1990:1). Six humanistic terms are identified here, with a debt to Buttimer for the larger portion of the list. They are: *humanitas*, *humanistic modes of knowing*, *humanities*, *humanitarianism*, *environmental humanism*, and *aesthetics*. All are connected by feedback loops.

Humanitas, that is, the nature of humanness, is seen as "diverse and recurrent themes about individuality and sociality, freedom and responsibility, rationality and hedonism, conservation and creativity" (Buttimer 1990:7). Geography has always been dependent on an understanding of societal imperatives to explain the nature of places and peoples. Postmodernists, especially deconstructionists, carry their analysis of humanness to the micro scale, turning to the attitudes and aspirations of individuals for understanding. In any event, human nature always plays a major role in geographical explanation.

Humanistic modes of knowing rely on the subjective—the inventive, intuitive, complex turns of the human mind. There is implicit (and sometimes explicit) rejection of the Cartesian world of objective reality. Geographers have often been caught up in the dichotomy of humanistic and objective ways of knowing. Because of their manners of research, this division has helped promulgate the separation between human geography and physical geography, the former more often and more overtly employing humanistic ways of knowing. There has been, especially in the writings of the postmodernists, an intensification of this tension that is expressed as one of theory and methodology but that places an onus on the practice of natural and physical science. Those committed to a geography of environmental interconnectedness, including the interplay of humans and the environment, appear in recent years to have been fighting a rear-guard battle to sustain traditions that trace back to von Humboldt, Karl Ritter, and some of the mid–twentieth century scholars of the regional and landscape schools (Marcus 1979; Bunske 1981). It is worth noting, however, that some postmodernists argue for the maintenance of appropriate positivist endeavor, especially as it relates to globalism and environmentalism (Harvey 1989).

The term *humanities* refers to "a range of knowledge fields whose central focus rested on the study of humanity, that is, history, literature, arts, rhetoric, and others" (Buttimer 1990:16). Geography has historically fluttered about the edges of the humanities. This positioning has usually depended upon the proclivities of individual geographers who move away from the mainstream of the earth science/social science channel. For example, occasional courses are taught under the so-called humanities distribution umbrella in American universities. Also, some geographers write from or toward a humanities perspective. Current exemplars are Yi-Fu Tuan and David Lowenthal (1975).

Humanitarianism refers to concern about the human condition and efforts to ameliorate and improve it. The numbers of geographers who are motivated in their work by humanitarian concerns cannot be counted, but their number must be appreciable across the spectrum of geographical activities and practices.

Any number of ideas have been posited under the rubric of *environmental humanism*. Its essences are the basic tenets of environmental unity; that is, that elements and processes of the earth environment are interrelated and interdependent, that change in one leads to change in all others, but with emphasis on the human factor in the global equation. A sequence of twentieth-century forays into environmental humanism would include the regional geography of the 1930s–1950s, the obsolete man-land tradition (replaced by people-environment), ecosystem studies, globalism, landscape studies born of the Sauer-Berkeley school, and the GAIA concept. The last (Lovelock 1979) enjoyed popular attention in the 1980s but is not, except for a heavy biological theme, all that different from concepts posited by centrist geographers for three centuries.

The *aesthetics* of humanism may not warrant separate classification insofar as aesthetics intersect all humanistic endeavor. They are, perhaps, most manifest in the realms of humanitas, humanistic modes of knowing, and the humanities. By any name, various aesthetic traditions cut across the fabric of geographical description and explanation. Bunske (1981), for example, has neatly interwoven the nature of Humboldt's pioneering works in the Andes on "visual translation of ideas" into the art of Frederick Church. Tuan (1989) has explored on a more philosophical level the aesthetic impulse implicit in observing and sensing the earth, while Eveanden (1981:150) has considered "the [unusual] decision to regard landscape beauty as a natural resource."

SCIENCE

The business of science is, at least superficially, less confusing. Science has come to its present state through a long and reasonably orderly evolution from the observations and classifications of Aristotle through the mathematical-astronomical science of Islam to Baconian scientific method. The pitfalls were numerous, especially those relating to religion, but the final product was a fair approximation of objective logic. Subsequently, scientific method was both refined and expanded by philosophers and students of science (for example, Dewey 1938; Northrop 1947; Kuhn 1970; Harvey 1973). The general methodology falls under the rubric of positivism or of the more rigorous logical positivism (Johnston 1986).

But in common currency, especially when seen as counterpoint to humanism, science is much more. It also is all the paraphernalia and products of science, technology, engineering, and medicine—computers, fast cars, pesticides, television sets, and erythromycin. To the lay person it is not so much the mode of reasoning that identifies science as it is the trappings: lasers, Star Wars, and the white lab coat.

The realities of science are certainly more complex than the public perceptions garnered from movies, Sunday supplements, and radio talk shows. On one hand, some scientists have approached their work with a blind dogma, suspicious of any innovation and unswerving in obedience to some sacred methodology. In physical geography, for example, intellectual progress in the first half of this century was seriously impeded by adherence to the Davisian landform model and the Koeppen climate classification. The fault lay not with the model creators, but with the inflexible disciples who spread these particular geographic gospels. While these examples are from the past, they remind us how painfully paradigm shifts occur—even in so-called logical science. On the other hand, not all logical positivists are inflexible, and much progress is generated by those with imagination and a willingness to dispute conventional scientific wisdom. Catastrophe theory and climatic teleconnections are recent examples of concepts that emerged quickly to impact physical geography.

ATTITUDES AND PRACTICE IN GEOGRAPHY

A complex, often convoluted, dispute is being pursued currently in geography. In simplified terms, it pits the social-scientific geographers of the logical positivist school against those loosely defined as postmodernists, but who have recent roots in any or all of phenomenology, neo-Marxism, structuralism, structuration, or deconstructionism. Some see this controversy as a return to humanism and the abdication of paradigm-constrained geography. Off to the side, but critically linked to the debate, are geographers who till the fields of religion, perception, aesthetics, and topics in the humanities such as poetry, literature, and art. The arguments for postmodern geography are legion and heterogeneous. Entrikin (1976) presented one of the best early assessments of humanistic geography approached from an existential, phenomenological perspective. A sampling of recent reviews and summaries of postmodern literature can be found in Ley (1980), Harvey (1989), Cloke, Philo, and Sadler (1991), and Relph (1991).

For a variety of reasons, the most salient being the need to apply scientific logic to much of their work, physical geographers have been absent from this controversy either as participants or as targets. But many of them have real stakes in the humanism-science issue; the physical geographer's raison d'être is often based on strong aesthetic and emotional drives, as often as not embedded in a strong environmental and humanistic ethic.

The science-humanism debate in contemporary American geography is not trivial. The outcomes will affect research pathways, the composition of academic departments, and the geographic employment pool for government and the private sector for a generation to come, just as the paradigm shift to logical positivism—which peaked during the employment binge of 1955–75—similarly overwhelmed the nature and practice of cultural and human geography. Given the forthcoming echo of the baby boom, which will coincide with the retirement of the last of the prepositivists and of the early generation of positivists, the disciplinary prize will be considerable.

GEOGRAPHIC INTERPRETATIONS OF THE HUMANITIES

It is both a strength and a weakness of geography that we see its manifestations in almost everything; spatial organization and interconnectedness are after all, universal. Thus, it is no surprise that some geographers have looked to art, literature, music, and film.

Perhaps the most common use of the humanities in geography is that of humanities-based materials in the classroom. The novel, for example, can illuminate a landscape or culture in ways alien to the textbook. Plot, attitude, character development, setting: all assume a special aura as seen through the author's eyes or through the eyes of the characters in the novel. It is one thing to study The Himalayan dilemma (Ives and Messerli 1989); it is another to visit Nepal by way of Peter Matthieson's *Snow Leopard* or Pico Iyer's *Video Night in Katmandu*. And how different physically similar landscapes appear in the dominant grays of Boris Pasternak's trans-Ural Russia, for example, and the dramatically colored spectacle of Jack London's Yukon.

Movies, too, illuminate geographical learning. How often are our visions of culture and place locked in the cinematic images of a *Fitzcarraldo*, a *To Kill a Mockingbird*, or an *Elvira Madigan*? As video tapes and disks proliferate, there will be increased use both at home and in the classroom of materials with both implicit and explicit geographic content. Concomitantly, more geographical truths and mistruths will be disseminated. Reality will be less certain as viewers increasingly struggle to filter it through the variety of cultural and perceptual barriers erected by filmmakers. Film cannot be expected to provide single, objective expressions of the geographies of place.

The humanities have also served as subjects for geographical studies. What the writer or painter conveys through art may reveal either geographical substance or abstraction. Interpretations of both explicit and implicit relationships between geography and the humanities increased in the 1980s and promise to be a favored topic of geographers through the 1990s. The *Geographical Review* has been the most prominent American journal publishing humanities-related scholarship.

Art has provided a fruitful topic for geographical analysis, especially as it involves nineteenth-century painters of landscape. In addition to the aforementioned connection between Alexander von Humboldt and Frederick Church, as well as the Western artist George Catlin (Bunske 1981), Ronald Rees (1976) has explored John Constable's style as a "naturalist painter" vis-à-vis his interest in empirical science. The works of John Constable, the painter of "everyday objects and activities of the English countryside," and of J.M.W. Turner, whose "abiding interest lay in the universals of nature, not the particulars," also have been interpreted by Rees (1982) as antipodal approaches to art and geographical description.

"Artist as Geographer: Richard Long's Earth Art" by William Romey (1987) provides yet another perspective: how one artist's use of landscape, field trips, and maps contributed to the emergence of the earth-art movement in the 1960s. Art is also often used to make a geographical point in textbooks. The patterned works of Escher are favored in geographical publications, and Muller and Oberlander (1984) open each chapter in their introductory physical geography textbook with an appropriate reproduction of works by either classical or modern masters.

The works of Yi-Fu Tuan (for example, 1974, 1989, 1990) have been modern guideposts for geography's concern with the humanities. His attention to moral and aesthetic values in understanding and using landscape has provided the platform from which geographic incursions into art and literature have been launched. A number of geographers have followed suit. Denis Cosgrove's "John Ruskin and the Geographical Imagination" (1979) interprets Ruskin's geographical sense, cutting across his works from art to architecture to prose. Wang (1990) casts the interplay of geography and ancient Chinese poetry in the context of Confucianism and Taoism, and Alexander (1986) illustrates

Dante's explanation of landform origins in the context of evolving fourteenth-century earth science. The setting of one of modern American literature's best-known romans à clef, Yoknapatawpha County, has received meritorious attention in two articles by Charles Aiken (1977, 1979); he scrutinizes the parallels between Yoknapatawpha County and Lafayette County, Mississippi, where author William Faulkner lived. The interweaving of imagination and reality provide intriguing insights into the Faulknerian South and the real South.

Douglas McManis (1978) turned to a more popular literature, the mystery-detective story, to observe how and why geography is used in that genre. Focusing on the stories of Agatha Christie and Dorothy Sayers, he illustrates how the authors used geography not only as settings for crime, but also, at times, as part of either the commission of the crime or its solution. As for the most famous of all detectives, the role of weather and climate in several of Sherlock Holmes's cases has been pinpointed by Randall Cerveny and Sandra Brazel (1989).

The role of maps in literature has also been explored. An excellent overview by Muehrcke and Muehrcke (1974) reveals a fascination for and use of maps across a remarkable spectrum of fictional styles and genres. The works of Antoine de Saint-Exupery, James Dickey, Cervantes, J.R.R. Tolkien, Robert Louis Stevenson, Joseph Heller, Mark Twain, and Robert Heinlein are only some of their examples. Cartography's role in humanism extends far beyond its place in the humanities, however; it also has a long-established familiarity with the tensions between humanism and science.

CARTOGRAPHY AS SCIENCE AND ART

The representational side of geography has long been caught in the dichotomy of humanism and science. Indeed, the traditional definition of cartography is "the art and science of mapmaking." Early mapmaking especially was as much art as science, but more recent developments of remote sensing and geographic information systems (GIS) are generally thought of as scientific. Representational subdisciplines within geography have nevertheless sustained a humanistic side. There are few other geographic specialties in which service to humanity (by way of the usefulness of the product) is so commonly invoked as the motivation for research and development. In theory, at least, a GIS system or topographic map will always have recognizable merit with respect to some human problem.

Addressing the humanistic side of historical cartography, Blakemore and Harley (1980) have shown how some historians of cartography evaluated what maps meant to the people of the period. Such studies bear a strong resemblance to those in the history of art. Although cartography has been generally influenced by humanism throughout its history, as surveying and representation became more precise and standardized, it has been more commonly associated with science and engineering. Historical study of the original meanings of maps

> corrects the tendency to measure early cartography against the more utilitarian objectives of much modern mapping. Many early maps were evocations of place as much as records of topography. . . . [The iconographic approach] encourages the interpretation of maps as a whole—as contemporaries would have viewed them—rather than as an arbitrary polarisation of decorative and scientific content (Blakemore and Harley 1980:81,85).

Such evocations are as applicable to modern day maps as to any in history: "The blue lines on the Ordnance Survey's Route Planning map will appear a celebration of the motor car before the oil ran out, just as the crowded city plan was perceived as a symbol of commercial wealth by the proud citizens of Venice in the fifteenth century" (Blakemore and Harley 1980:86).

In a full-scale treatment of a contemporary map, Wood and Fels analyze the state highway map of North Carolina in much the same fashion as historians of cartography have analyzed maps from earlier times:

> Map signs, and maps as signs, depend fundamentally on conventions, signify only in relation to other signs, and are never free of their cultural context or the motives of their makers. There is something about [the North Carolina map], something of veins and arteries seen through translucent skin, and if you stare at it long enough you can even convince yourself that blood is actually pulsing through them. (1986:54)

Wood and Fels interpret the various elements of the map as political tools and messengers of the myth of North Carolina. They suggest, for example, that the legend carries the message of the things for which North Carolina would like to be known and that its "scientific" function of relating symbol to referent is secondary at best.

The nonscientific face of maps is also manifest in the practice of collecting and displaying maps as a form of art. Contemporary as well as historical maps are favored as wall hangings and decorations. Also, the people involved in making maps are often drawn from artistic fields, especially those who are holistic practitioners of mapmaking, such as Richard Edes Harrison, Erwin Raisz, and Hal Shelton. Shelton, for example, began as an artist, later designed maps for the U.S. Geological Survey and commercial publishers, and finally, upon retirement, returned to landscape painting. His capstone work—a captivating, detailed rendition of the Grand Canyon for the Library of Congress—mixes the best of cartographic art and science.

The humanistic face of cartography is also reflected in sources of funding. The National Endowment for the Humanities, along with the National Science Foundation, is a major sponsor of the monumental History of Cartography project. Similarly, many of the state atlases published during the last two decades have drawn heavily upon funds from social and humanistic agencies.

If the art of cartography is significant, it is also true that a large share of modern cartography depends on science and technology. This scientific side is manifested across the full range of cartographic activity from the surveys upon which maps are based to methods of compilation, construction, and reproduction. Furthermore, science has been brought to bear in the study of perception and understanding of maps. For example, Eastman's (1985) investigation of the effects of graphic organization and memory organization on how people acquire information from maps, an experimental study of relationships between human user and graphic form, is but one of an array of such scientific efforts over recent decades.

Assessed against other fields of geography, it appears that cartography has an enviable record for mixing humanistic arts with the rigors of scientific practice. And even as advancing technology eliminates some human mapmakers, others emerge to reinforce the art that melds with scientific cartography.

PHYSICAL GEOGRAPHY AS HUMANISM

Human geographers typically perceive physical geographers as falling outside the realm of humanistic endeavor. As we have noted, the importance of scientific method in physical geographic scholarship distances physical geographers from the ongoing discussions of humanism. In truth, many physical geographers do occupy much or all of their research careers pursuing objective knowledge according to the tenets of rigorous scientific logic. The validity of their work in fact depends on scrupulous attention to what is real and measurable in the landscape, and they take pride in careful adherence to scientific methodology. Of course, not everyone believes this position is tenable. One imaginary post-modernist is purported to have said that "the 'scientific method' is a hoax. Scientists pretend to have an objective pipeline to reality, when in fact, every scientist has blinders and subjective filters that dictate not only the outcome of the experiment, but also the nature of the problem" (Bauer 1991:7). The above is conventional wisdom as expressed by most postmodernist human geographers who, having dismissed traditional scientific methodology, have thrown out the physical geography baby with that particular bathwater.

For all the loyalty to logical positivism on the part of physical geographers, the fact is that many practice various forms of humanism. There is no contradiction here; to practice humanism is not to be unscientific. C. P. Snow saw scientists as particularly aware of humanitarian needs:

> First, about the scientist's optimism. This is an accusation which has been made so often that it has become a platitude. It has been made by some of the acutest non-scientific minds of the day. But it depends upon a confusion between the individual experience and the social experience, between the individual condition of man and his social condition. Most of the scientists I have known well have felt—just as deeply as the non-scientists I have known well—that the individual condition of each of us is tragic.
>
> There is a moral trap which comes through the insight into man's loneliness: it tempts one to sit back . . . and let others go without a meal. As a group the scientists fall into that trap less than others. They are inclined to be impatient to see if something can be done: and inclined to think it can be done, until it is proved otherwise. That is their real optimism and it is an optimism the rest of us badly need. (1961:6–7)

These observations no doubt apply to many physical geographers. In fact, given the people-environment tradition in geography and the necessity of social science training in their degree programs, physical geographers may exhibit greater humanitarian interests than do some of their colleagues in the basic sciences. As for most people, however, scientists' attitudes are influenced by the milieu in which scientists operate. There is a conventional wisdom that pure (theoretical) scientists are intellectually removed from worldly concerns and conversely, that applied scientists and engineers ought to have greater empathy with the needs of society. Physical geographers should fall out accordingly along the intervening spectrum. There is, however, no clear evidence supporting those suppositions nor is it possible to divine each geographer's commitment to humanitarian good.

Motivations in physical geography are, just as in human geography, extraordinarily complex. It can be argued, for example, that modelers of global climate are motivated by self-interest and hidden agendas vis-à-vis future funding. This cynical appraisal is countered by the climatologists' concern for the future of society and environment; the implications of increased carbon dioxide and depleted ozone are no small matter. Similarly, biogeographers working in the Amazonian rainforest or North Slope tundra hold a stake in the future of fragile biomes that supersedes their individual research interests.

A recent rash of publications dealing with the highland-lowland ecosystem of the Himalayas is a case in point (for example, Byers 1987; Ives and Messerli 1989). There is currently a plethora of geographical field studies and planning proposals that range across geomorphology, hydrology and glaciology, biogeography, resource planning, agricultural geography, and population geography. Good research that is advancing the knowledge in those fields has been accomplished. But it would be a rare investigator, physical geographer or human geographer, who is not committed to relieving the human and environmental disasters that are overtaking the land, from overgrazed, overpopulated Himalayan uplands to storm- and flood-wracked deltas along the Indian Ocean.

Although physical geography's methodology follows the pathways of logical positivism, the routes to knowledge often remain tortuous and braided. Many nonscientists hold the mistaken belief that scientific research is always executed according to some fixed Baconian ritual. This is patently absurd. The research of physical geographers and other scientists does follow logical thinking, but humanist ways of knowing intrude upon the pattern. Intuition, inspiration, leaps of logic, and just plain guessing have helped complete more than one scientific jigsaw puzzle. Once in a while serendipity helps; however, it is the remarkable workings of the mind that still make for rigorous science. This was beautifully exemplified in James Watson's (1968) autobiographical narrative of events leading to the solution of the DNA molecule.

The reality, of course, is that this chapter is mistitled. The separation of physical geography from the rest of the discipline, while it might please certain narrow-minded physical geographers and a number of equally parochial postmodern human geographers, makes little sense. The physical geography-human geography dichotomy is a microcosm of the larger "two cultures" problem. Its manifestations are unfortunate insofar as two streams of intelligence have diverged from what is basically a spatially integrative discipline. Each intellectual stream is becoming unaware and unknowledgable of the other. As Snow said in addressing the problem at a larger scale: "This polarisation is sheer loss to us all. To us as people, and to our society. It is at the same time [a] practical and intellectual and creative loss. . . . The degree of incomprehension on both sides is the kind of joke which has gone sour" (1961:12).

GLOBALISM AND HUMANITARIANISM

Globalism, in an academic sense, has been around geography throughout its history. We have already seen that it is practiced under various conceptual umbrellas, such as environmental unity, ecosystem analysis, and human ecology. The themes of spatial organization, interconnectedness of processes and elements, and people-environment relationships have remained the basic stuff of geographical teaching and scholarship. Thus, a rather straightforward, perhaps naïve, globalism has been inherent in the discipline. However, significant shifts in perspective and understanding have occurred in recent decades: the one-Earth principle has attained wider support, and a humanitarianism theme has begun to permeate geographical work.

This theme, which involves complex, intertwining issues of environmentalism, social conscience, and political attitudes, plays differently for geographic specialties and, indeed, for individuals. The root causes are varied and not easily pinned down. In superficial definition they include: (1) the general "greening" of western society; (2) increased scholarly attention to society's disadvantaged, homeless, and outcasts; (3) greater grass-roots participation in humanitarian causes; (4) a philosophical shift in many political and academic circles to global thinking; and (5) newfound abilities, based in technology, to describe and study the earth at global scales. These changes are so far mainly indigenous to Western societies; they certainly carry a strong American flavor. The shifts also cross the spectrum of disciplines, with greatest visibility, not surprisingly, in the social and environmental sciences. Thus, while geographical thinking is significantly impacted, so too is the work of many other disciplines.

Again, it is important to note that the humanitarian concerns of geographers wear many faces, ranging from the quasi-political, postmodern scholarship of Michael Dear and Jennifer Wolch's *Landscapes of Despair: From Deinstitutionalization to Homelessness* (1987) to the latest U.S. National Academy of Science position paper on the rectitude of and social need for global climatic modeling. In any case, humanitarian motivations easily germinate global thinking. The irony is that most American geographers, while preaching and teaching global perspectives, have in the past thirty years largely abrogated extranational research and failed to sustain intercourse with geographers abroad except those in the English-speaking realm.

Globalism is really an attitude that geographers and others bring to their scholarship. Most of us, of necessity, operate at local or regional scales. Yet the extension of that work to global scales is implicit and, for most, brings with it humanitarian motives. Anne Buttimer sees globalism, particularly as expressed in the GAIA concept, as a challenge to geography and as a unifying theme for humanistic and other geographers: "The present global crisis, and the daring images projected by GAIA, may well be the catalysts for such an expansion of horizon. Neither the inherited divisions between intellectual and moral realms of discourse nor those academic fences between 'humanities,' 'divinities,' and 'natural science,' block creative imagination today" (1990:22).

GEOGRAPHERS IN THEIR ELEMENT

American-born geographers arrive at their disciplinary identity by innumerable paths. Many come from backgrounds in the social, biological, or physical sciences, and a few drift in from the humanities. Some come directly into geography, although that route is more common for foreign-born geographers working in the United States. But these are only academic hearths; most became geographers because of basic human drives (Buttimer's *humanitas*) that are integral to their lives and psyches.

Surely a majority of geographers, in their secret hearts, preserve that kernel of romance, adventure, and curiosity that led them to the discipline in the first place. Exploration and adventure are no longer fashionable in academic careers, but a few geographers have managed to partake of that suspect fruit while pursuing legitimate research objectives. These are the lucky ones, who have turned their aesthetic and emotional attachments—to exploration and travel, to mountain, desert, or rainforest landscapes, even to risk taking—into rewarding and needed research. Both human and physical geographers populate this realm. Their research may function in either humanistic or positivist modes, but they share the need to experience the world at first hand. Field work is integral to their research, and most would agree that their humanness is enriched by the field experiences that have become a necessity in their lives.

The proportion of geographers who regularly go into the field has steadily diminished in recent decades. This is partly because geographic field training has been gradually disappearing from American colleges and universities, partly because of increasing use of large data bases, and partly because so much can now be accomplished from a computer work station. Nevertheless, when the question is posed, "What makes a geographer?" the answer will reveal an aesthetic appreciation of earthscapes, fascination with cultural differences and similarities, commitment to environmental integrity, and above all, concern for human well-being.

Whether the geographer's focus is on urban places, village agriculture, or climate change, the emotional rewards accrue beyond specialization. Most still fantasize Richard Halliburton adventures, wonder at an Ansel Adams photograph, or revel in the spectacle of great urban places. Humanitarian concerns lie always beneath the surface, but the pure humanness of these experiences continues to bring joy to geographers' lives. If humanism and science in geography seem to have diverged into Snow's two cultures, the unifying themes of globalism and how people come to geography will remain to bridge the separation.

REFERENCES

AIKEN, CHARLES. 1977. Faulkner's Yoknapatawpha County: Geographical fact into fiction. *Geographical Review* 67:1–21.

———. 1979. Faulkner's Yoknapatawpha County: A place in the American South. *Geographical Review* 69:331–348.

ALEXANDER, DAVID. 1986. Dante and the form of the land. *Annals of the Association of American Geographers* 76:38–49.

BAUER, BERNARD. 1991. Disciplinary tectonics: The human-physical rift. *News: USC Geography Update* 5:6–8 (newsletter of the University of Southern California Department of Geography).

BLAKEMORE, M. J., and J. B. HARLEY. 1980. *Concepts in the history of cartography: A review and perspective.* Cartographica Monograph 26. Toronto: University of Toronto Press.

BUNSKE, EDMUNDS V. 1981. Humboldt and an aesthetic tradition in geography. *Geographical Review* 71:127–146.

BUTTIMER, ANNE. 1990. Geography, humanism, and global concern. *Annals of the Association of American Geographers* 80:1–33.

BYERS, A. 1987. Landscape change and man-accelerated soil loss: The case of the Sagarmatha (Mt. Everest) National Park, Nepal. *Mountain Research and Development* 7:209–216.

CERVENY, RANDALL S., and SANDRA W. BRAZEL. 1989. Sherlock Holmes and the weather. *Weatherwise* 42:80–81.

CLOKE, PAUL, CHRIS PHILO, and DAVID SADLER. 1991. *Approaching human geography: An introduction to contemporary theoretical debates.* New York: Guilford.

COSGROVE, DENIS E. 1979. John Ruskin and the geographical imagination. *Geographical Review* 69:43–62.

DEAR, MICHAEL J., and JENNIFER WOLCH. 1987. *Landscapes of despair: From deinstitutionalism to homelessness.* Cambridge: Polity.

DEWEY, JOHN. 1938. *Logic: The theory of inquiry.* New York: Henry Holt.

EASTMAN, J. RONALD. 1985. Graphic organization and memory structures for map learning. *Cartographica* 22, 1:1–20.

ENTRIKIN, J. NICHOLAS. 1976. Contemporary humanism in geography. *Annals of the Association of American Geographers* 66:615–631.

EVEANDEN, NEIL. 1981. The ambiguous landscape. *Geographical Review* 71:147–157.

HARVEY, DAVID. 1973. *Explanation in geography.* London: Edward Arnold.

———. 1989. *The condition of postmodernity: An enquiry into the origins of cultural change.* Oxford: Basil Blackwell.

IVES, JACK D., and BRUNO MESSERLI. 1989. *The Himalayan dilemma: Reconciling development and conservation.* London and New York: Routledge and United Nations University.

JOHNSTON, RONALD J. 1986. *Philosophy and human geography.* 2d ed. London: Edward Arnold.

KUHN, THOMAS S. 1970. *The structure of scientific revolutions* 2d ed. Chicago: University of Chicago Press.

LEY, DAVID. 1980. *Geography without man: A humanistic critique.* Research Paper 24. Oxford: University of Oxford, School of Geography.

LOVELOCK, JAMES. 1979. *GAIA: A new look at life on Earth.* Oxford: Oxford University Press.

LOWENTHAL, DAVID. 1975. Past time, present places: Landscape and meaning. *Geographical Review* 65:1–36.

MCMANIS, DOUGLAS R. 1978. Places for mysteries. *Geographical Review* 68:319–334.

MARCUS, MELVIN G. 1979. Coming full circle: Physical geography in the twentieth century. *Annals of the Association of American Geographers* 69:421–432.

MUEHRCKE, PHILIP C., and JULIANA O. MUEHRCKE. 1974. Maps in literature. *Geographical Review* 64:317–338.

MULLER, ROBERT, and THEODORE OBERLANDER. 1984. *Physical geography today.* New York: Random House.

NORTHROP, F. S. C. 1947. *Logic of the sciences and humanities.* New York: Macmillan.

REES, RONALD. 1976. John Constable and the art of geography. *Geographical Review* 66:59–72.

———. 1982. Constable, Turner, and views of nature in the nineteenth century. *Geographical Review* 72:253–269.

RELPH, EDWARD. 1991. Post-modern geography. *Canadian Geographer* 35:98–105.

ROMEY, WILLIAM D. 1987. Artist as geographer: Richard Long's earth art. *Professional Geographer* 39:450–456.

SNOW, C. P. 1961. *The two cultures and the scientific revolution.* New York: Cambridge University Press.

TUAN, YI-FU. 1974. *Topophilia: A study of environmental perception, attitudes, and values.* Englewood Cliffs, N.J.: Prentice-Hall.

———. 1989. Surface phenomena and aesthetic experience. *Annals of the Association of American Geographers* 79:233–241.

———. 1990. Realism and fantasy in art, history, and geography. *Annals of the Association of American Geographers* 80:435–446.

WATSON, JAMES D. 1968. *The double helix.* New York: Mentor.

WANG, XIAO-LUN. 1990. Geography and Chinese poetry. *Geographical Review* 80:43–55.

WOOD, DENIS, and JOHN FELS. 1986. Designs on signs: Myth and meaning in maps. *Cartographica* 23, 1:54–103.

Applications of Geographic

Concepts and Methods

Risa I. Palm and Anthony J. Brazel

··· ———————————————— ○ ———————————————— ···

WHAT IS APPLIED RESEARCH?

Applied research in any discipline is best understood in contrast with basic, or pure, research. In geography, basic research aims to develop new theory and methods that help explain the processes through which the spatial organization of physical or human environments evolves. In contrast, applied research uses existing geographic theory or techniques to understand and solve specific empirical problems. In this chapter, the term *applied research* will include those studies that apply geographic theory or methods to empirical or policy problems. Geographers employed in business, industry, and government make important contributions to applied research and are often involved in the implementation of their recommendations. But research need not be implemented for it to be considered applied (Frazier 1982:15). Nonimplementation does not detract from the relevance of the research, nor does it usually denote criticism of the scientific merit of the research (Gares 1989).

Geography originated as an applied discipline. Early geographers focused on estimating the size and shape of the earth, and they engaged in exploration on behalf of their royal sponsors. In contemporary geography, as in other basic sciences, the research agenda is set not only by individual scholars but also in response to current opportunities and directives. Basic research may be shielded from the direct interests of patrons, but applied research often serves the agendas of the private firms and government agencies that sponsor it. Basic research may also be influenced by patrons. The National Science Foundation in the United States, for example, has chosen to stimulate research on topics such as geographic information systems and global change by earmarking funds for those subjects.

WHY DO APPLIED RESEARCH IN GEOGRAPHY?

Three reasons for engaging in applied research are: to bridge the gap between researcher and subject or user by involving subjects in research design and implementation; to generate feedback that helps scholars refine techniques and theories; and to help solve pressing problems.

When understood as *praxis*, research that engages its subjects on a partnership basis, applied geography can be seen as an antidote to the alienation that results from positivist basic research that assumes detachment of researcher from subjects in the quest for objective truth. Considerable criticism has been leveled at classical social science approaches whose goal is to diagnose a problem by using survey questionnaires. Not only may surveys and interpretations be flawed, but the subjects who complete them often gain little or nothing for their participation in such projects. Whether research is cross-national or cross-cultural within the same nation, research techniques and projects should benefit the subjects as well as the scientists who profit by conducting the research. Projects that actively involve subjects may not be strictly and clinically replicable, but they offer several advantages.

> First, due to the active participation of subjects, there is a dialectic to the research that serves to unearth what may normally be hidden. Second, such research is a group process, not lonely reflection, and so reflects the actual processes through which individuals come to define and clarify their own view. Moreover, as research is most often tied to immediate practical action, through dialogue, participants are more able to see the relevance of the research and give it meaning in their own terms and situations (Hamnett et al. 1984:106)

Active participant involvement helps avert ethical problems and enables analysts to make recommendations with the confidence that subjects will support and understand changes that are suggested.

Applying geographical theories to real problems provides feedback for basic research. When empirical analysis shows how the effects of regional investment spread differ from those intended by development theorists, for example, the feedback helps clarify and correct theory (Pred 1977). Similarly, fatalities caused by dust blowing over highways gave rise to questions about the fundamental dynamics of airborne dust (Hyers and Marcus 1981; Nickling and Brazel 1984; Brazel and Brazel 1989). When geographers' recommendations regarding the locations of service facilities are implemented but fail to serve the target populations, geographers are forced to reexamine their recommendations and the theories upon which they are based.

Applications enrich the research base, but their primary purpose is to solve real problems. How much has geography contributed to resolving major economic, environmental, and social problems? Some geographers contend that the discipline has failed to develop methods and theory that will improve living conditions for the impoverished and deprived of the world (Zelinsky 1975); that applied geography often serves largely to impede needed change (Harvey 1974), or that problems it helps alleviate at one place reemerge "in another form elsewhere, thereby delaying the search for a cure and diverting needed intellectual resources into activities which, in the long-term global context, will be unproductive" (Johnston 1986a:19). For the most part, such criticisms have been directed toward basic research or to applications in which the notion of praxis (participant involvement) was absent. Applied geography indisputably includes work with narrow applicability to the companies or industries that underwrite it, thereby serving to reinforce the existing order. But geographers have also engaged in projects that involve subject populations and that produce recommendations for basic changes in the delivery of services or the organization of social and environmental systems.

GEOGRAPHY AND PUBLIC POLICY

Applied research addresses policies as well as problems, and when geographers focus on public policy, their work may be converted into legislation or recommendations for implementation if desired outcomes are known but the ways to obtain them are not. Translating research into policies and programs is complex and fraught with unanswered questions.

> Which research results are to be applied? When there are competing or conflicting research findings, as there almost assuredly will be, who will determine which are valid? Who, if anyone, will be responsible for prioritizing the problems which these findings address? By whom are they to be applied? To what purposes are they to be addressed . . . ? Who will benefit and who will be disadvantaged, or should these questions of vested interest be of concern when "public interest" is being served by the application of research results? To what extent can knowledge actually affect either practice or policy? (Nigg 1988:13)

Nigg's research on how to translate known methods of reducing seismic hazards in designing new buildings and retrofitting existing structures into legislation and reality led her to conclude that there is no simple way to accomplish such translations. Research findings are usually not in complete agreement on any complex issue. Furthermore, some people will profit and others will lose from any attempts to implement policies. Even the existence of rational solutions whose outcomes are highly probable or even certain is no guarantee they will be adopted and implemented, owing to the opposition they are certain to arouse.

Necessary conditions for translating basic research into public policy or private industry programs are: (1) the knowledge in question must be clear and understandable to users; (2) the knowledge must fit users' specific needs; (3) users should receive some obvious advantage from implementation; (4) users should be aware of advantages and disadvantages of implementation; (5) implementation must be feasible and practical; (6) the information source must be credible to users; and (7) implementation should be socially reinforced. The extent to which any or all of these desiderata are present will determine whether research insights can become policies and action programs (Palm 1983a; Nigg 1988). Additional characteristics of effectual applied research are that it be unbiased or that its biases be clearly stated, that the information source be available when answers are needed, and that the research contain specific proposals that can be implemented (Wilbanks 1982). Analysis of the differences between applications that succeed and those that meet with limited success or that fail demonstrates that complete agreement between the commissioning agency and the research team on the nature and methodology of the project and flexibility in research design and strategy are additional qualities that raise the probability that policy research will succeed (Glantz, Robinson, and Kranz 1985).

APPLICATIONS OF HUMAN GEOGRAPHIC RESEARCH

Time Geography and the Location of Public Facilities

Improving access to public facilities for populations that depend upon them requires that planners identify populations that need such facilities, site those facilities, provide transportation that will enable target groups to reach them, and ensure that they are open when clients need them. The elderly, for example, need access to health care and to shops and social centers (Golant 1976, 1984; Rowles 1978, 1986). Travel raises both symbolic and real problems for the elderly. Poor public transportation combined with the elderly's reduced strength and fears of leaving a familiar neighborhood to reach distant facilities in strange locations may place needed services beyond the physical and psychological range of oldsters. The elderly may be reluctant to move to locations near needed facilities because of the monetary and psychological costs of finding a new residence in a new area, and many elderly people with low incomes have few choices regarding housing under any circumstances. Geographers and planners are challenged to find areas within cities where concentrations of elderly individuals and needs for public facilities require intervention, and to site facilities and provide access to them in ways that assure adequate levels of service at minimal cost.

Another population with access problems amenable to geographic analysis is working parents with small children. Because women usually take primary responsibility for child care, several research projects have examined the ways time allocation affects women's access to employment, facilities, and opportunities in relation to their child care responsibilities and other household obligations (Tivers 1976; Palm and Pred 1978; Palm 1981a). Although such research may focus on access to resources and on roles within the household, it raises planning issues such as better access to day-care facilities and socioeconomic issues such as more flexibility in work hours without penalties for career advancement.

The perspective of time geography (see chapter 10) has provided especially helpful insights into the relationships between facilities and populations that need them. When a program of activities—the set of origins and destinations an individual needs to visit—is analyzed in space and time simultaneously, the constraints imposed by opening hours, the clustering or dispersal of facilities in space, and transportation alternatives are greatly clarified. Urban areas, neighborhoods, and regions can be compared with respect to the quality of daily life that a particular time geographic regime permits (Martensson 1978). Using this tool often reveals that seemingly minor changes in the temporal and spatial ordering of the environment can yield large improvements in access and quality of life (Carlstein 1978; Hägerstrand 1978; Lenntorp 1978).

Applied Geography and the Law

Geographers are increasingly called upon to provide expert testimony in courts or before public commissions on issues such as racial segregation and the implications of residential segregation for such policies as school busing. William A. V. Clark, for example, provided expert testimony to the U.S. Commission on Civil Rights on residential segregation in American cities (Clark 1986). He reviewed two measures of residential segregation (a dissimilarity index and the exposure index), noting changes in their values over time for 38 large metropolitan areas. He then assessed the contribution of discrimination to changes in the measures as opposed to the effects of affordable housing and social preferences, the availability of information about vacancies, and the influence of life-cycle changes on mobility. He concluded that discrimination is not the sole cause of racial separation, and that some causes "lie beyond the direct control of government or government-authorized bodies" (Clark 1986:122).

Morrill (1989) reached a similar conclusion in the study of the impacts of school busing he prepared for the Seattle School Board. Morrill prepared a time geographic plan that offered controlled choice—the movement of students who wished to study certain subjects and would attend different schools part of the time. Morrill concluded that mandatory busing was a less effective solution to school segregation than a plan that would make the integrated school an attractive voluntary alternative. "The costs of coercion are so high, so counterproductive in the long run, and the potential benefits of voluntary plans so promising that serious exploration of alternative techniques of integration deserves to be made" (Morrill 1989:353). Ultimately, the Seattle School Board adopted a version of Morrill's plan.

Examining issues such as residential segregation in a broader context can provide helpful insights into larger social processes, many of which bear upon geographic planning and practice. Recent decisions in which the U.S. Supreme Court has refused to interfere with the autonomy of suburban areas, thereby preserving de facto racial segregation in housing and schools, reflect the latent political balance and ideology of American society.

> Power, then, lies in the suburbs. It is bolstered by the decisions of the Supreme Court over the malapportionment cases, which have ensured that suburban representation is approximately equal to suburban population. From this, it is difficult to perceive any successful major challenge to present suburban practices. The legislatures will not act against suburban interests. Nor, it seems, will the Courts. (Johnston 1984:172)

Social organizations such as the courts act to preserve the power of the already powerful because justices and judges themselves operate within the constraints of societal values.

> Overall it is clear that the decisions taken have favored the status quo . . . the ideology into which they [the Supreme Court Justices and judges of other courts] have been socialized, and which they demonstrated to their patrons/constituents, strongly favors the status quo. The justices have been put there, and the texts that they have to interpret have been written, to provide stability, to mediate conflicts in such a way that the strength of the system is enhanced. (Johnston 1984:184)

Land-use change and land development processes give rise to many policy and practical questions amenable to geographic analysis such as legislative attempts to control growth, to reduce property tax variability, or to place restrictions on development that affect the distribution of certain activities. The literature on the impacts of taxation and land-use policies and regulations reports both successes and failures. Attempts to use taxation to improve mass transit in Los Angeles, for example, have largely failed (Adler 1986). Local and state legislation aroused the opposition of business and property groups outside the central business district. When the Los Angeles Metropolitan Transit Authority was established in 1951, it had limited power and financing authority. When the state established the Southern California Rapid Transit District (SCRTD) in 1964, the Los Angeles central business district was allocated only two representatives on an eleven-member board, and the district was granted neither sales nor property-taxing power to subsidize its operations. SCRTD's most recent successor, the Los Angeles County Transportation Commission, was created by the state legislature in 1976. It was granted the power to impose a sales tax in order to subsidize service, but that power has been eroded by a regional tax-sharing plan and by attempts by the City of Los Angeles to take over the commission's express bus operations. Downtown and suburban commercial interests are engaged in an intense jurisdictional struggle, complicated by side issues of serving transit dependent populations, enticing federal subsidies, and protecting the environment—all factors that have confounded the translation of legislation into effective practice.

Homelessness and Housing the Mentally Disabled

One of the most serious problems facing U.S. urban society is the homeless, many of whom are also mentally disabled. When the mentally disabled are deinstitutionalized and community mental health care is inadequate, some people are incarcerated "for crimes more indicative of their mental health disabilities than criminal intent" (Dear and Wolch 1987:174). Others become homeless, lacking access to even the lowliest single room in a downtown hotel. The addition of the mentally disabled to the numbers of those homeless for other reasons produces both an increase in the number of homeless individuals and a change in the age and gender composition of the group. Economic hardship and deinstitutionalization have created a situation in which "a single adverse event is sufficient to tip this marginalized population into homelessness" (Dear and Wolch 1987:199). Dear and Wolch suggest a process for assigning service-dependent individuals to particular communities within the metropolitan area, and for granting regulatory roles to regional governmental bodies (to allocate a fair share of the region's burden to each jurisdiction) and to local service agencies (to provide a place where community members can participate in decisions to provide services)—practical measures that would ameliorate a growing urban problem in America's cities.

Political Redistricting

Political districts should ideally contain roughly equivalent numbers of people, and they should not systematically disenfranchise racial or ethnic monitories.

Gerrymandering—drawing district boundaries to serve the interests of a political or ethnic group—is a spatial method of political control. Gerrymanders may ensure the reelection of a particular party or split an interest group to reduce its influence in a legislative body. Applied political geography usually suggests boundaries that would avoid gerrymandering, although like any kind of knowledge, it can also be used to create or strengthen gerrymanders. Geographic models and techniques of redistricting specify criteria for representation by population size, racial composition, and political composition that meet the tests of compactness, respect for previous boundaries, and conservative impacts on the previous system (Morrill 1981). Morrill (1973) developed a redistricting plan based on fundamental geographic principles that was adopted by the state of Washington when legislative districts were redrawn. Even if ideal plans are not adopted in their entirety, geographers can make great contributions by providing rational first approximations for redistricting that serve a base line from which deviations can be measured.

Applied Geography and Revolution

Another action-oriented approach to understanding the spatial organization of cities focuses on urban neighborhoods. Here, applied geography involves residents as active participants, and the goal of the effort goes beyond documentation to the promotion of change in the study area. William Bunge involved himself deeply in a Detroit neighborhood in order to understand the lives of its residents and to change its social order. Writing of himself in the third person, he "views himself as a fighter against the 'ins,' and for the 'outs,' a fighter who hopes for eventual personal and social peace. Yet Bunge is not out to eradicate the 'rotten rich,' only the institutions they support. Slum lordism, an enterprise over which even the rich balk, is one example" (Bunge 1971:136). Bunge's call to action is evident in his book's subtitle, *Geography of a Revolution*, as well as in its content.

> The book has had a human power of its own, and in a sense, it created the author, a geographer who has not only been writing about Fitzgerald, but living there, and is thus intricately involved with it. . . . The theme of the book was fighting slum processes, and a key lesson was the tremendous contribution of black people to the fight. . . . Fitzgerald started a revolution. The revolution is a revolution of the exploited against a ruling class, of black people against racism, and a revolution of all the people against a technology of death. (Bunge 1971:preface)

Bunge's attempt to foment revolution while producing urban regional geography is an extreme example of nonpositivist praxis that has not been equaled since in American geography.

Business and Industry

At the other end of the political spectrum is the long tradition of applying geographic expertise to problems that arise in business and industry. In contrast to the research reviewed above, this type of work is less likely to involve basic research or the development of new theory within geography. Furthermore, a great deal of this work is never published or is subject to proprietary claims (Epstein and Schell 1982). Commercial, industrial, and marketing geography have helped shape the nation's commercial and industrial landscapes, and marketing geography is an attractive option for many undergraduate majors.

Managers or entrepreneurs often ask geographers to evaluate potential sites for corporate expansion or to examine a given site to determine its most effective use. Clients requesting sites are usually retail and service firms seeking new opportunities who need places, whereas developers and real estate firms have places and want to know what to do with them. Both tasks involve research on population and housing characteristics, the location and nature of competition, site accessibility, and environmental characteristics, including both the physical and legal or regulatory environments. When geographers move from analysis to action in this realm, they may provide support for rezoning requests or for other modifications to the urban infrastructure and may be called to testify for or against proposed developments. In such instances "the marketing geographer serves as an analyst, an advisor, and an active agent for change" (Epstein and Schell 1982:275).

Patterns of Human Disease

Geographers address both the spatial and environmental aspects of human disease in order to understand its etiology and the likely courses of its spread and to recommend ways to intervene (see chapter 7). Lewis and Mayer (1988) have argued that geographers should view disease as a natural hazard. From this perspective, the threat of disease or the actual spread of disease could be analyzed in the same ways that preparedness for responses to extreme geophysical events are studied, that is, by analyzing how individual and structural responses either mitigate or exacerbate disease impacts. Other geographers have used cartographic techniques and analysis of spatial correlates to shed light on the environmental causes of disease as well on how infections move through space. The environmental factors associated with California encephalitis, for example, suggest that continued use of insecticides to control mosquitoes and thereby limit the spread of the disease is warranted (Pyle and Cook 1978). AIDS has been shown to expand geographically via a combination of hierarchical and contagious diffusion processes (Gould, Gorr, and Casetti 1988; Gould 1989), although reliance on spatial modeling and mapping that assumes passive receptivity of populations to AIDS must be supplemented by consideration of the complex relationships among behavior, host, disease agent, and environment peculiar to the disease (Shannon and Pyle 1989).

SOCIETY AND ENVIRONMENT

Environmental Law

Legislators often attempt to regulate the relationships between society and the environment to prevent misuse of the environment and to mitigate the worst effects of environmental variability. Analysis suggests that legislation that reinforces existing processes is most effective; legislation that runs counter to the existing political economy and the interests of major economic stakeholders is likely to be ineffective. The Coastal Barrier Resources Act of 1981 (16 U.S.C. Secs. 3501–10) imposed restrictions on the development of land susceptible to coastal flooding by identifying undeveloped coastal barriers where federal flood insurance and federal development subsidies would not be available for new construction or substantial improvements to existing dwellings. The proscribed areas include such popular resorts as Hilton Head, South Carolina, Marco Island and Amelia Island, Florida, Atlantic City, New Jersey, Galveston, Texas, and Scituate, Massachusetts. These areas are particularly susceptible to coastal flooding by hurricanes and are therefore especially dangerous for permanent or even seasonal occupancy. Although the legislation was intended to halt or at least slow construction in the designated areas, it increased rather than reduced exposure.

> Exposure of lives and investments to risk from natural hazards is increasing on most developed coastal barriers due to infilling of vacant land, intensification of already developed land, and rising property values of existing development. The federal government is continuing to contribute indirectly to this situation by providing the various benefits in developed areas that are now prohibited on undeveloped barriers. Furthermore, denial of such benefits in the CBRS [Coastal Barrier Resources System—the shoreline identified by the act as undeveloped] actually reinforces pressures for further development on developed coastal barriers. (Platt 1987:13)

Legislation that was intended to reduce the exposure of human life and property to hurricanes had the opposite effect.

The Alquist-Priolo Special Studies Zones Act enacted by the California Legislature in 1972 and modified in 1975 requires that surface fault rupture zones be identified and mapped, and that real estate agents disclose the locations of the zones to prospective property buyers. The purpose of the act was to establish a program to map the surface traces of active faults and to prevent further construction of large-scale public facilities or residential projects astride fault traces. It was also intended to inform prospective buyers of the location of these zones, presumably to increase awareness of earthquake hazards. Although the mapping program has been established and construction directly astride surface fault traces has stopped, the disclosure provisions intended to alert prospective buyers to the fact that existing structures were within the hazard zone has had little effect on development (Palm 1988). According to state and county officials, developers have responded by siting lifelines such as highways, pipelines, and utilities along fault traces, thereby transferring risk from the private homeowner to the public at large. Furthermore, disclosure of the locations of the fault zones has not affected home mortgage rates, appraisals, prices, or buyer demand (Palm 1983b). Home buyers frequently do not remember the disclosure, and it has virtually no impact on decisions to buy insurance or to take other mitigation measures (Palm 1981b). Real estate agents favor the mandatory disclosure because it has not adversely affected business and the signed disclosure offers them protection against subsequent claims if a house is destroyed or damaged by a major earthquake.

Societal Amplifications of Environmental Variability

As the examples of failures of legislation to produce intended consequences suggest, social practices and policies can amplify or reduce the impacts of environmental variability on individuals and households. Structuralist perspectives that examine how individuals and households function within social and political structures are especially helpful in highlighting the socially determined impacts of environmental variability:

> Drought does not necessarily "cause" famine, as much of the discourse surrounding the Sahelian famine of the 1970s implied. To make such an assumption is to make a major methodological error since it poses drought as a natural cause. . . . Much of what passes as natural hazards are not really natural at all; drought may be a catalyst or trigger mechanism in a sequence of events that lead to famine conditions, but the subsistence crisis itself is more a reflection of the structural ability of the socioeconomic system to cope with the unusual harshness of ecological conditions and their effects. To neglect this tendency is to resort to an ideology that sees natural disasters as "acts of God," placing responsibility upon a malevolent nature. (Watts 1983: 17)

For analysts such as Watts, famines associated with droughts are not natural phenomena; they are evidence that the social mechanisms that once provided a margin of subsistence security for peasants have been eroded by external influences from the global trade system (Watts 1987). Understanding a region's political economy and social order is essential to understanding the impacts of natural hazards and formulating effective palliatives and remedies for their socially amplified consequences.

Applications of Cartography and Related Technologies

Geographic methods, in the forms of cartography (manual and computerized), remote sensing, automated cartography, and geographic information systems (see chapters 4, 6, 7) have been harnessed for use by governments and industries throughout the world for purposes ranging from military operations to species protection. At the behest of the U.S. Department of Energy, a team of geographers used Landsat Thematic Mapper data to determine the sizes of the foraging areas needed by the wood stork. The information was then used to reserve adequate habitat for the endangered species in order to promote its recovery (Hodgson et al. 1988). Remote sensing applications have proved useful in measuring deforestation in tropical regions (Malingreau and Tucker 1987), in searching for mineral deposits in Brazil and Pakistan, and for monitoring crop production in order to identify regions where food will be needed (Gustafson 1982). Useful as such applications are, they, along with other forms of applied geography, have not escaped criticism from scholars concerned primarily with basic research who fear that university geography departments emphasize these skills at the expense of basic research and by "meeting demands of the private sector, which are predominantly concerned with short-term issues" (Johnston 1986:19).

APPLICATIONS OF RESEARCH IN PHYSICAL GEOGRAPHY

In most applied physical geography, human society is treated (1) as the target of a physical system, as in research on the climatic effects on human comfort or on flood protection strategies or (2) as a phenomenon that perturbs a physical process by creating urban heat islands, that changes climate by emitting carbon dioxide, or that disturbs a stream by channelizing its course. Human perturbation of physical systems may be intentional or inadvertent, and the societal component of applied research spans the continuum from local to global scales. The approaches and methods used in applied physical geography vary widely, but the pervading paradigm in most research is logical positivism—physical geographers believe a knowable physical world exists and that it is accessible via the scientific method. The goals of applications in physical geography are usually understanding physical systems as a prerequisite for ameliorating existing or prospective ecological problems.

In climatology and meteorology, numerical global circulation models' (GCMs) forecasts of future climate impacts on society resulting from carbon dioxide emissions are perceived by policymakers as critical features of national and international scientific agendas. The basic assumption underlying applications focused on the CO_2 issue is that even if answers are not yet definitive, the risks of not developing policies of adaptation, abatement, and control to ameliorate carbon dioxide effects are too high. A wait-and-see attitude would be "reckless rather than cautious" (Riebsame 1990:10). Trying to do something about carbon dioxide was not originally a priority among scientists, who are the first to acknowledge the limitations of climatic change research. Agenda setting, agenda abetting, and active lobbying brought this issue to public prominence (Ingram, Milward, and Laird 1990; Ingrams and Mintzer, 1990). The issue has become one of *mediarology* as much as meteorology (Schneider 1989). Consequently, many proposed policies related to the projected effects of carbon dioxide warming and ozone depletion are less than ideal examples of applied research. But policy analysis is required now, not when better models become available. Thus, the quality of applied research need not correlate with the likelihood of its potential implementation.

Activism led by a host of environmental groups, governments, and industries proposing major policy changes on a global basis lies at the heart of most climate change research applications. As often occurs internally in science, the carbon dioxide and ozone issues have attracted proponents and antagonists vying not only in the scientific literature, but through the mass media. In fall 1989, a half-hour carbon dioxide-climate change debate aired on Cable News Network's *Cross Fire* between geographer-climatologist Laurence S. Kalkstein of the University of Delaware and Patrick Michaels, Virginia's state climatologist. Many television programs, popular magazine articles, books, and newspaper accounts have publicized the greenhouse warming theory and the ozone hole controversy (Ingram, Milward, and Laird 1990). Scientists, who provide the factual substance, and networks of individuals and interest groups have helped frame the issue and move it into the political realm (*Time*, 30 Apr. and 21 May 1990). Few issues in applied physical geography reach this stage even though they may have equal merit. The carbon dioxide issue arose from a timely coalescence of events, policy entrepreneurs, and increases in scientific and technical knowledge (Ingram, Milward, and Laird 1990).

The threat of the short-term dire consequences of carbon dioxide and climate change has elevated climate dynamics and climate change to unparalleled heights in the history of climatology. During the 1970s climatology moved to the forefront as a significant scientific discipline (Mather et al. 1980), and climate came to be viewed as a major destabilizing force on world economies

(Hare 1979). Consequently, major national and international climate-related acts, programs, and policies were considered and implemented. The U.S. Climate Program Plan was initiated in 1977, to be followed by the National Climate Program Act in 1978 and the Global Climate Protection Act in 1987. These combined funding programs and policies stimulated basic and applied research in climatology, hydrology, geomorphology, and biogeography, and they emphasized scenario analyses of how climatic and other environmental processes affect different regions. Geographer-climatologists, for example, have played prominent roles in these programs.

The impacts of applied geography on policy also are evident in the Environmental Protection Agency's report to Congress, "The Potential Effects of Global Climate Change on the United States" (Smith and Tirpak 1988). One example is an extrapolation of mortality rates for forty-eight major U.S. cities under a scenario of a doubled carbon dioxide climate that concluded that summer mortality rates from heat-related causes would double whereas winter rates would decline by half or less (Kalkstein and Davis 1989). Key research issues are the complex and subtle relationships between human health, sociological and physiological factors, climatic region, and weather patterns. In general, the considerable contribution of geographers to the global warming and climate literature has emphasized (1) refining the science of climate change, (2) estimating the regional effects of climate change, and (3) developing response strategies (Kalkstein and Solomon, in press).

Applied climate research, exemplified by the carbon dioxide issue, poses a double ethical bind (Schneider 1989). On the one hand, scientists maintain their loyalty to the scientific method and its demands for full disclosure of scientific uncertainties. On the other hand, translation of a research finding into policy that will better society often demands advocacy and activism. Hedging on research findings may lead users of applied research results, who demand certainty, to mistrust them more than they should. Applied researchers try to find the delicate balance between fact and uncertainty when converting their findings into policies and implementation of programs.

Sorting out the relative influences of local and global forces at work in a given human or physical process is a challenging facet of geography. The issues that confront geographers who scale up small-area case studies they assume to be representative of larger regions and who address national and international cooperative efforts to ameliorate global problems are, in many respects, the opposite problems that face those engaged in applications of global climate change research. GCMs provide reasonably accurate signals for current climate. Scaling down such predictions to smaller regions to formulate response strategies can be risky. In climatology, statistical and model biases arise in spatially parameterizing or aggregating climate and environmental data over large areas, which may lead to inappropriate applications at local scales. The relative accuracy of the coarse resolution GCM spatial data for regional and local applications is currently being evaluated for North America (Brazel 1991) and other world regions (Kalkstein 1991). The goal of this research is to provide further guidance concerning applications of climate change data to studies of the effects of climate change.

In geomorphology, theoretical simulation modeling comparable to that found in meteorology and climatology at global and local scales is restricted because of limited spatial data and the uncertainty of many fundamental relationships. Nevertheless, there are myriad examples of applications and policy advocacy, because geomorphology and public policy are united by an understanding of changes resulting from human action that alter other components of natural systems (Coates 1982), as in the close ties between geomorphology and urban planning in the dry lands of the world (Cooke and Doornkamp 1974; Cooke et al. 1982). Research based on force and resistance concepts borrowed from physics and applied to stream morphology, sediment transfer in river systems, and movement of heavy metals in the environment have provided opportunities for field tests and theory development (Graf 1988). Research on aeolian dust and dust deposits provides many examples of useful applications in aeolian geomorphology (Pye 1987). Adverse effects of dust include soil erosion, air pollution, vegetation damage, visibility reduction and car accidents, disruption of communication, damage to industrial equipment, food and water contamination, disease spread, climate and weather modification, and radioactive dust fallout; dust can even influence the outcome of major political crises (Carter 1979). A fine and permeable line often exists between basic research and feedback from applications (for example, better understanding of turbulent flow and dust movement dynamics) that stimulates expansion and revision of existing theory.

Basic research in biogeography focuses on the spatial patterns of vegetation communities, paleoecology, bioclimatology, and ecosystem ecology, usually from an ecological perspective (Veblen 1989). Many applications occur in conservation, national park management, and reserve or preserve planning and management. They often revolve around issues of forest productivity and regeneration, tropical deforestation and agriculture, interactions between fauna and productive vegetation regions, fire regimes (chapter 3), maintenance of plant communities in urbanized regions, soil nutrient cycling, and vegetative forcing of climate change and responses to it.

An example of the convergence of concepts and theory, policy advocacy, development planning, and environmental assessment of a region is represented in the Himalayan dilemma. The notion that the Himalayan region is drifting inexorably into environmental super crisis and collapse has become widespread. Most development policies and foreign aid have been based on this theory of Himalayan environmental degradation. Projects have been undertaken with little realization of the many limiting administrative and financial control frameworks that exist in the region. Unwarranted generalizations are driving policy in the region, and as in many complex situations, quick fixes or ill-conceived policies will do more harm than good. Geographers Ives and Messerli have tried to strike a balance between activism and sound science. They remind U.N. administrators and heads of state of the need to maintain adequate links among scientific research, policy analysis, and policy-making because "uncertainty is a large element of the [Himalayan] problem" (1989:9).

Using the diverse methods and approaches described above, physical geographers will continue to investigate the relationships between human societies and the physical environment. They are applying and will likely continue to apply their knowledge to policy formulation for environmentally sensitive regions of the globe. Future efforts will be enhanced by expanding capabilities for monitoring earth systems in greater temporal and spatial detail. The current status of climate change research provides a model for other facets of applied physical geography.

APPLIED GEOGRAPHY

Geography is an applied discipline. Practicing geographers contribute directly to the betterment of society by helping solve its problems and indirectly by using the lessons drawn from applications to enrich basic research. Geography is based on a theoretical body of thought dealing with the spatial organization of society and the relationship between societies and their environments. The application of those theories lends strength to the discipline and constitutes one of the major justifications for geography's prominent role in university curricula. When recommendations for providing better spatial access to social services for the mentally disabled are implemented and the numbers of homeless individuals are reduced, geography and geographers have made a direct and measurable contribution to the well-being of those individuals and to society. When geographers measure rates of forest destruction and devise policies that slow or halt the process, they conserve an important part of the environment and contribute to the well-being of the planet. When geographers provide sound counsel for policy decisions, they help bridge the gap between science and the people who are sometimes its subjects, but who should always be its rulers.

REFERENCES

A baffling ozone policy. 1990. *Time*, 21 May, 20.

ADLER, S. 1986. The dynamics of transit innovation in Los Angeles. *Environment and Planning D: Society and Space* 4:321–335.

ALEXANDER, CHARLES P. 1990. A sizzling scientific debate. *Time*, 30 Apr., 84.

BRAZEL, A. J. 1991. Comparisons of RAND climatology and GCM outputs for North and Central America. In *Global comparisons of selected GCM control runs and observed climate data*, ed. L. Kalkstein, 1–63. Washington, D.C.: Office of Policy Analysis, Environmental Protection Agency.

BRAZEL, A. J., and S. W. BRAZEL. 1989. Applied and basic research at the Laboratory of Climatology and Office of the State Climatologist of Arizona. In *Papers and proceedings of applied geography conferences* 12:116–121.

BUNGE, WILLIAM. 1971. *Fitzgerald: Geography of a revolution*. Cambridge, Mass.: Schenkman.

CARLSTEIN, TOMMY. 1978. Innovation, time allocation, and time-space packing. In *Human activity and time geography*, ed. Tommy Carlstein, Don Parkes, and Nigel Thrift, 146–162. New York: John Wiley and Sons.

CARTER, J. 1979. How dust fouled up the hostage problem in Iran. Oval Office Tapes Service. Cited in Pye 1987:274.

CLARK, WILLIAM A. V. 1986. Residential segregation in American cities: A review and interpretation. *Population Research and Policy Review* 5:95–127.

———. 1987. Demographic change, attendance area adjustment, and school system impacts. *Population Research and Policy Review* 6:199–222.

COATES, D. R. 1982. Environmental geomorphology perspectives. In *Applied geography: selected perspectives*, ed. J. W. Frazier, 139–169. Englewood Cliffs, N.J.: Prentice-Hall.

COOKE, R. U., D. BRUNSDEN, J. C. DOORNKAMP, and D. K. C. JONES. 1982. *Urban geomorphology in drylands*. New York: Oxford University Press and United Nations University.

COOKE, R. U., and J. C. DOORNKAMP. 1974. *Geomorphology in environmental management*. Oxford: Clarendon.

DEAR, MICHAEL J., and JENNIFER R. WOLCH. 1987. *Landscapes of despair: From deinstitutionalization to homelessness*. Princeton: Princeton University Press.

EPSTEIN, BART J., and EILEEN SCHELL. 1982. Marketing geography: Problems and perspectives. In *Applied geography: Selected perspectives*, ed. John W. Frazier, 263–282. Englewood Cliffs, N.J.: Prentice-Hall.

FRAZIER, JOHN W. 1982. *Applied geography: Selected perspectives*. Englewood Cliffs, N.J.: Prentice-Hall.

GARES, PAUL A. 1989. Geographers and policy-making: Lessons learned from the failure of the New Jersey Dune Management Plan. *Professional Geographer* 41:20–29.

GLANTZ, MICHAEL H., JENNIFER ROBINSON, and MARIA E. KRENZ. 1985. Recent assessments. In *Climate impact assessment*, ed. R. W. Kates, J. H. Ausubel, and M. Berberian, 565–598. Chichester: John Wiley and Sons.

GOLANT, STEPHEN M. 1976. Housing and transportation problems of the urban elderly. In *Urban policymaking and metropolitan dynamics: A comparative geographical analysis*, ed. John S. Adams, 379–422. Cambridge Mass.: Ballinger.

————. 1984. *A place to grow old: The meaning of environment in old age*. New York: Columbia University Press.

GOULD, PETER R. 1989. Geographic dimensions of the AIDS epidemic. *Professional Geographer* 41:71–78.

GOULD, PETER R., WILPIN GORR, and EMILIO CASETTI. 1988. Understanding and predicting the AIDS epidemic in geographic space. Mimeo. University Park, Pa.: Penn State–Carnegie Mellon–Ohio State Consortium.

GRAF, WILLIAM L. 1988. *Fluvial processes in dryland rivers*. Berlin-Heidelberg: Springer-Verlag.

GUSTAFSON, GLEN C. 1982. Applications of aerial imaging technology to environmental problems. In *Applied geography: Selected perspectives*, ed. John W. Frazier, 170–196. Englewood Cliffs, N.J.: Prentice-Hall.

HÄGERSTRAND, TORSTEN. 1978. Survival and arena. In *Human activity and time geography*, ed. Tommy Carlstein, Don Parkes, and Nigel Thrift, 122–125. New York: John Wiley and Sons.

HAMNETT, MICHAEL P., DOUGLAS J. PORTER, AMARJIT SINGH, and KRISHNA KUMAR. 1984. *Ethics, politics, and international social science research: From critique to praxis*. Honolulu: East-West Center and the University of Hawaii Press.

HARE, F. KENNETH. 1979. Focus on climate. *Environmental Science and Technology* 13:156–159.

HARVEY, DAVID. 1974. What kind of geography for what kind of public policy? *Transactions of the Institute of British Geographers* 63:18–24.

HODGSON, MICHAEL E., JOHN R. JENSEN, HALKARD E. MACKAY, JR., and MALCOLM C. COULTER. 1988. Monitoring wood stork foraging habitat using remote sensing and geographic information systems. *Photogrammetric Engineering and Remote Sensing* 54:1601-1607.

HYERS, A. D., and M. G. MARCUS. 1981. Land use and desert dust hazards in central Arizona. In *Geological Society of America special paper 186*, ed. T. Pewe, 267–280. Washington, D.C.: U.S. Geological Survey.

INGRAM, H., H. B. MILWARD, and W. LAIRD. 1990. Scientists and agenda setting: Advocacy and global warming. Paper presented at the Western Political Science Association annual meeting, 23 March, Newport Beach, Calif.

INGRAM, H., and C. MINTZER. 1990. How atmospheric research changed the political climate. In *Global climate change: The meeting of science and policy*, ed. F. Burkhart, C. Hutchinson, M. Saint-Germain, and L. Beck. Issue Paper No. 1, 1–11. *Proceedings of understanding global change and Arizona: Boom, bust or another sunny day?* Tucson: University of Arizona.

IVES, JACK D., and BRUNO MESSERLI. 1989. *The Himalayan dilemma: Reconciling development and conservation*. London and New York: Routledge and United Nations University.

JOHNSTON, RONALD J. 1984. *Residential segregation, the state, and constitutional conflict in American urban areas*. London: Academic.

————. 1986a. Applied geography. In *The Dictionary of Human Geography*, ed.

Ronald J. Johnston, Derek Gregory, and David Smith, 17–20. 2d ed. Oxford: Basil Blackwell.

————. 1986b. *On human geography*. New York: Basil Blackwell.

KALKSTEIN, L. S., ed. 1991. *Global comparisons of selected GCM control runs and observed climate data*. Office of Policy Analysis monograph. Washington, D.C.: Environmental Protection Agency.

KALKSTEIN, L. S., and R. E. DAVIS. 1989. Weather and human mortality: An evaluation of demographic and interregional responses in the United States. *Annals of the Association of American Geographers* 79:44–64.

KALKSTEIN, L. S., and B. SOLOMON. n.d. Global warming and climate change: The roles of geographers. *Professional Geographer*. In press.

LENNTORP, Bo. 1978. A time-geographic simulation model of individual activity programmes. In *Human activity and time geography*, ed. Tommy Carlstein, Don Parkes, and Nigel Thrift, 162–180. New York: John Wiley and Sons.

LEWIS, NANCY D., and JONATHAN D. MAYER. 1988. Disease as natural hazard. *Progress in Human Geography* 12:15–33.

MALINGREAU, J. P., and C. J. TUCKER. 1987. The contribution of AVHRR data for measuring and understanding global processes: Large-scale deforestation in the Amazon Basin. In *Proceedings, IGARSS '87 Symposium*, 443–448. Ann Arbor.

MARTENSSON, SOLVEIG. 1978. Time allocation and daily living conditions: Comparing regions. In *Human activity and time geography*, ed. Tommy Carlstein, Don Parkes, and Nigel Thrift, 181–197. New York: John Wiley and Sons.

MATHER, J. R., R. T. FIELD, L. S. KALKSTEIN, and C. J. WILLMOTT. 1980. Climatology: The challenge for the eighties. *Professional Geographer* 32:285–292.

MORRILL, RICHARD L. 1973. Ideal and reality in reapportionment. *Annals of the Association of American Geographers* 63:463–477.

————. 1981. *Political redistricting and geographic theory*. Resource Publication 1981–1. Washington, D.C.: Association of American Geographers.

————. 1989. School busing and demographic change. *Urban Geography* 10: 336–354.

NICKLING, W. G., and A. J. BRAZEL. 1984. Temporal and spatial characteristics of Arizona dust storms. *Journal of Climatology* 4:645–660.

NIGG, JOANNE M. 1988. Frameworks for understanding knowledge dissemination and utilization: Applications for the National Earthquake Hazards Reduction Program. In *A review of earthquake research applications in the National Earthquake Hazards Reduction Program: 1977–1987, Proceedings of Conference XLI*, 13–33. Open File Report 88-13-A. Reston, Va.: U.S. Geological Survey.

PALM, RISA I. 1981a. The daily activities of women. In *Women and the social costs of economic development: Two Colorado case studies*, ed. Elizabeth Moen, Elise Boulding, Jane Lillydahl, and Risa Palm, 99–118. Boulder, Colo.: Westview.

————. 1981b. *Real estate agents and special studies zones disclosure*. Institute of Behavioral Science Monograph 32. Boulder: University of Colorado.

————. 1983a. Improving hazard awareness. In *Proceedings of conference: A work-*

shop on the 1886 Charleston, South Carolina, earthquake and its implications for today, 55–61. Open File Report 83–843. Reston, Va.: U.S. Geological Survey.

————. 1983b. *The response of lenders and appraisers to earthquake hazards*. Institute of Behavioral Science Monograph 38. Boulder: University of Colorado.

————. 1988. Alquist-Priolo legislation on active fault zones. In *A review of earthquake research applications in the National Earthquake Hazards Reduction Program: 1977–1987. Proceedings of Conference XLI*, 225–230. Open File Report 88-13-A. Reston, Va.: U.S. Geological Survey.

PALM, RISA I., and ALLAN PRED. 1978. The status of American women: A time-geographic view. In *An invitation to geography*, ed. David E. Lanegran and Risa I. Palm, 99–109. 2d ed. New York: McGraw-Hill.

PLATT, RUTHERFORD H. 1987. Overview. In *Cities on the beach: Management issues of developed coastal barriers*, ed. Rutherford H. Platt, Sheila G. Pelczarski, and Barbara K. R. Burbank, 1–14. Research Paper 224. Chicago: University of Chicago Department of Geography.

PRED, ALLAN R. 1977. *City-systems in advanced economies*. London: Hutchinson.

PYE, K. 1987. *Aeolian dust and dust deposits*. London: Academic.

PYLE, GERALD F. and ROBERT M. COOK. 1978. Environmental risk factors of California encephalitis in man. *Geographical Review* 68:157–170.

RIEBSAME, W. E. 1990. Anthropogenic climate change and a new paradigm of natural resource planning. *Professional Geographer* 42:1–12.

ROWLES, GRAHAM D. 1978. *Prisoners of space? Exploring the geographical experience of older people*. Boulder, Colo.: Westview.

————. 1986. The geography of aging and the aged: Toward an integrated perspective. *Progress in Human Geography* 10:511–539.

SCHNEIDER, S. H. 1989. *Global warming: Are we entering the greenhouse century?* San Francisco: Sierra Club.

SHANNON, GARY W., and GERALD F. PYLE. 1989. The origin and diffusion of AIDS: A view from medical geography. *Annals of the Association of American Geographers* 79:1–24.

SMITH, J. B., and D. TIRPAK, eds. 1988. *The potential effects of global climate change on the United States*. Draft Report to Congress, Vols. 1 and 2. Washington, D.C.: U.S. Environmental Protection Agency, Office of Policy, Planning, and Evaluation, Office of Research and Development.

TIVERS, J. 1976. Constraints on spatial activity patterns: Women with young children. Occasional Paper No. 6. London: Kings College, University of London, Department of Geography.

VEBLEN, THOMAS T. 1989. Biogeography. In *Geography in America*, ed. Gary L. Gaile and Cort J. Willmott, 28–46. Columbus, Ohio: Merrill.

WATTS, MICHAEL. 1983. *Silent violence: Food, famine, and peasantry in northern Nigeria*. Berkeley: University of California Press.

————. 1984. State, oil, and accumulation. *Society and Space* 2:403–428.

WILBANKS, THOMAS J. 1982. Location and energy policy. In *Applied geography: Selected perspectives*, ed. John W. Frazier, 219–232. Englewood Cliffs, N.J.: Prentice-Hall.

ZELINSKY, WILBUR. 1975. The demigod's dilemma. *Annals of the Association of American Geographers* 65:123–143.

The Peopling of

American Geography

Donald G. Janelle

Social and demographic profiles document the origins of the people who shape American geography. Most received formal training in the discipline since World War II. Regional dimensions of the likelihood of being a geographer, institutional settings for education in the discipline, and linkages among specializations reflect ongoing social changes in American geography. The growing participation of women, steps to increase the presence of minorities, and the immigration of geographers from other countries offer new paths for the discipline's development.

TIMING AND THE SOCIALIZATION OF GEOGRAPHERS

The education of the American geographers of the 1990s spans periods of intellectual socialization from the 1930s to the 1980s. The timing of one's training is critical to the path followed by aspiring geographers. It may even be a determinant of whether one enters the discipline at all. To illustrate the importance of timing and its linkage to social and institutional changes, compare John Fraser Hart's reports on manpower in geography for 1966 and 1972. In contrast to the 1950s, a period of limited opportunities for geography graduates, by 1966 Hart highlights severe shortages of geographers to fill teaching positions at all levels of an expanding educational system, an abundance of opportunities, and increasing graduate enrollments. The mid 1960s saw the arrival of the baby boom on university campuses. Financial support for military veterans also helped to swell the demand for education. The quantitative revolution and new technologies ushered in exceptional demands for geographers trained in quantitative techniques and remote sensing.

Meeting these demands required curriculum reform and teacher training. With financial support from the National Science Foundation (NSF), these requirements were met by the innovative efforts of the Commission on College Geography (1963–74) and the High School Geography Project (1963–70). In addition, amendments to the National Defense Education Act (NDEA) in 1964 bolstered student enrollments in graduate programs. The act's provision of fellowships and student loans gave special incentives to those interested in teaching at institutions of higher learning. NDEA also supported teacher-training institutes and encouraged the appointment of geographers to work with state education agencies. In response, student enrollments and geography programs expanded rapidly through the 1960s.

In 1972, six years after his first report, Hart presents the results of this expansionary era of geography in the United States. Thirty-nine geography departments granted 502 doctoral degrees between 1 July 1965 and 30 June 1970. This was an exciting turnaround for geography. However, by the end of this period, overinvestment led to a supply of geographers that exceeded demand. In later years, many new graduates left geography altogether or cultivated opportunities outside of academia. Academic institutions, under financial pressures, sought rationalization of programs. Severe cuts in financial support for higher education hurt geography on several campuses in the years to follow. For all of its benefits to the discipline, the NDEA ushered in a cyclic pattern of boom and bust in the demand for geographers.

By the 1980s most programs in geography had contracted to meet the realities of the marketplace. However, by the end of the decade, changes in the national economy and in technology were ushering in new demands and opening new horizons. Applied and business-oriented sectors of the discipline saw considerable expansion. This growth enhanced geography's linkages to scientific, public-agency, and business communities. In addition, public concern for the lack of geographical literacy posed opportunities for geography in elementary and secondary schools. A coalition of societies, including the National Geographic Society (generous with financial support and leadership), the National Council for Geographic Education, the American Geographical Society, and the Association of American Geographers, launched an aggressive effort to build a solid base for geography's future in America (see chapter 9).

Owing to expansions of microcomputer and communications technologies, the 1980s also fostered unprecedented changes in the methods of geography. Geographical Information Systems (GIS) captured the imagination of geographers and others concerned with managing and analyzing information on spatial patterns. Initiatives by the National Science Foundation to set up the National Center for Geographic Information and Analysis helped foster changes in the structure and development of university programs in the discipline. Again, the demand for geographers was on the rise.

This brief survey on timing and the socialization of geographers highlights the significance of public and private initiatives for attracting people into the discipline. Geography's status relates to society's prevailing but everchanging concerns about social and economic development. It is in its ability for adapting to new technological and social realities that geography commands recognition in scholarship, science, and application. This notion of changing social and technological milieux is central to interpreting the analyses that follow.

THE AAG MEMBERSHIP

These descriptive analyses will draw on a sizable, but not necessarily representative, mix of American geographers. They focus on U.S. resident members of the Association of American Geographers (AAG) in May 1989. Excluded are those who shun association with formal societies, those who do not see their interests served by the association, including many teachers below the university level, and many geographers in the private-sector economy. Uneven treatment of the interrelated aims of this chapter relates to its primary reliance on a single data source—the membership records of the AAG.

Of the 5,889 (grown by 1991 to about 6,200) members in the AAG in 1989, 5,021 met the principal criterion of U.S. residency. Excluded were institutional and corporate members who lacked clear identity with specific individuals, and departmental secretaries who, though contributing to the day-to-day operations of the discipline, are not geographers.

The Likelihood of Being a Geographer

In comparison with other occupations, the likelihood of being a geographer in the United States is very low. Figure 17.1 (*top*) shows the births of AAG members per 100,000 population by states, revealing sizable regional variations. Values range from highs of 7.6 in the District of Columbia and 6.47 in Illinois to lows of 0.63 and 0.66 in Alabama and South Carolina. High values occur in the northern tier of states of the Midwest and Great Plains (Wisconsin, Minnesota, and North Dakota), and in Utah, Nebraska, Massachusetts, and Delaware. Except for Maine and Vermont, values below 2 per 100,000 occur south of the Ohio River and in the block of southern states from New Mexico to Florida.

The current pattern of residence for AAG members (Fig. 17.1, *bottom*) shows unusually high concentrations in the Washington, D.C., area (District of Columbia, Virginia, Maryland, and Delaware). High densities in the national capital region reflect, in part, the importance of geographic expertise in addressing the knowledge and policy needs of the federal government.

It is not surprising that geographers respond to the same forces of migration

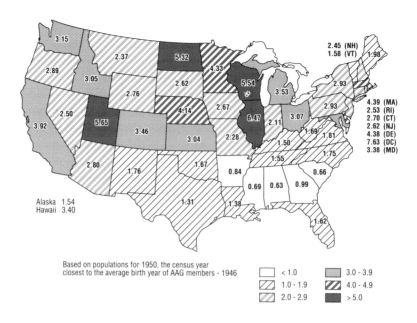

Based on populations for 1950, the census year
closest to the average birth year of AAG members - 1946

< 1.0	3.0 - 3.9
1.0 - 1.9	4.0 - 4.9
2.0 - 2.9	> 5.0

AAG Members per 100,000 Population by States

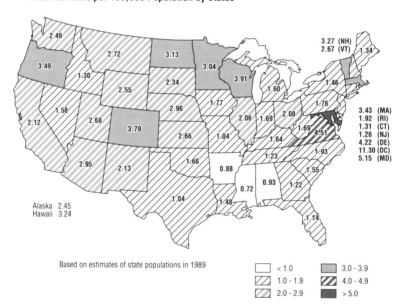

Based on estimates of state populations in 1989

< 1.0	3.0 - 3.9
1.0 - 1.9	4.0 - 4.9
2.0 - 2.9	> 5.0

Fig. 17.1. The Likelihood of Being a Geographer. Number of births and current residence of
AAG members per 100,000 population by states. Based on AAG membership in May 1989.

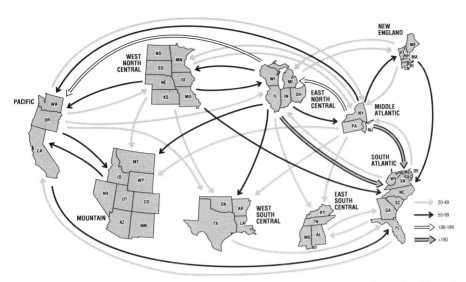

Fig. 17.2. Geographers on the Move. Migration from regions of birth to regions of residence in 1989. Only moves of 20 or more geographers are illustrated. Based on AAG membership in May 1989.

as does the population in general. The growth of the Sunbelt and American West is a case in point. Figure 17.2 documents the transfer of geographers from regions of birth to regions of residence in 1989. The South Atlantic region, with a strong emphasis on the Washington, D.C., area, is a dominant destination. The Middle Atlantic and East North Central states stand out as prominent source areas for geographers.

Table 17.1 offers additional insights into the general migratory patterns. South Atlantic and Pacific states have the highest retention rates. More than 50 percent of the geographers in these regions were born there. In contrast, geographers born in the West North Central and Mountain states show greater propensities to follow careers and educations in other areas. Less than 14 percent of geographers in the Mountain states were born there. Of the 711 geographers born outside the United States, the highest concentrations occur in the Pacific, East North Central, and South Atlantic areas. However, as a proportion of all geographers, their representation is highest in New England (19.2 percent). Do New England educational institutions and businesses show greater partiality than those of other areas for geographers of foreign origin?

In the AAG, which is largely academic in orientation, institutions of higher education play a leading role in motivating the interregional movement of members. For most, the selection of an undergraduate school starts a succession of events that leads to a career in academic geography.

Table 17.1. Geographers on the Move—Regional Migration

Region of birth	Region of residence in 1989									Total births by region	Percent living in region of birth in 1989
	PAC	MTN	WNC	WSC	ENC	ESC	NEW	MID	SOU		
Pacific	289	61	24	25	40	14	6	32	61	552	52.4
Mountain	42	50	10	18	11	4	4	4	19	162	30.9
West North Cent	62	44	134	41	66	8	11	25	61	452	29.6
West South Cent	17	13	19	81	11	6	4	12	30	193	42.0
East North Cent	123	62	72	50	426	34	33	60	194	1,054	40.4
East South Cent	14	10	5	12	14	40	2	4	28	129	31.0
New England	34	17	13	4	29	4	130	29	57	317	41.0
Middle Atlantic	76	45	41	31	102	25	63	281	205	869	32.3
South Atlantic	34	18	13	17	37	23	7	23	240	412	58.3

										Total
Total U.S.-born residents—1989	691	320	331	279	736	158	260	470	895	4,140
Puerto Rico							1	2	2	5
Outside U.S.	121	46	52	37	154	20	62	96	123	711
Total 1989 residents	812	366	383	316	890	178	323	568	1,020	4,856
Percent born in host region	35.6	13.7	35.0	25.6	47.9	22.5	40.2	49.5	23.5	40.4
Percent of foreign origin	14.9	12.6	13.6	11.7	17.3	11.2	19.2	16.9	12.1	14.6
Net internal migration	139	158	−121	86	−318	29	−57	−399	483	

SOURCE: *Calcuated by author for AAG membership (May 1989).*

Getting a Degree in Geography

The allocation of AAG members by six age cohorts helps to mirror the results of acculturation into geography during periods dominated by particular patterns of training and specialization. This allocation helps to describe changes in the productivity of universities and in the time required for securing degrees (figs. 17.3 and 17.4). For geographers born before 1925, for example, the average age for the receipt of the doctorate was 38.5 (fig. 17.3) and the average year of receipt was 1955 (fig. 17.4). The age for receiving the doctorate shows persistent decline in figure 17.4 for the next three succeeding age cohorts, to 31 years of age for the 1945–54 grouping. This shift is less extreme for the master's and bachelor's degrees. Military service in World War II, the Korean Conflict, and the Vietnam War may account for some of the variation. However, the practice of taking jobs while working on the doctorate was more common in the earlier periods. In addition, the discipline was under pressure in the 1960s to meet the demands of expanding educational systems.

The average age for receiving doctorate degrees by those in the 1955–64 cohort (26.9 years) will increase over time. Nearly 60 percent were still students in mid-1989 (Fig. 17.5). The percentages of this group with highest degrees at the doctor's, master's and bachelor's levels were 12, 29, and 39 percent, respectively.

The institutional source of one's degree is held by some as an indicator of status in the discipline, often conditioning the prospects for favorable jobs upon graduation. So too, institutions are eager to attract the strongest candidates for degrees and are picky about their origins. Along with other indicators of quality, particularly the levels of publication and citation of its faculty, the number of graduates provides a clue about a department's standing and influence in the

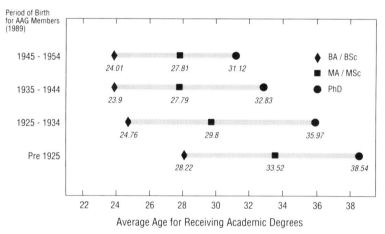

Fig. 17.3. Getting a Degree in Geography. Average age for recipients of degrees in geography, by period of birth. Based on AAG membership in May 1989.

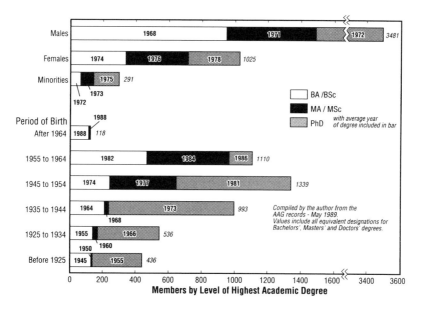

Fig. 17.4. Educational Characteristics of U.S.-Resident Members of the Association of American Geographers in May 1989. Males, females, minorities, and total members grouped by period of birth.

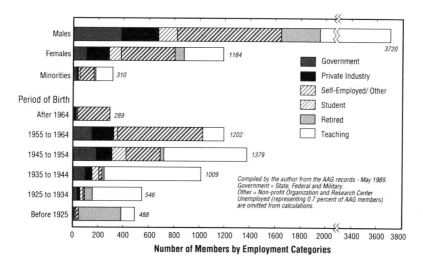

Fig. 17.5. Employment Characteristics of U.S.-Resident Members of the Association of American Geographers in May 1989. Males, females, minorities, and total members grouped by period of birth.

Table 17.2. Leading Ph.D. Universities: Number of Degrees Granted to AAG Members, by Period of Birth

Period of birth: Avg. year of Ph.D.:	Pre–1925 1955		1925–34 1966		1935–44 1973		1945–54 1981		1955–64 1986	
	Chicago	33	Wisconsin	27	Michigan St	36	UCLA	34	Ohio State	9
	Clark	33	Washington	26	UCLA	34	Ohio State	31	Illinois	7
	Michigan	28	Michigan	21	Minnesota	32	Wisconsin	30	UC Berkeley	6
	UC Berkeley	21	Syracuse	19	Wisconsin	28	Minnesota	29	Colorado	6
	Wisconsin	18	Michigan St	18	Chicago	27	Clark	28	Minnesota	6
	Columbia	15	Chicago	17	Illinois	27	Colorado	26	Louisiana	5
	Northwestern	14	Northwestern	17	Michigan	26	UC Berkeley	25	Washington	5
	Ohio State	12	UCLA	16	Washington	24	Michigan St	23	Clark	4
	UCLA	11	UC Berkeley	15	Georgia	23	Kansas	21	Indiana	4
	Maryland	10	Minnesota	15	Kansas	22	Washington	21	Kansas	4
	Nebraska	7	Clark	13	Oregon	22	Illinois	19	Rutgers	4
	Harvard	6	Illinois	12	Iowa	21	Syracuse	18	UCLA	4
	Illinois	6	Louisiana	12	UC Berkeley	20	Louisiana	15	9 univs.	3
	Washington	6			Northwestern	20	Penn State	14		
Total		220		228		362		334		64
Total Ph.D.s granted to age cohort		309		376		765		716		150
Number of univs. granting Ph.D.s to members		65		73		115		104		64
Percentage of Ph.D.s from: Above univs.		71.2		60.6		47.3		46.6		42.7
Non-U.S. universities		3.9		6.9		6.4		8.4		12.0

SOURCE: *Compiled by author for AAG membership (May 1989).*

discipline (see table 17.2). The focus in table 17.2 on AAG members, as opposed to total number of graduates, helps ground the listings on a strong indicator of commitment to the discipline. The groupings by age cohorts allow identification of changes in the productivity of departments for the post–World War II period.

The full doctorate degree count for the 1955–64 cohort is still in the making. However, individuals in the first four cohorts have, with minor exceptions, completed their formal educations. The separate tabulations reveal the persistence of some departments that took early leads in the discipline. These include, for example, Clark, University of California–Berkeley, Illinois, Louisiana, Ohio State, Syracuse, University of California–Los Angeles, Washington, and Wisconsin. Once prominent centers of innovation that have since declined are Chicago, Harvard, Michigan, and Northwestern. Finally, the rapid growth of major state universities is illustrated by Colorado, Kansas, Michigan State, and Minnesota.

The feature that stands out in the interpretation of table 17.2 is the concentration of degrees from relatively few universities. However, with the increasing number of doctoral departments and the growing significance of foreign institutions, the extent of concentration is dropping. Only fourteen universities accounted for 71 percent of all U.S.-resident doctorates for the pre-1925 cohort, but the same number accounted for only 47 percent of the 1935–44 and 1945–54 cohorts. In turn, the share of doctorates from non-U.S. institutions increased from 3.9 percent for the earliest cohort to 12 percent for the youngest group of doctorates.

Dean Rugg (1981) reviewed doctorate production in the period from 1893 to 1946, confirming the significance of the major Midwest universities. The most important of these, in order of graduates, were Chicago, Wisconsin, Michigan, and Nebraska. Because the Midwest accounted for 149 of 337 dissertations completed in that period, Rugg describes it as the hearth (the heartland) of American geography. The evidence (table 17.2) confirms the continued significance of institutions in the Midwest. However, the growth of geography in other regions has lessened the hegemony of Midwestern universities.

Student-mentor relationships reinforce the concentration of degree production by a few institutions. Research on this topic draws on a sparse body of literature, primarily by Bushong (1981, 1984). Focusing on the period 1907 to 1946, his genealogical perspective warrants review for insights on American geography as it entered the postwar era. Only eight individuals were responsible for 54 percent of all doctorates in the period. In order of the number of dissertations supervised, they were Charles C. Colby and W. Elmer Ekblaw with 25 each, Harlan H. Barrows, 24, Wallace W. Atwood, Sr., and Carl O. Sauer with 19 each, Clarence F. Jones, 18, Vernor C. Finch, 15, and Samuel

Van Valkenburg, 13. Tracing the descendants of these and other mentors of that period through four generations of academic offspring, Bushong's tabulations reveal that 56 percent of all doctorates granted through 1946 were descendant from only six mentors. These included Barrows, 104, Colby, 46, Atwood, Sr., 41, Rollin D. Salisbury, 31, and Ellen C. Semple and Finch, 28 each.

In more recent research, Bushong (1984) explores mentor roles among Latin American specialists in geography, for the years 1907 to 1981. Of the 152 scholars who directed the 379 successful dissertations on Latin American topics in the period, 27.4 percent were academic descendants of one individual—Sauer. Two departments, the University of California-Berkeley and the University of Florida, accounted for 19.5 percent of dissertations on Latin America. Such institutional and mentor concentrations play critical roles in the development of the discipline and its subspecialties.

Related areas of investigation concern the networks that link departments through the hiring of graduates as professors. Here, too, there is concentration. Thus, for the 1893–1946 period, 60 percent of the 176 staff members of the nation's 19 most significant doctorate departments came from institutions in the Midwest (Rugg 1981). An analysis by Sopher and Duncan suggested that such exchanges offer a means for discerning the peer status of academic units. Using the provocative analogy of Indian caste systems, they argued that "placing Ph.D.s gives prestige according to the rank of the taker." (1975:17). Their transaction analysis of 321 hirings among 51 departments from 1960 to 1974 confirmed an ordered directional movement of doctorates and a ranking among departments, which they interpreted as indicators of status in the discipline.

Aside from faculty, the exchange of students among universities offers another clue to the structure of the discipline. For the 1989 AAG membership, it is possible to define matrices of student flows among all universities. However, the number of universities involved is large. AAG members received bachelor's degrees from 661 different institutions, including 180 foreign schools. Master's degrees came from 360 universities, 114 of them outside the United States, and doctorates from 196 universities, 82 of them in other countries.

Analyses here are confined to the more productive institutions at each degree level. However, even this limited analysis yields comparatively sparse entries for the matrices, attesting to the broad diffusion of students from many sources to many destinations. Owing to the source of data, the analysis favors universities that are successful in encouraging graduates to be members of the AAG (table 17.3). No doubt tabulations based on total degrees granted would include other institutions in the rankings.

A simple correlation matrix of student exchanges is used to detect common patterns of institutional origins and destinations. Generalized by highest order of correlation, a linkage analysis (McQuitty 1957) clustered the universities (fig. 17.6). While linkage in a cluster does not mean that cluster members

Table 17.3. Number of Degrees Granted by Universities to U.S.-Resident Members of the AAG (May 1989)

Bachelor's degrees		Master's degrees		Doctoral degrees	
UCLA	95	Wisconsin	117	Wisconsin	106
Wisconsin	83	UC-Berkeley	94	UCLA	99
Minnesota	68	Chicago	89	Clark	97
Penn State	59	Minnesota	87	Chicago	91
Washington	57	Clark	79	Michigan	88
Clark	54	Illinois	77	UC-Berkeley	87
Michigan	50	UCLA	77	Minnesota	84
Michigan State	47	Penn State	65	Washington	82
Illinois	44	Ohio State	63	Michigan State	81
Colorado	42	Syracuse	63	Ohio State	76
Dartmouth	39	Michigan State	60	Illinois	71
Texas	39	Washington	60	Syracuse	61
Wayne State	38	Colorado	59	Kansas	57
SUNY-Buffalo	36	Michigan	59	Northwestern	57
Kent State	35	Columbia	56	Louisiana	53
Maryland	35	Indiana	56	Iowa	48
Syracuse	34	Northwestern	52	Columbia	45
Indiana	33	Kansas	41	Colorado	44
Ohio State	33	Georgia	40	Indiana	44
San Diego State	31	Rutgers	36	Penn State	39
Valparaiso	31	Louisiana	35	Nebraska	37
Massachusetts	29	Texas	34	North Carolina	36
Illinois State	27	Iowa	33	Oregon	35
Missouri	27	Oklahoma	33	Georgia	34
Oregon	27	Oregon	33	Tennessee	30
North Carolina	26	Arizona State	32	Johns Hopkins	29
Harvard	25	Nebraska	32	Maryland	28
Kansas	25	Oregon State	32	Rutgers	28
Calif. State-Chico	24	SUNY-Buffalo	31	Pittsburgh	27
Miami U.	24	Missouri	30	Oklahoma	26
Rutgers	24	North Carolina	29	Florida	24
SUNY-Albany	24	Arizona	28	Oregon State	23
		S. Illinois	28		
		Florida	27		
		Pittsburgh	26		
		Tennessee	26		
		Miami U.	25		
		Wayne State	23		

SOURCE: *AAG membership, May 1989.*

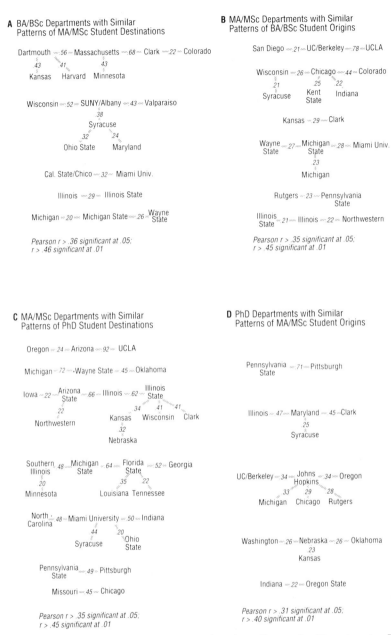

A BA/BSc Departments with Similar
Patterns of MA/MSc Student Destinations

Dartmouth ···56··· Massachusetts ···68··· Clark ···22··· Colorado
43   41   43
Kansas   Harvard   Minnesota

Wisconsin ···52··· SUNY/Albany ···43··· Valparaiso
38
Syracuse
32   24
Ohio State   Maryland

Cal. State/Chico ···32··· Miami Univ.

Illinois ···29··· Illinois State

Michigan ···20··· Michigan State ···26··· Wayne State

*Pearson r > .36 significant at .05;
r > .46 significant at .01*

B MA/MSc Departments with Similar
Patterns of BA/BSc Student Origins

San Diego ···21··· UC/Berkeley ···78··· UCLA

Wisconsin ···26··· Chicago ···44··· Colorado
21   25   22
Syracuse   Kent State   Indiana

Kansas ···29··· Clark

Wayne State ···27··· Michigan State ···28··· Miami Univ.
23
Michigan

Rutgers ···23··· Pennsylvania State

Illinois State ···21··· Illinois ···22··· Northwestern

*Pearson r > .35 significant at .05;
r > .45 significant at .01*

C MA/MSc Departments with Similar
Patterns of PhD Student Destinations

Oregon ···24··· Arizona ···92··· UCLA

Michigan ···72··· Wayne State ···45··· Oklahoma

Iowa ···22··· Arizona State ···66··· Illinois ···62··· Illinois State
22   34   41   41
Northwestern   Kansas   Wisconsin   Clark
32
Nebraska

Southern Illinois ···48··· Michigan State ···64··· Florida State ···52··· Georgia
20   35   22
Minnesota   Louisiana   Tennessee

North Carolina ···48··· Miami University ···50··· Indiana
44   20
Syracuse   Ohio State

Pennsylvania State ···49··· Pittsburgh

Missouri ···45··· Chicago

*Pearson r > .35 significant at .05;
r > .45 significant at .01*

D PhD Departments with Similar
Patterns of MA/MSc Student Origins

Pennsylvania State ···71··· Pittsburgh

Illinois ···47··· Maryland ···45··· Clark
25
Syracuse

UC/Berkeley ···34··· Johns Hopkins ···34··· Oregon
33   29   28
Michigan   Chicago   Rutgers

Washington ···26··· Nebraska ···26··· Oklahoma
23
Kansas

Indiana ···22··· Oregon State

*Pearson r > .31 significant at .05;
r > .40 significant at .01*

Fig. 17.6. Export-Import Similarities among American Geography Departments. Based on origins of academic degrees for members of the AAG in May 1989. Each university is linked with the university that has student origins/destinations most similar to its own, based on correlation coefficients.

necessarily share students with each other, it does show a tendency for them to derive degree candidates from the same set of universities and to send their students for higher degrees to similar institutions.

Although the correlation values are quite low, the groupings suggest pronounced regional forces in the derivation and allocation of students. Examples include Michigan-Michigan State-Wayne State, Berkeley-UCLA-San Diego, Pennsylvania State-Pittsburgh, and Illinois-Illinois State. Detecting the reasons for such groupings would require alternative data sets and different research questions. For instance, do differential tuition fees for in-state and out-of-state students condition the linkages among state-supported institutions? Similarity of departmental specializations, linkages among graduates (now faculty) with the same mentors, and commonality of programs may play roles in structuring the contact networks for different universities. The stability of these relationships over time are also of interest. Do they persist through changes in research or job opportunities? Do they survive major shifts in disciplinary structure brought about by technological or social forces? Principal indicators of such changes are choices of career and research specialization.

Specialization in Geography

Based on AAG membership in 1984, Goodchild and Janelle (1988) provide the most recent and complete analysis of specialization in American geography. One would not expect major shifts to occur in the span of only a few years; however, the introduction of new specialty groups and the prospect for analyzing changes in specialization patterns by age cohorts justify expanding on those analyses. The restructuring of specializations included new groups concerned with microcomputers, geographical information systems, hazards, students (abandoned in 1989), and energy and environment (replacing separate groups on environmental studies and energy). Analysis begins with a matrix of cross-memberships among the forty specialty groups. Although members could associate with three groups, only 50 percent did so. Twenty-three percent did not sign up for any specialty groups, 11 percent joined only one, and 16 percent opted for two affiliations.

Conversion of the data to a correlation matrix allows simple structural clustering of the specializations according to each group's highest correlation values (fig. 17.7). A high correlation between a pair of specialty groups implies a similar distribution of their members among other groupings. In other words, their members share common interests. The clusterings in figure 17.7 reveal a strong technical grouping, two general human geography clusters, and a historical-cultural grouping. The largest cluster, mostly urban in orientation, illustrates the interrelatedness of demographic, social, economic, and political concerns in geography. Physical geography specialties show a cohesive cluster, even though climate studies link more strongly to a general resource group.

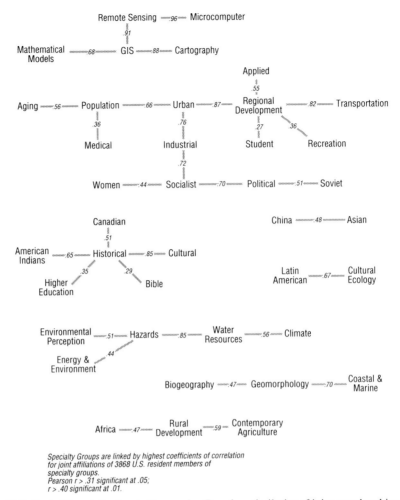

Fig. 17.7. Foci of Specialization in Geography. Based on similarity of joint memberships among AAG specialty groups in May 1989. Specialty groups are linked by highest coefficients of correlation for joint affiliations of 3,868 U.S.-resident members of specialty groups.

Aside from the distinct China and Asian linkage, regional interests retain, as in the 1984 analysis, strong associations with particular systematic foci. Thus, the Soviet Union and East Europe link with political geography, Canada with historical geography, Latin America with cultural ecology, and Africa with rural development.

The most prominent pattern is the exceptionally strong linkage among the discipline's primary technical subfields. In its brief existence, the GIS group has emerged as the largest specialty group in the AAG. It holds a central position among those interested in mathematical models, cartography, and remote sensing. Whereas all other groups had their highest correlations with other specialties, several had statistically significant correlations with the specialty groups in this cluster, particularly with GIS. In part, this represents genuine research and teaching linkages. However, many members may join the GIS group to learn about this comparatively new field.

Is There a Generation Gap in Geography?

The surge of interest in GIS is particularly strong among younger geographers. This interest could be symptomatic of a general age bias in the selection of disciplinary specializations. Consider the correlation coefficients of specialty group membership distributions for the six age categories. There is a surprisingly strong tendency for each to be most similar to its neighboring age cohorts and least similar to groups most distant in age. Does this apparent generation gap structure communication patterns among geographers?

Geographers born before 1925 share highest similarity in specialization with those born in 1925–34 (r = 0.83), with a rapid drop in similarity for each successively younger age category (fig. 17.8). Their correlation (r) with the youngest group, those born after 1964, is only 0.20. Since all age groupings are most similar in specialty affiliation to those closest in age, age is a factor in what geographers do and with whom they communicate.

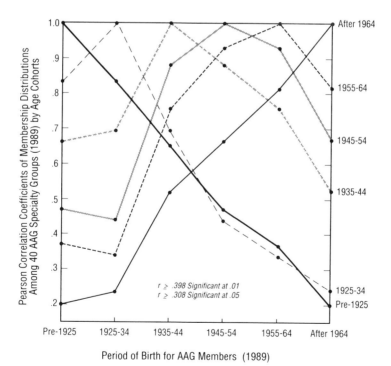

Fig. 17.8. Generation Gaps in American Geography. Based on correlation of membership in AAG specialty groups by period of birth. Data for May 1989.

In 1988, the AAG launched a direct assault against the generation gap. The Phoenix Group (named after the site of the association's annual meeting) was set up to bring together promising recent doctoral graduates with senior members of the discipline. A dozen senior geographers committed themselves to working with seventeen young scholars to help ease their transition into the profession.

The Phoenix Group sponsored workshops of interest to geographers who were just starting careers, and they examined their opportunities for participation in specialty groups and publication in major journals. This experiment, repeated with a new group of recent graduates in 1991 (the Miami Group), extends the mentor model to a broader level. It eases the socialization of junior scholars into the discipline and encourages supportive communication across the age spectrum.

Geographers as Representative of the American Population

Beginning in the 1960s, American geography has tried to achieve a membership that is more representative of the national population, with attention focused on minorities and women. Success in this effort is measured by their increasing proportionate representation among AAG members for successively more recent age cohorts (figs. 17.9 and fig. 17.10). This change reflects the dominant strategy of recruiting among students. There is movement toward a gender balance, but the attraction of minorities has barely kept pace with

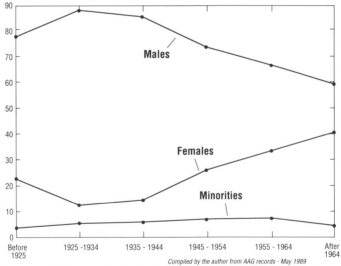

Compiled by the author from AAG records - May 1989

Fig. 17.9. Percentage of Membership in the Association of American Geographers by Period of Birth for Females, Males, and Minorities. Based on AAG U.S.-resident membership in May 1989.

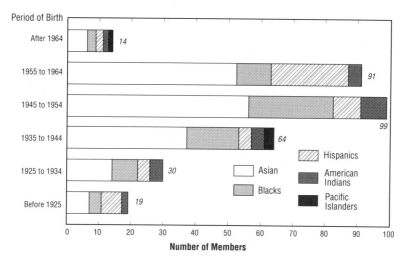

Fig. 17.10. Minorities in American Geography by Period of Birth. Based on AAG U.S.-resident membership in May 1989.

growth in the membership, and minorities remain far below their proportions in the general American population.

Women make up 24.7 percent of the total 1989 AAG membership living in the United States, and 28.7 percent of those born in the country. However, they exceeded these averages for the two most recent age groupings, accounting for 33.5 percent of those born between 1955 and 1964, and for 40.8 percent of those born after 1964 (fig. 17.9). Since women account for about 30 percent of all academic degrees granted in geography (AAG 1989), these values show a greater tendency for women to join the AAG than for men.

While progress in numerical representation is evident, the geography faculties of American universities remain decidedly male in composition. Surveys conducted between 1987 and 1989 show that women make up only 8.4 to 9.7 percent of the faculty. They hold only 5 percent of tenured positions, and are a mere 3 percent of full professors (Lee 1990). The AAG will need to keep its Committee on the Status of Women in Geography until there is much more evidence of women's integration into the full range of the discipline's activities, opportunities, and structures.

In contrast to a greater balance in gender representation, AAG membership of minority populations has shown only modest improvement. The proportion doubled, from 3.6 percent for the pre-1925 cohort to 7.4 percent for those born between 1955 and 1964 (fig. 17.10), but it remains far below the proportion of minorities in the American population. Indeed, there is a drop to 4.8 percent for the youngest cohort. It could be that minorities join the association at a later age, possibly resulting from low average discretionary incomes.

A report on the status of minorities in geography (Shrestha and Davis 1989) confirms low representation at all levels. From a survey of 161 institutions, only twenty-eight of 1,333 faculty were African American (2.1 percent), five were Hispanic, and two were Native American. Also, minority presence is low among the 9,026 students in the 161 institutions. African Americans made up 2.2 percent, Hispanics, 1.9 percent, and Native Americans, 0.4 percent. Asians, at 3.6 percent for students and 3.0 percent for faculty, fared better, but no account is available of those who migrated recently from other countries. It is unfortunate that this apparent lack of success conceals some innovative and successful efforts to attract and train minorities for careers in geography.

The Commission on Geography and Afro-America (COMGA), with support from the U.S. Office of Education, developed an active program to promote geography education and careers for African Americans, particularly in the South. Directed by Donald R. Deskins, Jr., it was active from 1969 to 1975. COMGA aided the development of African American institutions through fellowships for graduate students, faculty exchanges, information services, a placement program, and geographical research about African Americans in America. By 1971–72, 50 COMGA fellows were pursuing graduate degrees in 26 universities across the country. North Carolina Central University, a leading traditionally African American college with the country's largest minority geography department, was among the institutions that benefited significantly from COMGA initiatives. It was active in supporting the program (hosting summer workshops for training secondary school teachers) and has maintained an active undergraduate program since then. Unfortunately, financial support for this work ended in the mid-1970s.

A successor program to meet the needs of minority populations did not emerge until the late 1980s. Today, the association's Affirmative Action and Minority Status Committee monitors participation in the discipline, while a Committee on Minority Recruitment and Retention seeks to improve opportunities for minorities. A recent initiative is the Undergraduate Geography Training Program for Talented Minorities. It follows a successful offering of summer institutes for undergraduate minorities at New York's Hunter College (CUNY) starting in 1988. Its aim is to identify talented students from around the country for participation in special activities during their undergraduate years. Summer courses, summer research assistantships, and support for participation in AAG annual meetings expose them to intellectual challenges and career opportunities in geography. In hopes that they seek graduate training, a consortium of geography departments offers scholarships to those who complete the program. Persistent efforts of this nature may overcome the discipline's low profile among minorities.

Involvement in the discipline of a representative mix of the population by gender, ethnicity, and race draws on a richer set of perspectives for understanding the world than does a decidedly skewed mix. Though a commitment to social justice motivates interest in achieving this balance, geography benefits intellectually through direct access to the full range of people and experiences that comprise the world's humanity. Attempts to achieve this balance have yielded modest success and show promise of continued progress. In the meantime, the sociodemographic character of geographers and the practice of geography in the United States have borne dramatic transformations owing to migration from other countries. Among other factors, this migration has contributed to a growing internationalization of the discipline.

The Internationalization of American Geography

Numerical indicators and evidence of substantive changes in the content of the discipline illustrate the internationalization of American geography. Participation in the AAG by scholars from other countries and migration between these countries and the United States are of particular importance. These forms of connection broaden the range of communication in the discipline. They lead to the development of transnational networks of scholarship and application.

Nonresident members (664) represented nearly 12 percent of the AAG in 1989 (table 17.4). They are predominantly of British Commonwealth origin,

Table 17.4. Non–U.S.-Resident Members of the AAG, May 1989

Residence		Employment	
Canada	350	College/University	441
United Kingdom	50	Student	98
Japan	31	Retired	27
Australia	29	Government	26
Israel	21	Research Center	14
Saudi Arabia	17	Private Industry	13
Germany	15	Other Teaching	12
Korea	13	Self-employed	7
Hong Kong	11	Unemployed	3
The Netherlands	10	Nonprofit organiz.	2
Brazil	7	Other	9
Sweden	7	Missing information	12
Kenya	6	Total	664
Spain	6		
Switzerland	6	Males	546
India	5	Females	116
New Zealand	5		
Taiwan	5	Total Ph.D.s	447
Other	70	Ph.D.s from U.S. univs.	191
Total	664		

SOURCE: *Compiled by author from AAG membership records.*

**Table 17.5. U.S.-Born AAG Members Living in Other Countries,
May 1989**

Residence		Employment	
Canada	36	College/University	39
Australia	7	Student	11
England	6	Private Industry	3
Kenya	3	Government	3
The Netherlands	3	Self-employed	3
Brazil	2	Secondary school	2
France	2	Retired	2
Switzerland	2	Unemployed	2
Other	12	Other	4
		Missing information	4
Total	73	Total	73
Males	50	Total Ph.D.s	39
Females	23	Ph.D.s from U.S. univs.	31

SOURCE: *Compiled by author from AAG membership records.*

53 percent from neighboring Canada, mostly male (82 percent), and academic in orientation. More than 68 percent hold teaching positions, and 15 percent are students. Illustrating the importance of prior links with American geography, 43 percent of the doctorates were of American origin.

Only 73 of 664 nonresident members were born in the United States table 17.5). Half of these live in Canada, 32 percent are female, 56 percent hold teaching positions, and 15 percent are students. Nearly 80 percent of those with doctoral degrees received them from U.S. universities. A survey by Porteous and Dyck (1987) shows that 61 American-born scholars held faculty positions in Canadian universities in 1985. Thus, a significant proportion (possibly 40 percent) of Americans who work in Canada as academics are not members of the AAG.

It is, however, the increasing proportion of resident geographers from other countries that is of primary significance to the changing character of geography in the United States (table 17.6). More than 14 percent of American geographers were born in other countries. They come from all major regions of the world, but mostly from the United Kingdom (24 percent) and other British Commonwealth nations (25 percent). At 22 percent, female representation among immigrants is lower than for geographers born in the United States (29 percent). Although Europeans came primarily in the 1950s and 1960s, the most recent arrivals are from southern, eastern, and southeastern Asia. Of 171 geographers of Asian birth, 39 percent were students in 1989. They accounted for 35 percent of all foreign-born student members of the AAG.

Table 17.6. Foreign-Born Geographers in the United States, May 1989

Place of birth		Employment	
United Kingdom	171	College/University	409
Canada	66	Student	191
Germany	61	Private Industry	20
India	43	Government	20
China	38	Self-employed	8
Korea	18	Secondary school	4
The Netherlands	16	Retired	40
Hong Kong	15	Unemployed	9
USSR	14	Other	8
Australia	13	Missing information	2
Taiwan	13		
France	12	Males	552
Ireland	10	Females	159
New Zealand	10		
Bangladesh	9	Total Ph.D.s	443
Japan	9	U.S.	318
Other Asia	26	United Kingdom	47
Latin Amer./Carib.	43	Canada	30
Other Europe	70	Europe/USSR	30
SW. Asia/N. Africa	31	Australia/New Zealand	9
Africa	23	Other	9
Total	711		

SOURCE: *Compiled by author from AAG membership records.*

Academic degrees from U.S. institutions may be a principal factor in opening opportunities for long-term residence. Of those foreign-born geographers with doctorates, 72 percent earned them in American universities. An additional 19 percent of doctors, from universities in English-speaking British Commonwealth countries, would not have faced linguistic barriers to employment in the United States.

Ronald Abler (1989) suggests that recent migrants to the United States play an influential role in shaping the activity of the discipline. He documents a level of influence by British-trained and Commonwealth-trained academics that exceeds their numerical presence in American geography. For example, of 141 NSF research awards, 1986–88, 20 percent went to applicants who held their first academic degrees from UK-Commonwealth countries. One in six faculty of doctorate-granting departments for 1987–88 had similar origins. But, among the ten top doctoral departments 25 percent of 158 geographers had UK-Commonwealth origins.

It would be surprising if immigrant students, faculty and applied geographers did not alter the nature of research in American geography. The mix of cultural perspectives, the diversity of intellectual socialization, and the exposure to different national experiences offer fertile footing for innovation and insight. Citation analyses of the literature give important clues to such changes, brought about, in part, by the multinational population mix of American geography.

NETWORKS OF LINKAGE AND COMMUNICATION

The extent to which scholars reference the publications of others is now an accepted basis for measuring the commerce and status among authors, journals, disciplines, and institutions. This is possible through the services of the Science Citation Index and the Social Science Citation Index. In general, these sources of citation evidence focus on literature in the English language. Within that context, research by geographers provides a basis for detecting structural changes in the discipline.

Whitehand (1985) considered citations to human geographers. Of the thirty-two individuals he identified as "centurions" (authors whose work was cited at least one hundred times) in the period 1971–75, two-thirds were based in North America. One-third held doctorates from either the University of Chicago or the University of Washington. In addition, Whitehand observes that three-fourths of these scholars specialized in the spatial-analytic areas of human geography (for example, urban, economic, transportation, and quantitative methods).

Wrigley and Matthews (1987) updated the listing of centurions for the period 1981–85 and expanded the survey to include physical geographers. From 19 percent in the 1971–75 period, the number from British universities rose to 43 percent of 55 centurions. A rapid rise of newcomers in human geography was in contrast to the dominance in physical geography of older established leaders. There were no women on Whitehand's 1971–75 listing and only two (D. Massey and A. Buttimer) for 1981–85.

The most recent and most thorough citation analysis in geography is by Andrew R. Bodman (1991). His study covers 24,000 citations to the work of 2,706 geographers over five years, 1984–88. It is an exhaustive look at citations of academic geographers from Australia, Canada, the United Kingdom, and the United States. Bodman refers to those receiving at least one hundred citations in the period as the "master weavers."

Great Britain dominated the listing; 40 percent of the eighty-seven master weavers were on the faculties of British universities, and 60 percent earned their first academic degrees there. More than half of those with faculty positions in the United States were born in other countries. With the addition of Susan Hanson, women represented a scant 3.5 percent of the total. Relative to institutional concentration, Bodman notes that half of this elite group of authors received their doctorates from only eight universities (Cambridge, London, Bristol, Northwestern, University of California-Berkeley, Chicago, Washington, and Wisconsin).

Bodman's work exposes differences in the structural formation of physical and human geography. In human geography, though change is evolutionary, it occurs more quickly than in physical geography. For example, the 1984–88 citation leaders suggest a gradual relative decline of positivist spatial science. The primary area of growth is in the political economy approach to economic and social geography and to urban policy analysis. British-trained geographers on both sides of the Atlantic appear to be in command of this development. On the other hand, there is a revival of interest in cultural ecology and cultural geography, based largely on the success of scholars who subscribe to the Berkeley School.

In general, master weavers in human geography are younger than those in physical geography. In physical geography, the change in citation leadership occurs more slowly. Bodman describes this long-term dominance of influential figures (for example, Arthur Strahler) as indicative of a generational model of scientific change. Among the specializations of the field, American scholars are more prominent in biogeography, but British scholars dominate the citation leadership in geomorphology and hydrology.

Citation analyses also offer insights into a discipline's contacts and influence in other fields. For example, Bodman's analysis indicates stronger connections with other disciplines for physical geography than for human geography. In part this reflects the orientation of human geographers to the discipline's core journals. In contrast, physical geographers are more apt to publish research in the core journals of other disciplines. Overall, Bodman notes that 46.6 percent of citations to research by geographers occur in multidiscipline journals, 18.8 percent in the core journals of other disciplines, and the remainder in geography's core journals. Pasqualetti (1986) offers an interesting use of citation data to assess the influence of research by geographers in a specific policy area concerning nuclear power.

Though physical geographers accounted for only a quarter of Bodman's master weavers, they garnered more than 40 percent of the external citations. The strongest links were with geology and biology. For human geography, the greatest number of citations were in sociology journals. However, its greatest relative influence was in anthropology, mostly through the work of Karl W. Butzer.

SUMMARY AND DISCUSSION

The methodology and applications of geography in the United States reflect a close association with the origins of individuals who work and study here. As the mix of people changes, the nature of the discipline unfolds to embrace new perspectives and to tackle new problems. Cross-memberships among AAG specialty groups reveal a pronounced age-based generational pattern to activity in the discipline. Migration from other countries, particularly from British Commonwealth nations, has redirected some of the intellectual effort in American geography.

Strong mentor associations serve to reinforce the paths of intellectual activity, as seen in the continued strength of the Berkeley school of cultural ecology. Sustained mentor relationships also may provide a stabilizing force to counter successive age-based generational shifts in specialization patterns.

As chapters in this book attest, geographers are asking some of the same questions today that were asked in *American Geography: Inventory and Prospects*, the 1954 compendium on the status of geography in the United States. However, a comparison of these books points to dramatic shifts in thinking, in methods, and in the issues judged important. If nothing else, we are more self-conscious now, with concerns of how representative we are of the population (men, women, and minorities), more open about the ideological bases of our work, and more understanding of the interests we serve.

The regional distributions and migrations of geographers parallel the general population shifts within the nation. Nonetheless, the discipline responds to its own needs and opportunities. It is responsible for shaping its presence in different parts of the country. The leadership, commitment, and persistence of individuals and institutions shape the relative local and regional importance of geography. In part, they account for the place of geography in the Washington, D.C., area, its rapid expansion into the universities of the South, and its relative prominence in the Midwest and West.

Geography's presence in U.S. states is associated with strong university programs. However, improvements in geography at the elementary and secondary levels will have a profound influence on the future of the discipline in America. A discipline that reaches young people across the broad spectra of society early in their educational development will have the best chance of achieving a membership that is representative of that society. In turn, this achievement will affect its ability to serve society's needs. This brief survey on the peopling of American geography offers indirect support for this contention, particularly regarding responses to issues of gender, race, and ethnicity.

Concerns for women's issues, voiced seldom in the geography of the pre-1960s, gained prominence in recent years. Feminist views have shaped more inclusive images of human perspectives about our institutions, behaviors, and landscapes. On the other hand, the geographies of African Americans, Native Americans, Hispanics, and other minorities remain poorly defined. This reflects the continued weak presence of these peoples in the discipline. Aside from social motivations in seeing that all groups benefit from early exposure to geography, it is in the interest of scholarly/scientific work to tap the intellectual talent of these communities.

Policies and program initiatives to create a population mix in geography that is in balance with the nation's deserve continued attention by geographers. Current patterns in international migration favor the development of an unusually rich and varied base of human resources for American geography on the eve of the twenty-first century. However, as the issues that condition life in the nation exhibit connections with other parts of the world, we must consider how representative the international community of geographers is of its global population counterpart

NOTE

I thank the University of Western Ontario for a sabbatical leave and for research support in carrying out this study, and the Association of American Geographers for providing the opportunity. Gordon Shields and the UWO Department of Geography cartography unit deserve credit for the final graphics.

REFERENCES

ABLER, RONALD F. 1989. The Redcoats *keep* coming: The Britishization of American geography. Paper presented at the 85th annual meeting of the Association of American Geographers, Baltimore, Md.

ASSOCIATION OF AMERICAN GEOGRAPHERS. 1989. 1988 AAG survey of departments of geography in the United States and Canada. Washington, D.C.: AAG.

BODMAN, ANDREW. R. 1991. Weavers of influence: The structure of contemporary geographic research. *Transactions of the Institute of British Geographers* 16:21–37.

BUSHONG, ALLEN D. 1981. Geographers and their mentors: A genealogical view of American academic geography. In *The origins of academic geography in the United States*, ed. Brian W. Blouet, 193–219. Hamden Conn.: Archon.

———. 1984. Latin America as laboratory: Seventy-five years of doctoral research in Latin America by geographers in the United States. In *Latin America: Case studies*, ed. R. G. Boehm and S. Visser, 227–233. Dubuque, Ia: Kendall/Hunt.

GOODCHILD, MICHAEL F., and DONALD JANELLE. 1988. Specialization in the structure and organization of geography. *Annals of the Association of American Geographers* 78:1–28.

HART, JOHN FRASER. 1966. *Geographic manpower: A report on manpower in American geography*. Commission on College Geography Report 3. Washington, D.C.: Association of American Geographers.

————. 1972. *Manpower in geography: An updated report*. Commission on College Geography Report 11. Washington, D.C.: Association of American Geographers.

LEE, D. R. 1990. The status of women in geography: Things change, things remain the same. *Professional Geographer* 42:202–211.

McQUITTY, L. L. 1957. Elementary linkage analysis for isolating orthogonal and oblique types and typal relevancies. *Education and Psychological Measures* 17:207–229.

PASQUALETTI, M. J. 1986. The dissemination of geographical findings on nuclear power. *Transactions of the Institute of British Geographers*, n.s. 11:326–336.

PORTEOUS, J. D., and H. DYCK. 1987. How Canadian are Canadian geographers? *Canadian Geographer* 31:177–179.

RUGG, DEAN S. 1981. The Midwest as a hearth area in American academic geography. In *The origins of academic geography in the United States*, ed. Brian W. Blouet, 193–219. Hamden, Conn.: Archon.

SHRESTHA, NANDA R., and D. DAVIS, JR. 1989. Minorities in geography: Some disturbing facts and policy measures. *Professional Geographer* 41:410–421.

SOPHER, DAVID E., and J. S. DUNCAN. 1975. *Brahman and Untouchable: The transactional ranking of American geography departments*. Discussion Paper 10. Syracuse, N.Y.: Syracuse University Department of Geography.

WHITEHAND, J.W.R. 1985. Contributors to the recent development and influence of human geography: What citation analysis suggests. *Transactions of the Institute of British Geographers*, n.s. 10:222–234.

WRIGLEY, N., and S. A. MATTHEWS. 1987. Citation classics and citation levels in geography. *Area* 19:279–284.

Afterword

The Editors

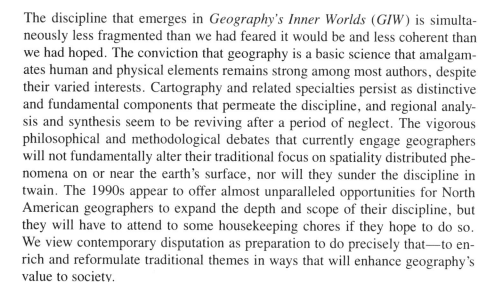

The discipline that emerges in *Geography's Inner Worlds* (*GIW*) is simultaneously less fragmented than we had feared it would be and less coherent than we had hoped. The conviction that geography is a basic science that amalgamates human and physical elements remains strong among most authors, despite their varied interests. Cartography and related specialties persist as distinctive and fundamental components that permeate the discipline, and regional analysis and synthesis seem to be reviving after a period of neglect. The vigorous philosophical and methodological debates that currently engage geographers will not fundamentally alter their traditional focus on spatiality distributed phenomena on or near the earth's surface, nor will they sunder the discipline in twain. The 1990s appear to offer almost unparalleled opportunities for North American geographers to expand the depth and scope of their discipline, but they will have to attend to some housekeeping chores if they hope to do so. We view contemporary disputation as preparation to do precisely that—to enrich and reformulate traditional themes in ways that will enhance geography's value to society.

RETROSPECT

The steering committee and the editors set out to produce an innovative and reasonably comprehensive overview of American geography in the 1990s that would appeal primarily to geographers, but that might also be of interest to nongeographers as an introduction to the discipline. In our judgement, that goal has been met. A diverse literature has been integrated under a set of headings that are common elements of all sciences, even if they are unorthodox rubrics under which to structure a discussion of contemporary geographic thought. But our authors have gone well beyond putting old wine into the editors' new bottles; *GIW*'s primary focus is on the present and what it portends for the future, although concepts and themes from the past pervade *GIW*, as befits a book treating a discipline that applies a comparatively stable set of ideas and techniques to a constantly changing world.

Ancestor and hero worship has been kept to a minimum, properly in our opinion. The methodological mainstays of the editors' intellectual youth and adolescence—demigods such as Richard Hartshorne, Carl Sauer, and even Brian Berry and Peter Haggett—have been debased. So have summary statements such as Ackerman's (NAS-NRC 1965) and the geography portion (Taaffe 1970) of the Behavioral and Social Sciences Survey (the BASS report). Protagonists in recent and ongoing conversations about the direction and scope of the discipline dominate citation lists, and the standard references that salted the debates of the past are scarcely in evidence. Also, the tight focus of the Ackerman and BASS reports stands in sharp contrast with the wide diversity of ideas and individuals recognized as geography's underpinnings in this volume. One of us went to so far as to suggest that *GIW* is postmodern in spite of its editors' inclinations.

Like earlier statements, *GIW* assumes that understanding the world demands an acquaintance with its geography. The core of geography is an abiding concern for the human and physical attributes of places and regions and with the spatial interactions that alter them. That traditional core permeates *Geography's Inner Worlds*, although varying in degree by chapter and section.

Unlike the Ackerman and BASS reports, which respectively gave physical geography cursory treatment and ignored it completely, *GIW* assumes that both human and physical geography are integral and complementary components of modern geography. Human geography's focus on features that are arrayed across a physically differentiated environmental complex distinguishes it from anthropology, economics, political science, and sociology. Physical geography's sensitivity to anthropogenic changes in physical systems distinguishes it from the biological sciences, geology, and meteorology. Both human and physical geography are further distinguished from cognate disciplines by their interest in how events at one place affect events and processes elsewhere.

Unlike the Ackerman and BASS reports, *Geography's Inner Worlds* bespeaks a renaissance of the regional geography that was in disfavor when they were written. Regionalism persists as a continuing heritage in geography, sometimes in traditional form, as outlined by Richardson in chapter 3, and other times in the postmodern neoregionalist form stressed by Pickles and Watts in chapter 12. Regional analysis and synthesis, albeit rooted in nontraditional epistemologies and philosophies, has again found favor among geography's junior scholars, and American geography will benefit greatly from that rebirth of geography's birthright.

Geography and cartography are still closely entwined two thousand years after their twin births. Whether mapping and maps are viewed as substance, as in chapter 4, as a means of data storage, analysis, and synthesis as they are examined in chapter 6, or as adjuncts to substantive exposition as they are used throughout *GIW*, geographers cannot escape or forsake them. Cartography, along with the related specialties of GIS and remote sensing, is a fundamental component of the discipline whether one considers its substantive content or its members' interests; cartography and related skills occupy a central position in the abstract intellectual space portrayed in chapter 18.

The quantities of spatially referenced data being captured are increasing rapidly. The need to store, analyze, and synthesize data in order to transform them into information is growing even more rapidly as more people crowd into a fixed amount of land and as detailed maps of people and of their activities become ever more critical components of intelligent land use. Given the growing applications of geographic information systems to manifold aspects of human and physical phenomena throughout North America and the rest of the world, cartography will continue to loom large on the discipline's research, teaching, and applications agendas in the years to come.

While we are more than satisfied with *GIW*, we have a list of desiderata that will inform any thoughts about a second edition. We wish we had done a better job of bridging the human geography–physical geography divide. We were unable to convince some human geographers to collaborate with physical geographers and thereby to give physical geography its due. We intended that both human and physical geography would be well represented in every chapter, either through the broad perspective of a single author or by the use of coauthors or collaborators. In some chapters, physical geography is a minor component or nearly absent despite the strenuous efforts of the editors. As a consequence, contemporary American physical geography is not as well represented in the volume as is human geography, and it is not as prominent as all three editors wished it to be at the outset. Capitalizing on the human-physical complementarity in geography should rank high on the discipline's agenda.

We also wish we had convinced all our authors to ground their discussions firmly in substantive research. Our intent was to produce a volume that focused, to use an old adage, on what American geographers *do* as a means of explaining what American geography *is*. Many authors accepted that schema, but some found it difficult to escape the customary academic practice of presenting ideas in the abstract and then citing empirical work that makes use of the ideas. Admittedly, moving from the empirical to the abstract is easier in some topics than in others, but we thought it should be possible throughout the book. There is a subtle but critical difference between the approaches; it differentiates discourse designed to teach people about the subject from that addressed primarily to other scholars. We hoped to explore worlds of American geography by providing exciting examples of how geographers address the world. But geographers find it difficult to break their habit of casting explanations of how they view the world in terms of what geography *is* rather than focusing on what geographers *do*.

Geography's abiding concern with primary data is more implicit than explicit in *GIW*. Geographers mistrust secondary data. Despite their commitment to technologies such as remote sensing and GIS that enable practitioners and scholars to manipulate data for places they have never seen, geographers remain equally committed to the ground truth that can be achieved only by going and by looking. Geographers present a far more opulent array of field trips in conjunction with their annual meetings than do adherents of any other discipline. When they visit new places, they look at different features and photograph different things than do tourists. A geographer who takes a slide of a monument is most likely a specialist in tourism who is more interested in the number of people visiting the site than in the attraction itself. Fieldwork fell into disfavor in human geography from the 1960s to the 1980s, but it seems to be a major component of the new philosophical viewpoints that are revitalizing regional geography, and it remained an important part of physical geography throughout the period.

Although the editors did not intend it, geographic practice has been slighted in *GIW*. Almost 20 percent of the AAG's members work in commerce, industry, and government, yet we neglected to engage a practitioner to present their viewpoint. That failure reflects an unfortunate gap in North America between those who profess and those who practice. Applications of the discipline's theories and techniques to practical problems strengthen geography. They enrich basic research and enhance geography's reputation in society. Geography lacks a natural constituency in American society such as economics enjoys in the business community and political science in local, state, and national politicians. If geographers are ever to develop a broad base of support they will have to create it themselves, and they will do so more through practice than they will by professing. Practice and policy formulation will offer especially rich opportunities for geographers in the 1990s, and a revision of this volume would have to attend to those facets of the discipline.

Overall, we are greatly pleased with the process that produced *Geography's Inner Worlds* and proud of the product. In spite of its shortcomings, we believe *GIW* represents American geography reasonably well. The two metaphors we evoked for the volume in the preface and first chapter were a field trip and fragments from an ongoing conversation. The field trip has been a reconnaissance rather than a full exploration, but we think it has assessed the rough outlines of the current geographical landscape faithfully. The snatches of the conversation are sufficient to give the flavor and the salient points of the ongoing debate in the United States about how to conduct geographic inquiry.

Obviously, although *Geography's Inner Worlds* focuses on pervasive themes in contemporary American geography, its content would more accurately be described as being based on the work of geographers resident in the United States and Canada who draw their intellectual sustenance from throughout the world. Few of the roots of the American geography summarized in 1954 in *American Geography: Inventory and Prospect* were native stock. Most were cuttings from Germany and France or hybrids produced by grafting New World ideas onto Old World root stock. Four decades later, considerably more of the intellectual sustenance of geography in North America appears to have originated in the United Kingdom and the Commonwealth nations than was true in the mid-1950s. The United States owes a great and growing debt to the British and Commonwealth geographers who now constitute a significant fraction of the American and Canadian professorates and a disproportionate percentage of the discipline's intellectual leadership (chapter 17). There is little in American geography that is distinctly American, and much that is multinational in flavor. Increasingly, the North Atlantic English-speaking community engages in an intertwined, if not a single geographical conversation.

As it was and remains diverse in origin, the American geography abstracted in *GIW* remains wonderfully diverse in substance, method, and philosophy. Compared with other disciplines, geography has always been tolerant of variety, and often enthusiastic about multiplicity. With the possible exception of the period before 1920 when environmental determinism achieved a measure of intellectual hegemony, American geography has never been a normal, paradigmatic science of the kind postulated by Thomas Kuhn (1970). No single group within geography has succeeded in marginalizing all others. Competing and contending substantive interests, methodological preferences, and philosophical schools have ceaselessly enlivened the discipline since the 1920s.

American geography was postmodern long before the term was invented. It has historically been eclectic and self-contradictory in many respects. It has often playfully delighted in juxtaposing dissonant substantive interests and intellectual traditions. American geography will prosper in the 1990s not so much because geography will change radically, but rather because the discipline's social and intellectual environments have now evolved to where geography has long been. Geographers are at last professing and practicing a postmodern discipline in a postmodern era.

PROSPECT

Geography's Inner Worlds appears at a critical and exciting juncture. Geography in the United States stands on the threshold of a period of rapid expansion and wider acceptance. Major nationwide programs to improve school geography and endorsement of the discipline by President George Bush and Secretary of Education Lamar Alexander as one of the five key components of secondary school curricula (*America 2000* 1991) have brought new prominence to geography and unprecedented interest in the discipline. The ABC television network has appointed Harm J. de Blij of Georgetown University as its geography editor. He appears on its top-rated morning news program several times a month to provide a geographic interpretation of current events (de Blij 1990). Although originally surprised at these developments, American geographers increasingly recognize that they offer exceptional opportunities for disciplinary growth during a period in which decline or stagnation in support for the academic enterprise is likely to be the rule.

But the process that produced *GIW* has highlighted three broad challenges American geographers will have to overcome in order to capitalize on current support for the discipline's revitalization. As we read what our authors have said, a basis for the coherence the discipline needs certainly exists within American geography. But some of the difficulties we encountered in trying to lead our authors and ourselves to that common ground reveal impediments that could hamper widespread dissemination of the geographical perspective. To prosper, American geographers would be wise to engineer the synergies that remain latent in the discipline's human and physical components and in the division between those who profess and those who practice. To accomplish those two tasks, the discipline will have to avoid the kinds of overreaction that have characterized its last two major restructurings.

Geographers committed to human capital, structurationist, and postmodern perspectives are indifferent to physical geography at best and hostile at worst. We accept their contention that concepts such as nature and natural resources are anything but natural; they are indeed human constructs freighted with ideological baggage. We agree that the respective abilities of different societies to cooperate with or manipulate environments are products of political and power relationships. At the same time, we feel constrained to note that the logical end of that line of reasoning is a cultural determinism that would be as extreme as environmental determinism—geography's hoary bogey man. No matter how much geographers' thinking about the physical environment is a human artifact, physical environments cannot be defined or wished out of existence; they exist, they must be reckoned with, and a positivist approach yields superior understanding of their operations for many purposes.

As noted in chapter 16 (which fell to the editors by default) the physical geography–human geography dichotomy is a microcosm of the larger two cultures problem noted by Snow (1961). Perhaps human geographers are indifferent or hostile to physical geography because they misunderstand its fundamentally human and humanistic origins. If humanism is a "stance on life and the world shared by people in many walks of life" as Buttimer (1990) asserts, most physical geographers would consider themselves to be humanists. Climatologists are, as Marcus observes, deeply concerned "for the future of society and environment; the implications of increased carbon dioxide and depleted ozone are no small matter. Similarly, biogeographers working in the Amazonian rain forest or North Slope tundra, hold a stake in the future of fragile biomes that supersede their individual research interests" (p. 336). Aside from the basic curiosity that animates many human as well as physical geographers, a driving motivation for physical scientists generally and physical geographers specifically is to make the world a better place—the same end that excites the postmoderns.

A basis for a coherent, synergistic disciplinary stance with respect to human and physical geography exists in the motivations human geographers and physical geographers share, namely that of improving the human condition. Separating physical geography from the rest of the discipline might satisfy a few narrow-minded physical geographers and a number of equally parochial postmodern human geographers. But widening the existing divide would be a tragic mistake that would seriously weaken the discipline at precisely the time American society needs a synthetic, coherent view of how humanity uses and abuses its physical environment. Human geographers cannot hope to meet that need without the insights of physical geographers. Incorporating positivist views of physical systems into the broader context of postmodern human geography strikes us as no different than citing data gathered on positivist bases and no different than using fundamentally positivist analytical techniques to arrive at conclusions nested in postmodern philosophical and theoretical frameworks.

One inner world of geography we sought to describe in *GIW* consists of the empirical subject matter persistently studied by geographers: the ideas and methods they consistently bring to those investigations and the broader conceptual formulations or theories that structure their selections of data and methods—in short, how geographers address the world. Subject, method, and theory serve to make any discipline distinct from others. Another inner world is composed of the conversation that occurs among geographers about which topics, tools, and paradigms ought to be studied and employed at any given place or time. As we have mentioned, many geographers find it difficult to separate the conversation among geographers about how to extend and improve the discipline from the one that tells society what geographers know about their outer world.

The tension between the editors' and authors' perspectives on which inner world should be given priority in *GIW* emerged often during the editing process. Most authors are conditioned to tell who dunnit first and what she or he done second. The editors were committed to stressing what geographers know about the world; who produced the knowing was of secondary importance to us. When we edited draft chapters, we consistently changed "Prufrock (1990) observed the land is more expensive in the centers of cities than on their peripheries" to "Land is more expensive in the centers of cities than on their peripheries (Prufrock 1990)." Many authors just as consistently revised the editors' changes back to the original, thereby emphasizing who dunnit. A process-product confusion is probably an inevitable feature of any disciplinary conversation. Furthermore, a volume on geography's core ideas written primarily for geographers might reasonably be cast primarily in who dunnit terms. Yet the editors cannot shake their conviction that geographers' persistent habit of structuring their narratives to focus attention first on the actors' names—and only secondarily on the drama they are playing out—hampers the discipline's ability to speak to what should be its broad constituency.

Without question, an inner-focused conversation about what geographers do with relevant data, methods, and theory must be prosecuted vigorously. A specialty that fails to engage in ongoing debate about its fundamental assumptions and procedures would not be a discipline. On the other hand, preoccupation with the inner-focused conversation impedes relations with the worlds outside geography. Geography enjoys no ready-made constituency in society. Its prospective constituency consists of people who are innately curious about places, peoples, and interactions among them, and of bureaucrats and business people. That audience is far less interested in what geography is than in the useful knowledge geographers can marshall to satisfy its curiosity, solve practical problems such as finding suitable sites for solid waste disposal, and inform policy formulation for issues such as national management plans for wetlands.

The urgent needs to reintegrate human and physical geography and to find the language and narrative forms that will engage the attention of American society will be pursued within a broader reformulation of the discipline's agenda that has been underway for some years. A complicated series of debates in contemporary American geography finds physical and positivist human geographers at odds with a postmodern group whose thinking is based in combinations of phenomenology, human capital approaches, structurationist views, and deconstruction. The former view the world as ultimately knowable through positivistic methods. The latter espouse the view that the world is knowable only by attending primarily to the political and social relationships positivists neglect. As Pickles and Watts put it, "Geographers have begun to realize more clearly how their concepts, theoretical frameworks, methodologies, categories, and language arise out of a particular historical and spatial conjuncture—modernity. Behind this recognition is also acceptance of the need to rethink many of the approaches they use to deal with the world" (p. 320).

As emphasized in chapter 16, the debate is far from trivial. The shift to logical positivism and spatial analysis after 1955 fundamentally altered the way geography was professed and practiced. Coming as it did at the beginning of an employment boom in academia, the shift set the dominant tenor of the discipline for a generation. Cultural and regional geography were denigrated, and they languished, much to the ultimate detriment of the discipline. American geography now finds itself in the early stages of another employment boom caused by the simultaneous retirement of the large cohort of geographers who completed doctorates in the years after World War II, the echo of the postwar baby boom, and the discipline's growing popularity. As in the 1950s, the outcomes of current debates over disciplinary priorities will govern undergraduate and graduate curricula, research agendas, and funding priorities through the teens and twenties of the twenty-first century.

Twice in the twentieth century—at an interval of about 35 years—geographers have thrown out a baby with its bath water. It would be tragic if they were to commit that mistake yet again. When American geographers rejected environmental determinism in the 1920s, they simultaneously abandoned the search for theory (grand and petite) and the quantitative methods they associated with the discredited view and some of its prominent advocates, such as Ellsworth Huntington. When a subsequent generation embarked in reaction on a quest for theory and up-to-date methods in the 1950s, it abandoned cultural geography and regional studies. Consequently, American geography in the 1990s finds itself bereft of scholars who possess a deep understanding of overseas areas at precisely the time the country desperately needs the insights seasoned regional geographers could provide. Is American geography destined to reel off on another tangent thirty-five years later, leaving in the lurch traditions it will discover it needs and wants thirty-five year hence?

The answer to that question lies in the hands and the wisdom of geographers who completed their doctorates within the last fifteen years. A major philosophical and methodological shift *is* underway in geography, and there can be little doubt as to its eventual outcome—youth will triumph. Intellectual changes do not result from conversions of individuals from old ideas to new ones, even in nonparadigmatic specialties such as geography. Disciplines evolve because of births and deaths. Adherents of what was once a dominant view eventually become a minority, and then nonentities, as they are replaced by younger scholars reared in different intellectual traditions. Wide swings in disciplinary perspective and philosophy may be inevitable; wholesale rejection of the traditional may be prerequisite to innovation. But we hope that is not so in this instance, and we hope that the next generation of geographers will profit from the mistakes made by its two predecessors.

One basis for our hope resides in our optimism that those on both sides of the debate will realize the degree to which geography could be enriched by modifying and incorporating existing traditions into new perspectives, a process that appears to be underway. No *ism* is as pure in application as it appears in methodological discussion. Ideal models of research presented in methodological treatises (whether those of positivism or of postmodernism) are caricatures of what actually happens when real research needs to be done; substantive problems clarify the philosophical mind wonderfully. Scholars and practitioners grappling with a real problem use whatever ideas and tools advance their understanding of the puzzle at hand, regardless of their provenance; scientists will guess and humanists will count when doing so yields insights that cannot be obtained in other ways.

New perspectives that attend carefully to gender relationships, political power, and social structure will broaden the ways the discipline can understand places, regions, and global systems, and they should excite all geographers, regardless of age or specialty. If those deeply committed to the new perspectives prefer to think of traditional approaches as quaint but occasionally helpful ways of addressing problems, so be it. The critical matter is that all parties avoid the wholesale rejection of each others' viewpoints that occurred in geography after 1920 and again after 1955.

A second reason to hope that American geography's young scholars will not make the same mistakes as their elders lies in a global shift from analysis toward synthesis, and correspondingly, from emphasis on explanation to a greater concern with understanding. Analysis has dominated American science, including geography, since World War II. Synthesis has been more than neglected—it has been nonexistent as a serious consideration. Thousands of analytical works and pieces devoted to analytical techniques have been written for every substantive attempt at synthesis in the last 50 years. One would search in vain for a treatise on synthetic technique in geography. In an era characterized by progressively greater specialization, scientists and geographers lost sight of the fact that analysis and even rigorous scientific explanation are not ends in themselves. In the coming decades, societies will be more interested in how diverse phenomena interact with each other than they have been in the past. Understanding will consequently be valued more than explanation. As a discipline, geography is well positioned to respond to that kind of intellectual and social need. Those who profess and practice geography will better respond to those needs if they remain open and eclectic.

If the hunches we have outlined in this Afterword are correct, *Geography's Inner Worlds* represents neither the last gasp of logical positivism nor the triumph of the postmodern view. We confess considerable trepidation on that point. *American Geography: Inventory and Prospect* was published in 1954 on the eve of what has often been termed the quantitative revolution. Geographers who voted with their research and teaching specialties decided it was mostly retrospect. Whether *GIW* suffers the same fate is in the laps of the gods that watch over geography, and more specifically, as we noted earlier, in the hands of a new generation that will hit its stride in the 1990s and the early years of the next century. We certainly hope that *GIW* is a milepost rather than a turning point. We reiterate our conviction that the kind of wholesale shift in emphasis that occurred in the 1920s and the 1950s—for which geography is ripe—would be unfortunate.

More than any other discipline, geography dwells at the juncture of the humanities, the sciences, and the social sciences. Geographers have historically drawn from and engaged in all three kinds of intellectual activity. In the process, they have created a wonderfully diverse tradition that embraces a spectrum ranging from literary and even poetic evocations of places to highly abstract mathematical models of natural systems. We believe that the ends of the geographical universe are not as distant from each other as many of our geographical brothers and sisters are wont to believe, and we thank our authors for helping us find unity and coherence where we had not previously recognized it. We are convinced that if geographers bend their efforts toward enhancing and enriching that diversity in a synthetic framework, their work will be highly valued and generously supported by American society in the years ahead. A coherent, synthetic, global discipline focused on human use of the earth lies within the grasp of American geographers if they possess the will, the good will, the imagination, and the boldness to create it.

REFERENCES

America 2000. 1991. Washington, D.C.: U.S.G.P.O.

BUTTIMER, ANNE. 1990. Geography, Humanism, and Global Concern. *Annals of the Association of American Geographers* 80:1–33.

DE BLIJ, HARM J. 1990. Geography on GOOD MORNING AMERICA. *Focus*, Winter 1990.

JAMES, PRESTON E., and JONES, CLARENCE F. 1954. *American Geography: Inventory and Prospect*. Syracuse: Syracuse University Press.

KUHN, THOMAS S. 1970. *The Structure of Scientific Revolutions* 2d ed. Chicago: University of Chicago Press.

NATIONAL ACADEMY OF SCIENCES-NATIONAL RESEARCH COUNCIL. 1965. *The Science of Geography*. Publication 1277. Report of the Ad Hoc Committee on Geography.

SNOW, C. P. 1961. *The Two Cultures and the Scientific Revolution*. New York: Cambridge University Press.

TAAFFE, EDWARD J., ed. 1970. *Geography*. Englewood Cliffs, N.J.: Prentice-Hall.

Index